D1528461

Electrical Insulation in Power Systems

POWER ENGINEERING

Series Editor

H. Lee Willis

*ABB Power T&D Company Inc.
Cary, North Carolina*

1. Power Distribution Planning Reference Book, *H. Lee Willis*
2. Transmission Network Protection: Theory and Practice, *Y. G. Paithankar*
3. Electrical Insulation in Power Systems, *N. H. Malik, A. A. Al-Arainy, and M. I. Qureshi*
4. Electrical Power Equipment Maintenance and Testing, *Paul Gill*
5. Protective Relaying: Principles and Applications, Second Edition, *J. Lewis Blackburn*

ADDITIONAL VOLUMES IN PREPARATION

Electrical Insulation in Power Systems

N. H. Malik
A. A. Al-Arainy
M. I. Qureshi
*King Saud University
Riyadh, Saudi Arabia*

MARCEL DEKKER, INC.   NEW YORK · BASEL · HONG KONG

Library of Congress Cataloging-in-Publication Data

Electrical insulation in power systems / by Nazar Hussain Malik, A. A. Al-Arainy, Mohammad Iqbal Qureshi.
    p. cm. -- (Power engineering ; 3)
    Includes bibliographical references and index.
    ISBN 0-8247-0106-2 (hardcover)
    1. Electric insulators and insulation. 2. Electric power systems--Protection. 3. Dielectrics. I. Malik, Nazar Hussain. II. Al-Arainy, A. A. III. Qureshi, Mohammad Iqbal. IV. Series
TK3401.E42 1997
621.319'37--dc21   97-34427
                                                                                                   CIP

The publisher offers discounts on this book when ordered in bulk quantities. For more information, write to Special Sales/Professional Marketing at the address below.

This book is printed on acid-free paper.

Copyright © 1998 by MARCEL DEKKER, INC. All Rights Reserved.

Neither this book nor any part may be reproduced or transmitted in any form or by any means, electronic or mechanical, including photocopying, microfilming, and recording, or by any information storage and retrieval system, without permission in writing from the publisher.

MARCEL DEKKER, INC.
270 Madison Avenue, New York, New York 10016
http://www.dekker.com

Current printing (last digit):
10 9 8 7 6 5 4 3 2 1

PRINTED IN THE UNITED STATES OF AMERICA

Series Introduction

Power engineering is the oldest and most traditional of the various areas within electrical engineering, yet no other facet of modern technology is currently undergoing a more dramatic revolution in both technology and industry structure. As these changes take place, many of the traditional and fundamental areas of power engineering become more important than ever. Certainly one such area is insulation. Always an essential element for satisfactory power system operation, its correct interpretation and application becomes more critical at a time when deregulated power grids are utilizing ever higher voltage levels to move greater amounts of power over longer distances.

Electrical Insulation in Power Systems, by Drs. Malik, Al-Arainy, and Qureshi, is a particularly useful book, because it combines a very comprehensive coverage of insulation methods with a consistent attention to detail and practical application. All of the insulation approaches used in modern power systems are presented in a straightforward and thorough manner, including vacuum, air, gas, liquid, solid and composite dielectric technologies. Within each area, the authors address the traditional fundamentals completely, and provide a good discussion of recent developments and their applications. Equally important are the final three chapters of the book, which provide a very cogent and well-organized presentation on testing and diagnostic procedures and the interpretation of their results.

As the editor of the Power Engineering Series, I am proud to include *Electrical Insulation in Power Systems* among this important group of books. Like all the books in this series, this volume presents modern power technology in a context of proven and practical application. It is useful as a reference book, as a text in an advanced power systems curriculum, or for self-study and tutorial application. The Power Engineering series will eventually include books covering the entire field of power engineering, in all of its specialties and subgenres, all aimed at providing practicing power engineers with the knowledge and techniques they need to meet the challenges of the electric industry in the 21st century.

H. Lee Willis

Preface

Electrical insulation is the backbone of all modern power system networks. Different types of dielectrics that constitute these insulating elements are subjected to a variety of stresses during their life span. A thorough knowledge of their fundamental properties is therefore essential for their appropriate and optimal design as well as their pre- and postinstallation testing and reliable operation.

The rapidly growing demand for electrical energy in the world today necessitates its transmission at extra high voltage levels. Electrical power engineers must efficiently tackle the complicated problems of insulation and overvoltages arising in power networks. To train and better equip these engineers now and for the future, it is necessary that they be provided with a relevant specialized background. At present, only a few books are available that address this problem, particularly when compared with the number of those published in other areas of electrical engineering. Even among these few, no single book covers the entire range of essential topics. Some are almost two decades old, and others are limited to monographs on specific individual topics such as switchgear, corona, insulating oils and gas insulated systems (GIS). There is therefore a great need for a book that covers in a single volume the topics that are pertinent to the electrical engineer working with high voltage insulation.

Electrical Insulation in Power Systems is an attempt to fill this gap. It is expected to be of considerable importance not only for engineers involved with power utilities and the insulation industry, but also for senior undergraduate and postgraduate students. We are associated with the subject of high voltage insulation as researchers and as teachers, at both the undergraduate and graduate levels, of specially tailored courses for practicing engineers working in industry and electric utilities. Most of the material presented in this book is the outcome of lectures prepared for university and industry audiences. The book covers in detail the fundamental properties of dielectrics and their desirable properties for pertinent power applications as well as the means of assessing the performance of materials and equipment in the laboratory and in service. This book will certainly enable the reader to understand high voltage insulation better and apply that knowledge more effectively.

This book comprises 12 chapters. Each chapter deals with a specific aspect of the subject and can be read as a self-contained unit with inclusive references. In Chapter 1, the reader is introduced to the importance and development of high voltage insulation technology through a brief description of important individual elements of modern power system networks together with a general perspective of the important properties of insulating materials. Chapter 2 is devoted to basic gas discharge mechanisms; Chapter 3 describes the air clearances necessary for transmission network and substation equipment, plus the factors that conspicuously control these clearances. No networks exist that do not employ sections of SF_6 GIS. Chapter 4 describes the fundamental mechanisms that control breakdown in SF_6 gas and offers a detailed account of SF_6 circuit breakers and GIS techniques generally encountered in modern power networks. Chapters 5 and 6 describe the importance and classification of liquid and solid dielectrics. We deal with their electrical properties in detail and then cover their conduction and breakdown aspects. Vacuum as a dielectric is described in Chapter 7, whereas Chapter 8 is concerned with composite insulation. Polymeric and oil-filled cables are an integral part of a power network, so Chapter 9 has been devoted solely to the treatment of this topic. Chapter 10 contains a discussion of the methods for generation and measurement of high test voltages.

With the advent of digital/opto electronics and microprocessors, most of the measurement techniques in almost every branch of technology have been modified. This applies to the high voltage measurements and diagnostic techniques as well. In the past decade, a variety of such techniques have been introduced and today many power stations are equipped with this equipment. Until now, details of such methods were available only in scattered form in the literature. Chapter 11 is fully devoted to a detailed survey of these techniques.

It is very important, for design engineers as well as practicing engineers, to test the ability of an insulation system or apparatus to meet the guaranteed specifications. For this purpose, testing prior to and after installation is generally necessary. Chapter 12 deals with high voltage testing techniques, including their internationally accepted classifications, test procedures and standards. Newcomers to this field as well as experienced designers of equipment and practicing engineers will find this vital information presented in a systematic way to make them aware of the latest techniques in their chosen field.

We welcome constructive suggestions for future editions of this book.

N. H. Malik
A. A. Al-Arainy
M. I. Qureshi

Contents

Series Introduction	H. Lee Willis	*iii*
Preface		*v*

1 Introduction to Electrical Insulation in Power Systems — 1

1.1	Introduction	1
1.2	Properties of Dielectrics	2
1.3	Classification of Insulating Materials	5
1.4	Applications of Insulating Materials	8
1.5	Electric Fields	10
1.6	Design Parameters of High Voltage Equipment	18
	References	19

2 Gas Dielectrics — 21

2.1	Introduction	21
2.2	Gas Behavior Under Zero Electric Field	21
2.3	Generation of Charged Particles	23
2.4	Deionization Processes	28
2.5	Uniform Field Gas Breakdown	29
2.6	Nonuniform Field Gas Breakdown	36
2.7	Time to Breakdown	42
2.8	Discharges Under Nanosecond Pulse Voltages	43
2.9	Gap-Type Discharge	44
2.10	Choice of Dielectric Gases	46
	References	48

3 Air Insulation — 49

3.1	Introduction	49
3.2	Air Insulation Applications and Modeling	49
3.3	Voltage Stresses	51
3.4	Impulse Breakdown Probability	51
3.5	Breakdown Voltage Characteristics	52
3.6	Volt Time Curve and Insulation Coordination	65
3.7	Phase to Phase Breakdown Characteristics	69
3.8	Arc Discharge	71
3.9	Undesirable Effects of Corona	72
3.10	Television Interference	80
	References	80

4 SF_6 Insulation — 83

4.1	Introduction	83
4.2	Basic Properties of SF_6 Gas	83
4.3	Breakdown Processes in SF_6	85
4.4	Uniform Field Breakdown	86
4.5	Nonuniform Field Breakdown	87
4.6	Estimation of Minimum Discharge Voltages	90
4.7	Factors Affecting Discharge Voltages	91
4.8	Arc Interruption in SF_6	96
4.9	Gas Insulated Switchgear	98
4.10	Compressed Gas Insulated Cables	107
4.11	Other Applications of SF_6	108
4.12	SF_6 Gas Handling	108
	References	109

5 Liquid Dielectrics — 111

5.1	Introduction	111
5.2	Classification of Insulating Oils	112
5.3	Essential Characteristics of Insulating Oils	119
5.4	Streaming Electrification	123
5.5	Reconditioning of Insulating Oils	124
5.6	Electric Conduction in Insulating Liquids	125
5.7	Breakdown in Insulating Liquids	129
	References	143

6 Solid Dielectrics — 147

6.1	Introduction	147

6.2	Solid Insulating Materials	148
6.3	Dielectric Loss in Solid Insulating Materials	167
6.4	Breakdown in Solid Insulation	173
	References	186

7 Vacuum Dielectrics 188

7.1	Introduction	188
7.2	Prebreakdown Electron Emission in Vacuum	189
7.3	Factors Affecting Breakdown Voltage in Vacuum	192
7.4	Breakdown Mechanisms	195
7.5	Arc Interruption in Vacuums	198
7.6	Vacuum Circuit Breaker	202
	References	206

8 Composite Dielectrics 209

8.1	Introduction	209
8.2	Dielectric Properties of Composites	210
8.3	Edge Breakdown	213
8.4	Cavity Breakdown	214
8.5	Breakdown Due to Surface Erosion and Tracking	217
8.6	Chemical and Electrochemical Deterioration and Breakdown	218
8.7	Materials of Outdoor Insulators	220
8.8	Oil-Impregnated Insulation	224
8.9	Flexible Laminates	234
	References	238

9 High Voltage Cables 241

9.1	Introduction	241
9.2	Cable Materials	241
9.3	Types of Cables	244
9.4	Cable Constants	245
9.5	Electric Stress in Cables	249
9.6	Cable Losses	250
9.7	Cable Ampacity	251
9.8	Partial Discharges in Cables	253
9.9	Treeing in Cables	255
9.10	Cable Aging and Life Estimation	263
9.11	Cable Accessories	266
9.12	Cable Fault Location	270
9.13	Recent Advances in Cable Technology	272

		References	273

10 Generation and Measurement of Testing Voltages — 276

- 10.1 Introduction — 276
- 10.2 High Voltage DC Generation — 276
- 10.3 High Voltage AC Generation — 279
- 10.4 High Voltage Impulse Generation — 282
- 10.5 Nanosecond Pulse Generation — 288
- 10.6 Spark Gaps as a Voltage Measuring Device — 289
- 10.7 Potential Dividers for High Voltage Measurement — 294
- 10.8 Other High Voltage Measuring Devices — 295
- 10.9 Measurement of Corona and Gap Discharge Currents — 302
- References — 304

11 New Measurement and Diagnostic Technologies — 306

- 11.1 Introduction — 306
- 11.2 Digital Impulse Recorders — 306
- 11.3 Digital Techniques in HV Tests — 311
- 11.4 Testing Automation — 315
- 11.5 Electric Field Measurements — 316
- 11.6 Electro-Optic Sensors — 317
- 11.7 Magneto-Optic Sensors — 325
- 11.8 Measurements of Very Fast Transients in GIS — 328
- 11.9 Space Charge Measurement Techniques — 329
- 11.10 Electro-Optical Imaging Techniques — 335
- References — 339

12 Insulation Testing — 342

- 12.1 Objectives of Testing — 342
- 12.2 HV Test Classification — 343
- 12.3 Test Voltages — 345
- 12.4 Test Procedures and Standards — 349
- 12.5 Testing of HV Measuring Devices — 351
- 12.6 Partial Discharge Test — 355
- 12.7 Dielectric Loss Test — 358
- 12.8 Testing of HV Apparatus — 361
- 12.9 Electrostatic Hazards — 375
- References — 378

Index — *381*

1
Introduction to Electrical Insulation in Power Systems

1.1 INTRODUCTION

The economic development and social welfare of any modern society depends upon the availability of a cheap and reliable supply of electrical energy. Extensive networks of electrical power installations have been built in industrialized countries and are being constructed in developing countries at an ever-increasing rate. The major function of such power systems is to generate, transport and distribute electrical energy over large geographical areas in an economical manner while ensuring a high degree of reliability and quality of supply.

The transmission of large amounts of electrical power over long distances is best accomplished by using high voltage (HV), extra high voltage (EHV) or ultra high voltage (UHV) power lines (see Table 1.1 for voltage classification). Thus, high voltage equipment is the backbone of modern power systems. Besides generation, transmission and distribution of electrical energy, high voltages are also extensively used for many industrial, scientific and engineering applications such as:

1. Electrostatic precipitators for the removal of dust from flue gases
2. Atomization of liquids, paint spraying and pesticide spraying
3. Ozone generation for water and sewage treatment
4. X-ray generators and particle accelerators
5. High-power lasers and ion beams
6. Plasma sources for semiconductor manufacture
7. Superconducting magnet coils

Table 1.1 Voltage Classifications (V = Rms, Line to Line Voltage)

Voltage class	Voltage range
Low voltage (LV)	$V \leq 1\text{kV}$
Medium high voltage (MHV)	$1 \text{ kV} < V \leq 70 \text{ kV}$
High voltage (HV)	$110 \text{ kV} \leq V \leq 230 \text{ kV}$
Extra high voltage (EHV)	$275 \text{ kV} \leq V \leq 800 \text{ kV}$
Ultra high voltage (UHV)	$1000 \text{ kV} \leq V$

In all such applications, the insulation of the high voltage conductor is of primary importance. For proper design and safe and reliable operation of the insulation system, knowledge of the physical and chemical phenomena which determine the dielectric properties of the insulating material is very important. In addition, the basic processes which lead to degradation and failure of such materials and appropriate diagnostic techniques are of prime importance since any such failure can cause temporary or permanent damage to the system, thereby influencing its reliability and cost. Considering the high cost and comparatively long life span (20–40 years) of high voltage equipment, every effort should be made to select the most appropriate materials as well as the design, installation and operational parameters for such apparatus. This book attempts to provide an overview of different aspects of electrical insulation as practiced mainly in high voltage power systems.

This chapter outlines some basic definitions and fundamental concepts which are essential for proper understanding of the electrical insulation behavior. It further provides a brief overview of general categories of insulating materials, a short description of their applications and some desirable properties of various classes of dielectric materials. Subsequent chapters provide the detailed properties, applications, failure modes and diagnostic techniques used for evaluating and testing different insulating materials.

1.2 PROPERTIES OF DIELECTRICS

There are several properties of a dielectric which are of practical importance for an engineer. The most important of these properties are briefly defined here.

1.2.1 DC Conductivity

DC conductivity, σ, is defined as $\sigma = J/E$ where J is current density (in A/m^2) resulting from the application of a direct electric stress E (in V/m). It is related to the bulk resistivity ρ of the dielectric by $\sigma = 1/\rho$ and is calculated from measured values of the insulation resistance. Alternatively, if ρ and geometry are known, insulation resistance can be calculated. Insulation resistance is used as an indication of conduction behavior of insulating materials in many practical applications, such as in hi-pot testing. For most insulating materials, σ depends upon the material purity, its temperature T and electric stress E. It generally increases as the ionic impurities in the insulation system increase. Similarly σ also tends to increase with T and E in most cases following a relationship of the type

$$\sigma(T) = Ae^{-E/kT} \tag{1.1}$$

where k is the Boltzman constant, A is a constant, and $\sigma(T)$ is the value of σ at temperature T. In addition, due to polarization effects, σ also depends upon time of application of the stress. It influences the power losses in a dielectric and controls the electric stress distribution under direct voltage applications.

1.2.2 Dielectric Permittivity

Dielectric permittivity, relative permittivity or dielectric constant, ε_r, of an insulating material is defined as $\varepsilon_r = C/C_o$, where C is the capacitance between two parallel plates having the space between them filled with the insulating material under discussion, and C_o is the capacitance for the same parallel plates when these are separated by vacuum. Generally ε_r is not a fixed parameter but depends upon temperature, frequency and molecular structure of the insulating material.

1.2.3 Complex Permittivity, Loss Angle and Dissipation Factor

In order to determine the response of dielectrics to alternating voltages, it is traditional to model the dielectric by a parallel RC network. Such a network along with its phasor diagram is shown in Figure 1.1. Here R represents the lossy part of the dielectric taking account of losses resulting from electronic and ionic conductivity, dipole orientation and space charge polarization, etc., and C is the capacitance in the presence of the dielectric as defined earlier. If an AC voltage, $v = \sqrt{2}\, V \sin \omega t$, is applied then the capacitive component of current is $I_C = j\omega CV$ while the resistive component of current is $I_R = -jI \tan \delta$. Since loss angle δ is usually very small,

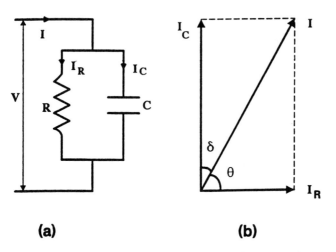

Figure 1.1 (a) Parallel equivalent circuit of a dielectric material and (b) corresponding phasor diagram.

$I_C \approx I$ and $I_R = -jI_C \tan \delta$. Hence, total current $I = I_R + I_C$ can be expressed as:

$$I = j\omega C_o V(\varepsilon_r - j\varepsilon_r \tan \delta) = j\omega C_o V\varepsilon^* \qquad (1.2)$$

where ε^* = complex relative permittivity having a real part equal to the dielectric constant, ε_r, and an imaginary part equal to the loss factor, $\varepsilon_r \tan \delta$. The loss factor differentiates the losses in one dielectric material from those in the other one. Tan δ is commonly known as loss tangent or dissipation factor or sometimes as power factor (cos θ) of the dielectric. It usually depends upon frequency and may also be influenced by the applied electric stress as well as the temperature. The power loss in the dielectric is given as:

$$\text{Power loss} = \omega C V^2 \tan \delta = \omega C_o V^2 \varepsilon_r \tan \delta \qquad (1.3)$$

Furthermore σ and tan δ are related as:

$$\tan \delta = \frac{\sigma}{\omega \varepsilon_o \varepsilon_r} \qquad (1.4)$$

where $\varepsilon_o = 8.85 \times 10^{-12}$ F/m is the permittivity of free space or vacuum.

1.2.4 Polarization

Unlike conductors where free electrons are easily available, most of the electrons in insulating materials are bound and not free to move. Under the influence of an applied electric field, the resulting electrostatic forces create some level of polarization forming dipoles. It is this electronic polarization which results in relative permittivity of more than 1 for most dielectric materials. In some crystalline dielectric materials, relative displacement occurs between positive and negative ions such as in Na^+ and Cl^- producing atomic polarization. In another mechanism which operates in organic substances including many polymers, permanent molecular dipoles are reoriented in electric field. The last type of mechanism to be considered is interfacial polarization which is observed for heterogeneous materials. In this case, mobile conduction charges are held up at some boundary within the dielectric (see chapter 6, section 6.3 and chapter 8, section 8.2.1 for more details). Electrolytic capacitors where charge is held up at the electrode surface double layer is an example of this type of polarization [1].

1.2.5 Dielectric Strength

This is defined as the maximum value of applied electric field at which a dielectric material, stressed in a homogeneous field electrode system, breaks down and loses its insulating property. It is given in V/m. The breakdown strength of most dielectrics depends upon the purity of material, time and method of voltage application, type of applied stress as well as other experimental and environmental parameters. Although in many cases, it may be difficult to assign a unique value of dielectric strength to a given material, a range of values can be found and used for application purposes. Although by definition dielectric strength refers to a uniform field system, in many applications breakdown strength under inhomogeneous field conditions needs to be defined and is sometimes referred to as the nonuniform field dielectric strength.

1.3 CLASSIFICATION OF INSULATING MATERIALS

There are virtually hundreds of insulating materials which are used in the electrical power industry. All such materials can broadly be classified into different categories: gases, liquids, solids, vacuum and composites. Some of the materials commonly used under each category and their desirable properties are summarized next.

1.3.1 Gases

In normal state most gases are good insulators. Consequently, overhead lines and open air circuit breakers using air insulation are in service since the early days of the electrical power industry. More recently, metalclad switchgear and gas-insulated cables filled with compressed sulfur hexafluoride gas, SF_6, have made their entry into the electric power systems. In special applications, as for instance in Van de Graaf accelerators or in measurement capacitors, other gases or mixtures of SF_6 with gases such as N_2, O_2, CO_2, air, and N_2O are also used.

An ideal gaseous insulator should be cheap, chemically and thermally stable, and should not form toxic, corrosive or flammable products under prolonged electrical stress. It should have good heat transfer and arc quenching properties, and exhibit low condensation temperature even when pressurized. Most importantly, it should have high uniform and nonuniform field dielectric strengths under DC, AC and impulse voltages and should not suffer any loss of these values under prolonged use.

1.3.2 Vacuum

The absence of any residual gas in the interelectrode gap space results in a vacuum medium which has excellent insulating and arc quenching properties. A true vacuum is very difficult to achieve and residual gas pressure of the order of 10^{-9} to 10^{-12} bar may exist in vacuum insulated equipment. In such equipment, material, shape and surface finish of electrodes, residual gas pressure and contaminating particles are important factors. Vacuum insulated medium voltage switches and circuit breakers are being used more frequently.

1.3.3 Liquids

An ideal liquid insulant should have high values of dielectric strength, volume resistivity, specific heat, thermal conductivity and flash point plus low values of loss factor, viscosity, pour point and density. Furthermore, the liquid should be noncorrosive, nonflammable, nontoxic and chemically stable having good arc quenching as well as gas-absorbing properties [2].

No single liquid is available which possess all of these properties, and compromises usually have to be made. Mineral oil having alkanes, cycloalkanes and aromatics as the main constituents has been used since the last century. Another class of liquids which have been used for transformers and capacitors are chlorinated aromatics or askrals, also called PCBs. They exhibit excellent flame resistance and very good electrical properties and are derived from benzene or from biphenyl. In the 1970s, it was found that these compounds exhibit a considerable health hazard and most countries have now legally banned the production and use of these liquids. Many, new liquids have since been developed which are without adverse ecological effects and also possess properties which are comparable with those of PCBs. These include silicone oil, synthetic hydrocarbons and fluorinated hydrocarbons.

1.3.4 Solids

An ideal solid dielectric must have some of the properties mentioned earlier for gases or liquids. In addition, it should have good mechanical and bonding properties. Inorganic as well as organic solid insulating materials are widely used in electrical power components. The most important inorganic materials are ceramic and glasses which are used to manufacture insulators, bushings and other high voltage components. The most prominent organic materials are thermosetting epoxy resins or thermoplastic materials such as polyvinylchloride (PVC), polyethylene (PE) or cross linked polyethylene (XLPE). Thermoplastic materials are mainly used for manufacture of extruded dielectric power cables. Kraft paper, natural rubber, ethylene polypropylene rubber (EPR), silicon rubber and polypropylene are some of the other solid dielectric materials which are widely used.

1.3.5 Composites

In many engineering applications, more than one class of insulating materials are used together, giving rise to a composite or a hybrid type of insulation system. Examples of such systems employing solid/gas insulation are transmission line insulators and solid spacers used in gas insulated switchgear (GIS). In solid/gas composites, the solid/gas interface usually represents the weakest link and has to be carefully designed. Similarly, in vacuum insulated systems, the interface of solid insulating spacer and vacuum proves to be a weak link. Examples of solid/liquid composite insulations are oil impregnated paper tapes used in high voltage cables, transformers, capacitors and bushings. Similarly oil impregnated, metallized plastic films used in power capacitors also belong to this category.

In the applications of composites, it is important to ensure that both components of the composite should be chemically stable and not react with each other under combined thermal, mechanical and electrical stresses over the expected life of the equipment and should have nearly equal dielectric constants. Furthermore, the liquid insulant should not absorb any impurities from the solid which may adversely affect its resistivity, dielectric strength, loss factor and other properties.

1.4 APPLICATIONS OF INSULATING MATERIALS

An electric power system has many high voltage components such as generators, transformers, circuit breakers, cables, bushings, overhead lines, surge arresters, GIS, capacitors, protective gaps and rotating machines. All such components need proper insulation. There are four principal areas where insulation must be applied [3]. They are (1) between coils and earth (phase to earth insulation), (2) between coils of different phases (phase to phase insulation), (3) between turns in a coil (inter-turn insulation) and (4) between coils of the same phase (inter-coil insulation). A brief introduction to the use of insulating materials in major high voltage system components is provided next.

1.4.1 Transformers

Present high voltage power transformers use enameled conductors, paper, glass or thermoplastic insulating tape, pressboard, glass fabric, porcelain and mineral or silicone oil. The windings are insulated by tape, held in place over the iron core by pieces of pressboard, glass fabric or porcelain, and impregnated with an insulating fluid which also acts as the cooling medium. Various designs of windings and oil cooling medium are employed [4,5]. In small power transformers as well as in current and voltage measurement transformers, the insulating materials used are thermosetting resins, insulating tapes, SF_6 gas, etc. In such cases, pressurized SF_6 gas provides insulating as well as cooling functions. Fire resistant transformers use insulating fluids such as high flash point mineral oil, chlorofluorocarbons or perchloroethylene.

1.4.2 Circuit Breakers

High voltage circuit breakers use ceramics, epoxy resins, epoxy resin bonded glass fiber, polyester resins, vulcanized fiber, synthetic resin bonded paper, SF_6 gas, air, vacuum and mineral oil. The air, oil, SF_6 gas or vacuum serves as the main insulation and arc quenching medium whereas ceramic or epoxy resin parts are used for mechanical support, bus bar insulation and arc chamber segments, etc. In low voltage breakers, synthetic resin moldings are used to carry the metallic parts.

1.4.3 Power Cables

The insulating materials used in power cables are paper or plastic tape, thermoplastic materials (such as PE, XLPE or PVC), silicon rubber, EPR, thermosetting resins, SF_6 gas and mineral oil. In oil filled cables, the inner conductor is insulated by lapped paper tape and impregnated with mineral oil. In polymeric insulated cables, the conductor and the insulating materials are extruded jointly and then insulation is cured and crosslinked. In gas-insulated cables, the inner conductor is held concentrically in a metallic tube by insulating spacers made of thermosetting resins and the tube is filled with pressurized SF_6 gas. Low voltage cables employing PVC, PE or XLPE insulation are normally without the other screen.

1.4.4 Bushings

Bushings are made of porcelain, glass, thermosetting cast resin, air, SF_6 gas, paper tape and oil, etc., and are constructed such that the feed-through conductor is insulated by paper tape and oil and is housed in a porcelain tube that enters the enclosure of the high voltage equipment. Two types of construction are normally used resulting in noncondenser and condenser graded bushings. Condenser graded bushings are used for rated voltages of over 50 kV, whereas noncondenser bushings are preferred for lower voltage applications. The paper tape used in bushings is usually resin bonded paper, oil impregnated paper or resin impregnated paper.

1.4.5 Overhead Lines

Overhead power lines use porcelain, glass, thermosetting resin, and air as the main insulation where the conductors are suspended via insulator chains from towers. Insulators are made out of porcelain or hard glass. Plastic insulator chains employing fiber glass and cast resins are also being used more in recent years. Room-temperature vulcanized rubber (RTV) coating is also being employed to improve the ceramic insulator's performance in polluted environments.

1.4.6 Gas Insulated Switchgear

Gas insulated switchgear (or GIS) use SF_6 gas, thermosetting resins and porcelain as the main insulating materials. In GIS construction, different components such as bus bars, interrupters and earthing switches are located in adjacent cylindrical compartments which are air tight, sealed and contain compressed SF_6 gas as the insulation medium. The inner live conductors are separated, at regular intervals, from the grounded enclosure by insulating spacers made of epoxy resins.

1.4.7 Surge Arrester and Protective Gaps

Lightning or surge arresters and protective gaps consisting mainly of air, SF_6, porcelain and metal oxide resistors are used to limit the transient overvoltages caused by lightning or switching actions in high voltage systems. In the simplest form, air insulated rod-rod chopping gaps are used across bushings, insulator chains, cable terminations or live conductors, etc. Alternatively, nonlinear resistors made of metal oxide (such as zinc oxide) with or without series spark gaps are used. In most recent designs, metal oxide surge arresters (MOAs) are used without any series spark gap. Ceramic or porcelain housing is used for mechanical support and for protection from the exposure to the environment.

1.4.8 Power Capacitors

Modern power capacitors consist of metallized polypropylene film, aluminum foil and polypropylene film, or metallized paper electrodes and polypropylene or other film and the impregnation fluid. The metallized foils represent the capacitance, the fluid minimizes the voids, increases the dielectric strength and sometimes the capacitance via its dielectric constant. The fluids used in recent years are isopropylbiphenyl, phenylxylylethane and silicone liquid impregnant. The whole assembly is housed inside a container with appropriate terminals or bushings for external connections.

1.4.9 Rotating Machines

Generators use mica tape system on conductors impregnated with either an epoxy or polyester resin. Other insulation materials normally used in rotating machines are polyvinyl acetal, polyester enamel or bonded fiber glass for inter-turn insulation; bakelized fabric, epoxy fiber glass, mica glass sheet, epoxy impregnated mica paper and varnished glass for inter-coil or phase-to-earth insulation; and bakelized fabric or epoxy fiber glass strips for slot closure. The impregnation treatment normally consists of alkyd phenolic estermide or epoxy based varnishes.

1.5 ELECTRIC FIELDS

Proper design of any high voltage device requires a complete knowledge of the electric field distribution and methods to control this field. Furthermore, for an understanding of the insulation failure modes, some knowledge of the electric field concepts is a prerequisite. A very brief introduc-

Introduction

tion to this subject is provided here; more details can be found in the references cited.

The electric field intensity E at any location in an electrostatic field is related to force F experienced by a charge q as F = qE. Moreover the electric flux density D associated with E is given as D = εE where ε = $\varepsilon_o \varepsilon_r$ is the absolute permittivity of the medium in which the electric field exists. If the medium is free of any space charge, the electric field is obtained from the solution of the Laplace equation:

$$\nabla^2 \phi = 0 \qquad (1.5)$$

where the operator ∇^2 is called the laplacian and ϕ is the potential which is related to E and path ℓ through which the charge is moved by

$$\phi = -\int E \cdot d\ell \qquad (1.6)$$

If the field medium has a space charge of density ρ_s, then the field is governed by the solution of the Poisson's equation:

$$\nabla^2 \phi = \frac{-\rho_s}{\varepsilon} \qquad (1.7)$$

When the medium under discussion is gaseous or vacuum, $\varepsilon = \varepsilon_o$.

1.5.1 Field Distribution Types

Broadly speaking, the field distribution in a region may be classified as homogeneous (or uniform) and nonhomogeneous (or nonuniform). In a homogeneous field, E is the same throughout the field region; whereas in a nonhomogeneous field, E is different at different points in the region. In the absence of space charges, the electric stress, E, in a nonuniform field gap usually obtains the maximum value at the surface of the conductor which has the smallest radius of curvature and achieves the minimum value at the conductor having the largest radius of curvature or the earth. In this case, the field is nonhomogeneous as well as asymmetrical. Most of the practical HV components used in electric power systems have nonhomogeneous and asymmetrical field distributions. Uniform or approximate uniform field distributions exist between two infinite parallel plates or two spheres of equal diameters with a gap spacing which is smaller than the sphere radius. Sphere electrodes are frequently used for high voltage measurements and in impulse voltage generation circuits. Similarly "profiled" parallel plates of finite sizes are also used to simulate homogeneous fields. In some gaps, which produce nonhomogeneous fields, the field along the gap axis may be symmetrical towards both electrodes with respect to the gap center. Examples of such nonuniform symmetrical fields are fields produced by similar diameter rod-rod or sphere-sphere gaps (with large distance between the spheres). When one of the electrodes of such symmetrical field gaps is earthed, the field symmetry may be disturbed due to the earthing effect. Consequently, the high voltage electrode has somewhat higher electric stress than the grounded electrode.

1.5.2 Methods of Field Estimation

In simple physical systems such as a single conductor above ground, two parallel conductors above ground, two equal diameter spheres, two infinitely long parallel plates, coaxial cylinders, and concentric spheres, it is possible to find an analytical field solution. However, for most high voltage components, the physical systems are so complex that it is extremely difficult to find an analytical field solution. In such cases, numerical methods are employed for electric field calculations. The existing methods include the finite difference method, the finite element method, the Monte Carlo method, the moment method, the method of images, the charge simulation method, the surface charge simulation method and combinations of these methods. Details of some of these methods have been discussed in the literature [5-10] and will not be repeated here. For complex field problems, experimental modelling using electrolytic tank, semiconducting paper or resistive mesh analog are also useful. Computer software packages are now available to carry out most of the field calculations.

1.5.3 Field Enhancement Factor

Whereas any designer of the high voltage apparatus must have a complete knowledge of the electric field distribution, for a user of the system the knowledge of the maximum value of the electric field E_{max} to which the insulation is likely to be subjected and the location of such a maximum gradient point is generally sufficient. Consequently, the concept of field enhancement factor or simply field factor f is of considerable use. This factor is defined as:

$$f = \frac{E_{max}}{E_{av}} \qquad (1.8)$$

where E_{av} is the average field in the gap and is equal to the applied potential difference divided by the gap separation between the electrodes. Values of f for most geometries of practical interest to a power system engineer are

Introduction

summarized in Table 1.2. These equations can either be derived analytically or are empirical equations derived from numerically computed stress values for a single homogenous dielectric medium [1,5,6,11,12,15–17]. From these equations one can easily estimate the maximum stress E_{max} to which a given dielectric material may be subjected when a voltage V is applied

Table 1.2 Field Enhancement Factor for Some Common Configurations

Configuration	f value	Comment
(a) Parallel plates	1	Uniform field
(b) Concentric spheres	R/r	R = outer sphere radius
(c) Coaxial cylinders	$(R - r)/(r \ln R/r)$	R = outer cylinder radius
(d) Hyperbolic point-plane	$2x/\ln(1 + 4x)$	
(e) Equal diameter spheres	$0.25[B + \sqrt{B^2 + 8}]$	Symmetrical voltages and B = x + 1
(f) Equal diameter spheres	$0.943 + 0.458x + 0.121x^2$	$x \leq 2$ and one sphere earthed
(g) Sphere-plane	$x + 0.55$	$x > 2$ and plane earthed
(h) Equal diameter parallel cylinders	$A/\left(2 \ln \left[1 + \dfrac{x}{2} + \dfrac{A}{2}\right]\right)$	$A = \sqrt{x^2 + 4x}$
(i) Conductor-plane	$2 + 0.32x$	$x \leq 5$
	$2.642 + 0.216x - 0.0002x^2$	$5 \leq x \leq 2000$
(j) Cylinder surrounded by a torus	$0.433 + 0.307 (R/r) + 0.0095 (R/r)^2$	R = torus radius, r = cylinder and torus cross-sectional radius
(k) Hemispherical rod-plane	$0.85 (1 + x)$	$x \leq 3$
	$\dfrac{0.45x \ln (6x)}{\ln (x)}$	$3 \leq x \leq 500$
(l) Bipolar DC line with 2 subconductors	$d \left(\dfrac{1}{r} + \dfrac{1}{D}\right)/\ln (4h^2/rD)$	d = distance between bundle centers
(m) 3-phase line with 2 subconductors	$d/\left(2r \ln \left(\dfrac{d}{\sqrt{rD}}\right)\right)$	d = distance between adjacent phases
(n) Two hyperboloidal points	$\dfrac{x\sqrt{1 + 2/x}}{\tan h^{-1} \sqrt{\dfrac{x}{x + 2}}}$	

d = gap length, r = high voltage electrode radius, x = d/r, h = height above ground, and D = bundle diameter.

across the two conductors with a gap spacing of d since $E_{max} = f \cdot V/d$. Instead of field enhancement factor f, sometimes the field utilization factor, $\mu_f = 1/f$, is used, which gives more insight about the effective use of the dielectric space since a larger value of μ_f represents a more compact equipment.

In a multidielectric media, the field computations become complicated, since in addition to the Laplace or Poisson equations, the boundary conditions (see section 1.5.6) must also be satisfied at the interface of the two dielectrics. Therefore, in such cases, except for some very simple configurations, numerical computations are essential. General values of f for such cases, therefore, cannot be given.

1.5.4 Field Intensification at Protrusions

Protrusions at the electrode surfaces are created during manufacture, installation or operation of many high voltage components. At the tip of such a protrusion, the field lines converge. Consequently, the microscope field at this tip, $E_p = fE$, becomes greater than the macroscopic gap field E. The field enhancement factor f in this case depends upon the shape and size of the protrusion. Equations for stress concentration have been derived for hyperbolic, spheroidal and ellipsoidal points. The stress enhancement factor at a conducting sphere surface in a uniform field (either an isolated sphere or a spherical boss on a conducting plane surface) is around 3. For a conducting ellipsoidal boss on a conducting plane or an isolated conducting ellipsoid in a uniform field f is given by [11]:

$$f = \frac{2n^3}{m \ln\left(\frac{m+n}{m-n}\right) - 2n} \tag{1.9}$$

where m is the ratio of the major to minor axis of the ellipsoid and $n = \sqrt{(m^2 - 1)}$. For a hyperbolic point to plane geometry, factor f is as given in Table 1.2. For needle-like geometries which have $\lambda > 10$, f can be approximated by [12]:

$$f = \frac{\lambda^2}{\ln \lambda - 0.3} \tag{1.10}$$

where λ = protrusion height/protrusion base radius. For some other geometries, the value of f depends upon the protrusion shape, its base height (h) as well as its radius of curvature or its base radius (r). As a limiting case, if h = r, i.e., a hemispherical protrusion, the value of f is approximately 3. The protrusions and surface defects play a prominent role in the

Introduction

initiation of partial discharges and ultimate breakdown of air and SF_6 insulation, polymeric insulated cables and vacuum insulated equipment, etc. Figure 1.2 shows values of f as a function of protrusion parameters for a number of protrusion configurations [7,20].

1.5.5 Field at the Interface of Composites

At the interface of two different dielectric materials (A and B) having dielectric constants of ε_A and ε_B, the tangential electric field and the normal flux density must be continuous. Therefore, in reference to Figure 1.3, the following conditions must be satisfied:

$$E_{At} = E_{Bt} \tag{1.11}$$

$$\varepsilon_A E_{An} = \varepsilon_B E_{Bn} \tag{1.12}$$

If there is a surface charge at the interface, equation (1.12) is modified as follows:

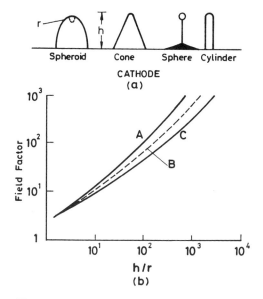

Figure 1.2 Typical protrusion geometries are shown in (a) and the resultant field enhancement factors for various values of h/r are shown in (b). Curve A is for a sphere or for a cylinder. Curve B is for a cone. Curve C is for a spheroid. (From Ref. 20.)

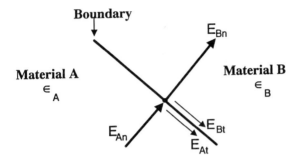

Figure 1.3 Boundary between two dielectrics.

$$\varepsilon_A E_{An} - \varepsilon_B E_{Bn} = \rho_s \quad (1.13)$$

where ρ_s is the surface charge density. It is important to note that for spacers and insulators used in air or SF_6 (A = spacer, B = air or SF_6), ε_A is typically 2 to 4 whereas $\varepsilon_B = 1$. Therefore, application of equations (1.12) and (1.13) show that the field is distorted at the solid/gas interface and the net field on the air side of the interface becomes larger, which makes such an interface the weakest link in the system. For DC voltage applications, the insulating materials can get charged. Similarly, there may be charging of insulator due to corona or other types of discharges giving rise to a surface charge density which can enhance the total surface field on the gas side. Similar arguments apply for voids or gas cavities inside solid or liquid insulating materials as discussed next. Therefore, each of such solid/gas, solid/vacuum, solid/liquid or solid/solid interfaces needs careful consideration.

1.5.6 Field Inside Cavities

Cavities can be generated inside solid or liquid dielectrics during manufacture or operation. Such cavities are usually spaces of lower density material or a gas pocket in a surrounding solid or liquid dielectric. Such gas spaces experience higher electric stress than the bulk liquid or solid media as a result of their lower dielectric constant. A cavity whose diameter perpendicular to the electric field is appreciably larger than its depth in the direction of the electric field, and when its depth is small compared to the total insulation spacing, experiences an electric field E_c under AC voltages of:

Introduction

$$E_c = \frac{\varepsilon_d}{\varepsilon_c} E_d \qquad (1.14)$$

where ε_d and ε_c are the relative permittivities of the dielectric and the cavity respectively, and E_d is the electric stress in the dielectric medium. For a gas cavity whose diameter is quite small in proportion to its depth in the electric field direction, the electric field in the cavity approaches nearly equal to that in the surrounding medium, i.e., $E_c = E_d$. For a spherical shaped gas cavity having cavity depth equal to cavity diameter, E_c is given as:

$$E_c = \left(\frac{3\varepsilon_d}{\varepsilon_c + 2\varepsilon_d}\right) E_d \qquad (1.15)$$

The electric stress in cylindrical cavities of various depths and diameters is given in [13]. For prolate and oblate spheroidal cavities, the exact electric stress in the cavities can be calculated [14].

The stress distribution under DC voltages will be dependent on time. As the voltage is raised, the initial distribution is determined as for AC voltages. However, it subsequently assumes a distribution dependent on the surface and volume resistivities of the cavity and the volume resistivity of the dielectric in series with the cavity. Consequently, the mean stress in the cavity under DC conditions can be determined from the calculations described above for AC voltages and by considering the relative permittivity of the solid dielectric to be very large, i.e., ε_d tends to ∞ [15]. Hence generally large stress values occur inside cavities under DC voltages than under AC applied voltages. Insulation is sometimes subjected to both AC and DC voltages simultaneously, or the DC voltage can contain an AC ripple component. In such a case, the cavity stress could be estimated by superposition principle. For fast impulse voltages, the stress distribution in cavities can be found by using the AC method discussed above.

1.5.7 Fields at Free Particles

In liquids and gaseous as well as vacuum insulation systems, free conducting or insulating particles may be introduced during manufacture, installation or operation. These particles can acquire some charge as a result of various mechanisms and may drift in the insulating medium. The electric stress at the ends of such particles can be enhanced, and consequently these particles can trigger breakdown of the insulation medium. The field enhancement factor for cylindrical metallic particles of radius r and length a is given as [18]:

$$f = \left[2 + \frac{a}{r}\right]\left[1 - \frac{a}{d}\right] \tag{1.16}$$

where d = gap separation. For free spherical metallic particles, the field factors are given in Figure 1.2. In the presence of ionic or electronic space charges and for particles of significant resistivity, the field factors are significantly influenced by the experimental parameters.

1.5.8 Electric Stress Control

Electric stress values have to be controlled in the design of high voltage equipment since a higher value of electric stress may trigger or accelerate the degradation and failure of the insulation. Thus, electrical stresses are controlled in cable terminations, high voltage bushings, potential transformers, etc. Methods have been suggested and employed with the aim of optimizing the stresses throughout different parts of the high voltage equipment in order to arrive at the most economical design. In addition, corona-free connections are desired during the testing of some high voltage equipment, thereby limiting the stress to values below critical values. Thus electrical stress control method are an important area and its details are available elsewhere [3,5,6,8,10,11].

1.6 DESIGN PARAMETERS OF HIGH VOLTAGE EQUIPMENT

Insulating materials play a critical role in the design and performance of high voltage power equipment. These materials must not only meet the dielectric requirements, but must also meet all other performance specifications including mechanical and thermal requirements, reliability, cost, ease of manufacture as well as environmental concerns. Table 1.3 summarizes the typical operating electric stress values of various types of power equipment and compares the mechanical forces experienced in operation, the relative complexity of the insulation and relative automation used in manufacture [19]. It is interesting to note that the highest operating electrical stresses are for capacitors where there is the simplest insulation configuration employing nearly uniform field distribution, highest manufacturing automation and where mechanical forces are fairly low. The design electrical stresses values given in Table 1.3 are determined by equipment test or basic impulse level (BIL) requirements. Generally, for economic reasons, the operating electric stress increases with system voltage whereas the design electric stress decreases with system voltage. Thus,

Introduction

Table 1.3 Relative Dielectric/Design Parameters of Typical High Voltage Power Equipment

Equipment	Operating stress (kV_{rms}/cm)	Design stress (kV_{peak}/cm)	Mechanical force (relative)	Insulation complexity (relative)	Manufacture automation (relative)
Generators	25	130	1.0	1.0	0.2
Transformers	15	115	1.0	0.9	0.3
SF_6 equipment	40	180	0.2	0.3	<0.1
Capacitors	600–1000	2000–3000	0.1	0.2	0.8

Source: Ref. 19.

as the system voltage increases, the equipment operates closer to its limits and presence of any defects can have serious consequences for the equipment's life expectancy.

REFERENCES

1. A. Bradwell (ed.), *Electrical insulation*, Peter Peregrinus Ltd., London, England, 1983.
2. A.C. Wilson, *Insulating Liquids—Their Uses, Manufacture and Properties*, Peter Peregrinus Ltd., London, England, 1980.
3. M.S. Naidu and V. Kamaraju, *High Voltage Engineering*, Tata McGraw-Hill, New Delhi, India, 1982.
4. A. White, "Design of High Voltage Power Transformers" in *High Voltage Engineering and Testing* by H.M. Rayan (ed.), Peter Peregrinus Ltd., London, England, 1994.
5. T.J. Gallagher and A.J. Pearmain, *High Voltage Measurement, Testing and Design*, John Wiley, New York, 1983.
6. M. Khalifa (ed.), *High Voltage Engineering: Theory and Practice*, Marcel Dekker, Inc., New York, 1990.
7. J.M. Meek and J.D. Craggs (eds.), *Electrical Breakdown in Gases*, John Wiley, New York, 1978.
8. E. Kuffel and W.S. Zaengl, *High Voltage Engineering Fundamentals*, Pergamon Press, New York, 1984.
9. M. Chari and P. Silvestor (eds.), *Finite Elements in Electrical and Magnetic Field Problems*, John Wiley, New York, 1980.
10. N.H. Malik, IEEE Trans. on Elect. Insul., Vol. 24, No. 1, pp. 3–20, 1989.
11. H. Bateman, *Partial Differential Equations of Mathematical Physics*, Cambridge University Press, New York, 1944.

12. R.V. Latham, *High Voltage Vacuum Insulation: The Physical Basis*, Academic Press, San Diego, 1981.
13. H.C. Hall and R.M. Russek, IEE Proc., Vol. 101, p. 47, 1954.
14. C.J. Bottcher, *Theory of Electric Polarization*, Elsevier Publishing Co., Amsterdam, The Netherlands, pp. 52–54, 1952.
15. R. Bastnikas (ed.), *Engineering Dielectrics—Vol. 1: Corona Measurement and Interpretation*, ASTM Press, Philadelphia, 1979.
16. Y. Qiu, IEEE Trans. on Elect. Insul., Vol. 21, No. 4, pp. 673–675, 1986.
17. L.L. Alston (ed.), *High Voltage Technology*, Oxford University Press, New York, 1968.
18. H.C. Miller, J. of Appl. Phys., Vol. 38, No. 11, pp. 4501–4504, 1967.
19. A.H. Cookson, IEEE Elect. Insul. Magazine, Vol. 6, No. 6, pp. 7–10, 1990.
20. P.A. Chatterton, Proc. Physical Society London, Vol. 88, pp. 231, 1966.

2
Gas Dielectrics

2.1 INTRODUCTION

Gases are the simplest and the most widely used dielectrics. In order to utilize these dielectrics efficiently, it is necessary to know their electrical behavior, especially the physical processes which lead to ionization and breakdown under different practical electrode systems. Each gas will breakdown at a certain electric stress. The breakdown voltage is defined as the peak value of the applied voltage at the instant of a spark discharge. This chapter highlights the basic mechanisms of gas breakdown in order to provide the base for understanding its engineering implications.

2.2 GAS BEHAVIOR UNDER ZERO ELECTRIC FIELD

In the absence of an electric field, a gas obeys Boyle's law, i.e.

$$Pv = C \tag{2.1}$$

where P and v are gas pressure and volume, respectively, and C is a constant which depends on the absolute temperature T and the mass m. However, v varies with T according to Gay Lussac's law, i.e.

$$\frac{v}{v_o} = \frac{T}{T_o} \tag{2.2}$$

where v_o and T_o are the initial values of volume and temperature, respectively. From equations (2.1) and (2.2) it follows that

$$Pv = n_k RT \tag{2.3}$$

where n_K is the number of kilomoles of gas and R is the universal gas constant, i.e., 8314 J/°K. Assume N_0 is the number of gas molecules in a mole (Avogadro's number) = 6.02×10^{23} molecules/mole, and N' is the number of total molecules in the gas which is equal to Nv, where N is the gas number density. If n_K in equation (2.3) is replaced by N'/N_0, and N' by Nv, it can be written as:

$$Pv = \frac{N'}{N_o} RT = \frac{Nv}{N_o} RT \tag{2.4}$$

Thus,

$$P = NkT \tag{2.5}$$

where $R/N_o = k$ = Boltzmann constant = 1.3806×10^{-23} K.

Using classical mechanics, with some assumptions, the kinetic energy of a gas can be related to its thermal energy by:

$$\frac{1}{2} m\nu^2 = \frac{3}{2} kT \tag{2.6}$$

where ν is the molecular velocity. In the absence of an electric field, the gas molecules will have random velocities ranging from zero to infinity. Maxwell derived the distribution function of molecular velocities and proved that this function is unique for a fixed gas temperature. Figure 2.1 shows this function, which can be expressed as [1]:

$$f(\nu_r) = \frac{4}{\sqrt{\pi}} \nu_r^2 \exp(-\nu_r^2) \tag{2.7}$$

where the relative velocity ν_r is given as $\nu_r = \nu/\nu_p$ and ν_p is the most probable velocity.

The average velocity $\bar{\nu}$ and the effective or rms velocity ν_{ef} can be related to ν_p as [1]:

$$\bar{\nu} = 1.128 \, \nu_p \tag{2.8}$$

$$\nu_{ef} = 1.224 \, \nu_p \tag{2.9}$$

Since the gas molecules move randomly they will collide with each other and with the walls of the container. The distance a particle traverses be-

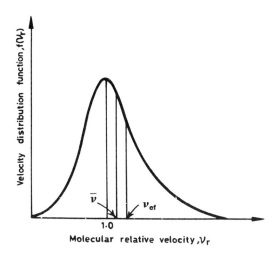

Figure 2.1 Maxwell distribution function for molecular velocities. (From Ref. 1 © Wiley, 1971.)

tween two successive collisions is called its free path (λ). Obviously free path values differ greatly, and hence the concept of mean free path ($\bar{\lambda}$) is more practical. If there are N_p pairs of particles of radii r_1 and r_2 then:

$$\bar{\lambda} = \frac{1}{\pi(r_1 + r_2)^2 N_p} = \frac{1}{\delta} \qquad (2.10)$$

where δ is the effective collision cross-section. For a certain gas, $\bar{\lambda}$ is directly proportional to gas temperature and inversely proportional to gas pressure. Table 2.1 shows values of $\bar{\lambda}$ for some selected gases.

2.3 GENERATION OF CHARGED PARTICLES

If a voltage is applied across two electrodes separated by an insulating gas, the conduction current will flow only if there are charged particles in the gas media. Such particles may include:

1. An electron $\bar{e}$ (negative charge)
2. A positive ion (neutral atom missing one electron) $A^+ = A - \bar{e}$
3. A negative ion (neutral atom with one excess electron) $A^- = A + \bar{e}$

Table 2.1 Values of the Mean Free Path $\bar{\lambda}$ of Some Gas Molecules Calculated from the Kinetic Theory of Gases at T = 288 K and P = 1013 mb

Gas	Molecular weight	$\bar{\lambda}$ (10^{-8} m)	Diameter (10^{-10} m)
H_2	2.016	11.77	2.74
He	4.002	18.62	2.18
H_2O	18.000	4.18	4.60
Ne	20.180	13.22	2.59
N_2	28.020	6.28	3.75
O_2	32.000	6.78	3.61
Ar	39.940	6.66	3.64
CO_2	44.000	4.19	4.59
Kr	82.900	5.12	4.16
Xe	130.200	3.76	4.85

These particles can be generated through various processes. Figure 2.2 shows the main processes which result in the generation of such particles in a gas discharge. Before discussing some of these processes, the Bohr's theories concerning the atomic structure and the energy levels are briefly summarized as follows.

1. The electrons can exist only in discrete stable orbits around the nucleus without radiating any energy. These stable orbits are located at a distance r from the center of the nucleus, where:

$$r = \frac{qh}{2\pi m_e \nu_e} \qquad (2.11)$$

where q = quantum number (an integer), h = Plancks constant = 6.6257 × 10^{-34} J·s, ν_e = electron velocity, and m_e = electron mass.

2. When the energy of an atom changes from a value W_1 to a lower value W_2, the excess energy is emitted as a quantum of radiation (photon) whose frequency (f_p) is related to Plancks constant by:

$$hf_p = W_1 - W_2 \qquad (2.12)$$

Generation of free electrons can result from ionization of neutral atoms or from detachment of negative ions. The ionization process needs a specified minimum amount of energy. If the energy absorbed by an atom is below this specified amount, it may lead to excitation, where the electron will not leave the atom but go to a higher energy level (or an outer orbit). Normally, excited states are not stable and the atom may absorb more energy to become ionized or it may go back to its original stable state by

Gas Dielectrics

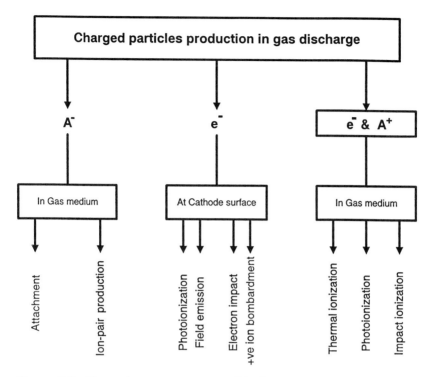

Figure 2.2 The main processes responsible for the production of the charged particles in a gas discharge.

radiating the excess energy. A brief description of the main ionization processes is given below.

2.3.1 Ionization by Collision (Impact Ionization)

When an electric field is applied between two electrodes and a free electron is present in the gap, the electron gains energy as it travels in the field towards the anode. While it traverses, it also collides with neutral gas molecules. If the electron energy is less than the ionization energy of the gas atom or molecule, an elastic collision may result. Alternatively, the gas atom may be excited. However, when the electron energy is larger than the ionization energy of the gas, an inelastic collision may occur, causing the ionization of the gas atom or molecule. In such inelastic collisions, each impact produces a positive ion and an extra electron. The positive ion is attracted to the cathode. Now these two electrons will gain energy from the field and may ionize further gas molecules. Depending upon the physical conditions, this process will either reach a stable condition, where a certain number of charged particles are generated and some conduction current flows, or the ionization process progresses to avalanche and then breakdown (see section 2.5). The following equations outline the ionization progression by this process:

$$A + \bar{e} \rightarrow A^+ + \bar{e} + \bar{e} \quad \text{or}$$
$$A + 1\bar{e} \rightarrow 2\bar{e} + A^+$$
$$2A + 2\bar{e} \rightarrow 4\bar{e} + 2A^+$$
$$4A + 4\bar{e} \rightarrow 8\bar{e} + 4A^+ \text{ and so on} \quad (2.13)$$

2.3.2 Photoionization (Ionization by Radiation)

An excited atom generally has a lifetime in the range of 10^{-7} to 10^{-9} s. When it returns to its ground state, it emits extra energy as a photon. Such low-energy photons may lead to ionization as follows:

$$A^* \rightarrow A + hf_p$$
$$B + hf_p \rightarrow B^+ + \bar{e} \quad (2.14)$$

Here A^* represents the excited state of an atom A and hf_p is the photon energy which is assumed to be more than that the ionization energy of atom B. External sources such as x-rays, nuclear radiation or cosmic rays may cause photoionization from much deeper energy levels within a molecule. Photoionization is an important process, especially in the breakdown of gas mixtures composed of the rare gases whose excited states can have long lifetimes.

2.3.3 Thermal Ionization

A sufficient increase in gas temperature will cause its particles to move faster, and consequently may cause ionization on collision between gas atoms or molecules. Thermal energy (W_t) can cause ionization by itself at high temperature like in flames and arcs as follows:

$$A + W_t \rightarrow A^+ + \bar{e} \quad (2.15)$$

2.3.4 Electron Detachment

An electron may detach from negative ion as follows:

$$\bar{A} \rightarrow A + \bar{e} \quad (2.16)$$

Although the number of charged particles do not increase in this process,

Gas Dielectrics

detachment can be considered as an ionization process since in this process the slowly moving negative ions are converted to fast-moving electrons. Since the kinetic energy is directly proportional to mass and to the square of the speed, the lighter but fast-moving electrons will have much higher kinetic energy compared to the heavier but slow-moving negative ions. Therefore electrons can cause further ionization more effectively than the negative ions.

2.3.5 Cathode Processes

The charged particles can also be supplied from the electrodes especially the cathode. At normal state, the electrons are bound to the solid electrode by electrostatic forces between electrons and ions in the lattice. For the electron to leave the cathode, a minimum specified energy, known as work function, is required where its value depends on the material. The source of the energy required can be one or combination of the following.

Positive Ion and Excited Atom Bombardment

When a positive ion has an impact on the cathode, an electron is released provided the impact energy is equal to or more than twice the cathode work function. At least two electrons will be released; one will neutralize the positive ion and the other will be ejected to the gas medium. An electron might also be emitted as a result of the bombardment of cathode by neutral excited atoms or molecules.

Photoemission

If the energy of a photon striking the cathode surface is higher than the cathode work function, an electron may be ejected from the cathode.

Thermionic Emission

Raising the cathode temperature to a very high value (around 2000 K) will lead to some electrons leaving its surface since the violent thermal lattice vibrations will provide the electrons with the required energy. The thermionic emission process has been widely used since the early days of electronics.

Field Emission

A high electrostatic field may overcome the binding force between electrons and protons and lead to the liberation of one or more electrons from the cathode. This takes place when the electric field value is of the order

10^7–10^9 V/cm. Most of power system components do not operate at such high stress values. However, conditions for field emission can exist at electrode protrusions and microdefects.

2.4 DEIONIZATION PROCESSES

Deionization is the process by which the number of charged particles in a gas volume, especially the electrons, decreases. Since these processes oppose the ionization, in some applications they are desirable, e.g., to prevent the avalanche growth or to quench an arc. The main deionization processes are briefly described here.

2.4.1 Diffusion

Charged particles move from the region which has a higher concentration to the region which has a lower concentration. The general diffusion equation is given as:

$$J = -D\nabla n \qquad (2.17)$$

where J = the rate of charge flow, ∇n = charge concentration gradient, and D = diffusion constant, which is expressed as [1]:

$$D = \frac{\bar{\lambda} \cdot \bar{\nu}}{3} \qquad (2.18)$$

For electrons D will be three orders of magnitude higher than that of ions due to their higher mean velocity $\bar{\nu}$. When time is taken into account, the rate of change of ion density n is given as [1]:

$$\frac{\partial n}{\partial t} = -\nabla \cdot \bar{J} = D \cdot \nabla^2 n \qquad (2.19)$$

Solution of equation (2.19) will give the concentration of ions (n) at any time and at any point. If this equation is solved for the case of diffusion from a cylindrical concentration, the average displacement (r_d) will be given as [1]:

$$r_d = \sqrt{4Dt} \qquad (2.20)$$

2.4.2 Recombination

Positive and negative ions combine to form neutral atom as:

$$A^+ + B^- \rightarrow AB + hf_p \qquad (2.21)$$

Gas Dielectrics

The rate of recombination is directly proportional to the concentration of both positive n_+ and negative n_- ions, i.e.

$$\frac{dn_+}{dt} = \frac{dn_-}{dt} = -\rho n_+ n_- \qquad (2.22)$$

where ρ is the recombination constant. Since in general $n_+ = n_- = n$, it follows that:

$$\frac{dn}{dt} = -\rho n^2 \qquad (2.23)$$

If this equation is integrated with respect to time, the density of charged particles at any given time will be:

$$n(t) = \frac{n_o}{1 + n_o \rho t} \qquad (2.24)$$

where n_o is the initial concentration.

2.4.3 Electron Attachment

Some electronegative gases such as O_2, CO_2 and SF_6 attach slow moving free electrons to neutral gas molecules and form heavy negative ions. Hence electron attachment, which constitutes a deionization process, can be written as $A + \bar{e} \rightarrow \bar{A}$. This process is opposite to the detachment process described in section 2.3.

2.5 UNIFORM FIELD GAS BREAKDOWN

The application of a specific electric field E across a dielectric gas will result, at first, in its ionization. As the time and/or the applied field increases, the ionization may lead to other physical processes, as shown in Figure 2.3. The exponential growth of ionization usually leads to avalanche formation. Such avalanches may, in turn, result in the formation of streamer, leader, spark, arc or plasma. The electric field can be uniform or nonuniform depending on the electrode arrangement. In case of a uniform field gap, the electric stress is the same everywhere and hence the ionization and deionization parameters are constant. Therefore, the physics of uniform field breakdown can be understood more easily. The two most widely accepted theories of breakdown are the Townsend theory (1910) and the streamer theory (1940). These are briefly described next.

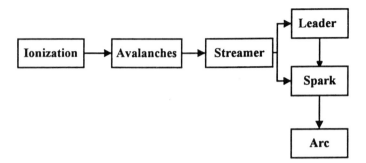

Figure 2.3 The various possible discharge process in a gaseous insulation.

2.5.1 Townsend Theory of Breakdown

Townsend investigated the ionization and breakdown under DC voltage conditions, and proposed a theory to explain the experimental observations. Considering the circuit of Figure 2.4, he assumed that n_o electrons are being emitted from the cathode per second by the ultraviolet light. The Townsend's first ionization coefficient α is defined as the number of ionizing collisions made by an electron in per unit distance as it travels in the direction of the applied field. It depends on gas pressure and electric field. The number of electrons at a distance x from the cathode will be n_x and is given by:

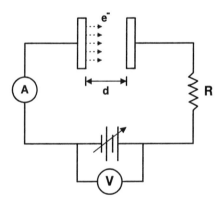

Figure 2.4 Experimental circuit for the study of Townsend discharge.

Gas Dielectrics

$$\frac{dn_x}{dx} = \alpha\, n_x \quad (2.25)$$

and hence

$$n_x = n_o \exp(\alpha x) \quad (2.26)$$

Similarly, n_d, the number of electrons reaching the anode placed at distance d, will be given by:

$$n_d = n_o \exp(\alpha d) \quad (2.27)$$

The number of new electrons created on the average by each primary electron leaving the cathode is:

$$\frac{n_d - n_o}{n_o} = \exp(\alpha d) - 1 \quad (2.28)$$

The average current in the gap, which is numerically proportional to the number of electrons traveling per second, will be:

$$I = I_0 \exp(\alpha d) \quad (2.29)$$

where I_0 is the initial current at the cathode. The above constitutes a single avalanche process. During the amplification of electrons in the field by α process, additional electrons are being liberated in the gap by other (secondary) processes as well. The secondary electrons thus produced create their own avalanches. Secondary processes include positive ion bombardment on cathode and photo-ionization and detachment. Townsend second ionization coefficient γ is defined as the net number of secondary electrons produced per primary electron leaving the cathode. γ is a function of E/P as well as electrode material, etc. The influence of secondary process on the current growth can be considered as follows. Let n_s = number of secondary electrons produced at cathode per second, and $n_t = (n_o + n_s)$ = total number of electrons leaving the cathode per second. Therefore, the total number of electrons reaching the anode becomes

$$n_d = n_t \exp(\alpha d) = (n_o + n_s) \exp(\alpha d) \quad (2.30)$$

However, by definition

$$n_s = \gamma(n_d - n_t) \quad (2.31)$$

Thus,

$$n_s = \gamma[n_d - (n_o + n_s)] \quad (2.32)$$

Rearranging equations 2.30 and 2.32, we get:

$$n_d = \frac{n_o \exp(\alpha d)}{1 - \gamma[\exp(\alpha d) - 1]} \quad (2.33)$$

Thus, current growth in the presence of α and γ processes is given as:

$$I = \frac{I_0 \exp(\alpha d)}{1 - \gamma[\exp(\alpha d) - 1]} \quad (2.34)$$

At breakdown $I = \infty$, since the current is only limited by the resistance of the external circuit. This condition is called Townsend breakdown criterion and can be written as:

$$\gamma[\exp(\alpha d) - 1] = 1 \quad (2.35)$$

Normally $\exp(\alpha d) \gg 1$, therefore the above equation becomes:

$$\gamma \exp(\alpha d) = 1 \quad (2.36)$$

Since α and γ are dependent on E and P, thus for a certain value of d there will be a value of E and hence V which will satisfy the Townsend breakdown criterion. The voltage V which satisfies the breakdown criterion is called the sparkover or breakdown voltage V_s and the corresponding distance d is called the sparking distance.

For electronegative gases where electron attachment takes place in addition to α and γ processes, the attachment coefficient is also considered in the current growth equation. Consequently, Townsend breakdown criterion is also modified for such cases [1]. Townsend mechanism explains breakdown phenomena only at low pressures corresponding to Pd ≤ 1.45 bar-cm for air. For gaps with larger Pd values or for breakdown under fast surge voltages, usually the streamer theory of gas breakdown applies.

2.5.2 Streamer Theory of Breakdown

Townsend theory fails to explain some experimental observations such as: the zigzag and branched paths of the spark channel, and short breakdown times when gaps are overstressed or have large Pd values. Due to these and other limitations, streamer theory was proposed. This theory can be briefly summarized qualitatively as follows.

1. In the uniform field gap of Figure 2.5, an electron avalanche consisting of fast moving electrons and slow moving positive ions is generated as discussed earlier. Thus, a space charge field opposing the main field is formed. Soon the electrons will be absorbed by the anode leaving behind an accumulation of positive ions. Excitation of atoms also takes place as a result of collisions during the primary avalanche and photons are emitted from the excited atoms.

Gas Dielectrics

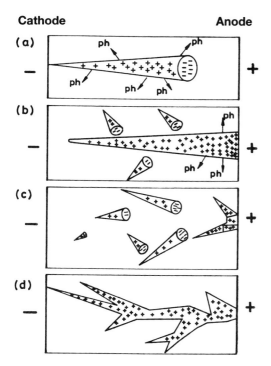

Figure 2.5 The development of avalanche to a streamer in a uniform field (+ = positive ions, − = electrons, ph = photons emitted from the avalanche).

2. These photons will be absorbed by gas atoms, and photoelectrons are produced in the gas at various distances from the avalanche (Figure 2.5a). If the space charge field mentioned earlier is of the same order of magnitude as the original applied field, then a second generation of auxiliary avalanches will be started by some of the most suitably located photoelectrons.

3. As auxiliary avalanches are formed, further photons and consequently photoelectrons are produced. Thus new third generation auxiliary avalanches will be created (Figure 2.5b). The electrons do not follow the original field lines any longer because space charge field distorts the original applied field. Besides, many avalanches may be created almost simultaneously. This is the cause of the observed branching and the zigzag paths of breakdown channels (Figure 2.5c). The auxiliary avalanches will be continuously absorbed by the primary avalanche and positive ions space charge will grow towards the cathode. Hence the ionized channel extends from anode to cathode. The ionized channel is called a streamer. The streamer tip forms branches that grow as a result of the incoming ava-

lanches (Figure 2.5b). Those electrons at the tips will soon be absorbed by the streamer and move in the channel towards the anode by virtue of a potential gradient within the streamer channel.

4. The propagation of one streamer tip continues while the others stop advancing due to the lack of avalanches feeding into them as shown in Figure 2.5c. If this process continues, a final streamer channel will be formed between the anode and the cathode, causing a complete breakdown. This channel will be similar to the one sketched in Figure 2.5d with numerous "incomplete" branches.

Once a streamer is formed, it usually leads to breakdown quickly. Mathematically an empirical streamer breakdown criterion for uniform field gaps can be formulated as:

$$\int_0^d \alpha \cdot dx = n_c \qquad (2.37)$$

where n_c is the critical number of electrons or ions in an avalanche when it transforms into a streamer. Usually it is believed that $n_c \approx 10^8$ for air and other gases.

2.5.3 Paschen Law

Based on the Townsend breakdown criterion, a relation between breakdown voltage V_s and the product of pressure and gap spacing can be established. This relation can be deduced as follows.

Both α and γ depend upon electric field E and gas pressure P. Therefore we can write

$$\frac{\alpha}{P} = f_1\left(\frac{E}{P}\right); \ \gamma = f_2\left(\frac{E}{P}\right) \qquad (2.38)$$

and since in a uniform field, $E = \frac{V}{d}$,

$$\frac{\alpha}{P} = f_1\left(\frac{V}{Pd}\right); \ \gamma = f_2\left(\frac{V}{Pd}\right) \qquad (2.39)$$

Substituting equation (2.39) in equation (2.35) leads to:

$$f_2\left(\frac{V}{Pd}\right)\left[\exp\left\{Pd\ f_1\left(\frac{V}{Pd}\right)\right\} - 1\right] = 1 \qquad (2.40)$$

There is only one value of V for a particular Pd value which satisfies equation (2.40). This value of V is the breakdown voltage V_s which can be written as:

Gas Dielectrics

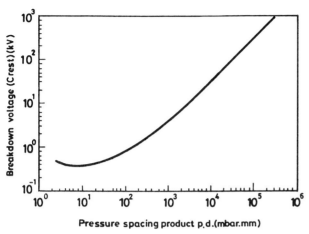

Figure 2.6 Paschen curve for air.

$$V_s = f(Pd) \tag{2.41}$$

Equation (2.41) shows that for a particular gas, the breakdown voltage is a unique function of the product of pressure and gap length. This relation is known as the Paschen law. The Paschen curve for air is shown in Figure 2.6, which shows a minimum value of V_s around a particular value of the product Pd. Table 2.2 shows minimum V_s values ($V_{s\ min}$) for some gases along with the corresponding Pd values [2].

Table 2.2 Minimum Sparking Potential for Various Gases

Gas	$V_{s\ min}$ (V)	Pd at $V_{s\ min}$ (Pa-cm)
Air	327	75.6
Ar	137	120
H_2	273	153.3
He	156	533
CO_2	420	68
N_2	251	89
N_2O	418	66.65
O_2	450	93.3
SO_2	457	93.3
H_2S	414	80

2.6 NONUNIFORM FIELD GAS BREAKDOWN

Any gas can withstand a certain electrical stress, and when the stress exceeds this value, a discharge will ensue. Since the stress in a uniform field gap is equal everywhere, discharge in such gaps usually takes the form of a complete breakdown. However, in non-uniform field gaps, the discharge will take place only in the areas where the stress is higher than the dielectric strength of the gas. This is known as a partial discharge (PD), and when it occurs at electrodes in air or other gases it is called "corona." In high voltage systems, often it is not economical to design equipment that is free of corona at nominal working voltages. The knowledge of corona onset voltage and the physical damage caused by corona is therefore important.

2.6.1 Corona Inception

Corona onset field is the critical stress value at the conductor surface corresponding to the corona inception. For power frequency applications, the following relations are generally applicable for corona inception.

1. For a single conductor above ground, the critical field (E_c) which will cause stable corona on the conductor is given as:

$$E_c = 30 \, m_s RAD \left(1 + \frac{0.301}{\sqrt{RAD \cdot r}}\right) (kV_{peak}/cm) \qquad (2.42)$$

where r = radius of the conductor, RAD = relative air density and m_s = surface irregularity factor (m_s = 1 for smooth conductors whereas for rough conductors its value is less than 1). The corona onset voltage V_c in this case can be written as:

$$V_c = E_c r \ln\left(\frac{2H}{r}\right) \qquad (2.43)$$

where H is the height of the conductor above ground.

2. For coaxial cylinders of inner and outer radii, r_1 and r_2, respectively, E_c and V_c are given as:

$$E_c = 31 \, m_s \, RAD \left(1 + \frac{0.308}{\sqrt{r_1 \cdot RAD}}\right) (kV_{peak}/cm) \qquad (2.44)$$

$$V_c = E_c \, r_1 \ln\left(\frac{r_2}{r_1}\right) \qquad (2.45)$$

In the above equations, relative air density (RAD) at pressure P (mbar) and temperature T (K) is given as:

Gas Dielectrics 37

$$\text{RAD} = \left(\frac{P}{1013}\right)\left(\frac{293}{T}\right) \qquad (2.46)$$

2.6.2 Corona Discharges

Corona discharges are best investigated using a sphere (or rod) to plane electrode configuration where the sphere (or rod) radius is chosen according to the field non-uniformity desired. Depending on the applied voltage and the shape of the electrode, there are six possible modes of corona: three for the positive DC or +ve half cycle of the AC, and three for the negative DC or −ve half cycle of the AC voltage. Table 2.3 and Figure 2.7 illustrate the occurrence of various corona modes in air, as they appear with an increase in the applied voltage.

The voltage values for positive and negative corona onset are approximately similar, however the transition between different modes occurs at different voltages for the two polarities. Depending on the rod radius and the interelectrode gap spacing, one or more of the above-mentioned modes may be absent for some electrode arrangements. In addition, two modes may occur simultaneously in some cases. Brief descriptions of different corona modes are given next.

Negative Corona

Trichel Pulses
As the voltage is raised up to the critical field intensity, electron avalanches are formed and propagate towards the anode leaving behind positive ions. When the electrons enter the low field region, they form negative ions. Thus, a space charge field is formed as shown in Figure 2.8. Here E_o is the applied electrostatic field and E_s is the field due to space charges. The discharge stops when the effective field ($E_o - E_s$) in the vicinity of the sharp electrode drops below the critical field value. After the space charges clear the gap, the field in the vicinity of the rod recovers and the process

Table 2.3 Corona Modes

Negative corona	Positive corona
Trichel pulses (TP)	Onset pulses (OP)
−ve pulseless glow (−ve G)	+ve Hermstein glow (+ve G)
−ve prebreakdown streamer (−ve st)	+ve prebreakdown streamer (+ve st)
Complete breakdown	Complete breakdown

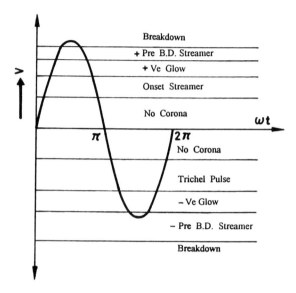

Figure 2.7 AC cycle with the various possible corona modes.

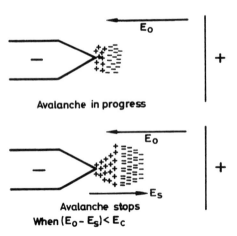

Figure 2.8 Negative corona avalanche.

Gas Dielectrics

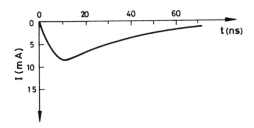

Figure 2.9 Negative corona current (Trichel pulse).

is repeated again. This leads to the formation of corona current pulses as shown in Figure 2.9. Upon raising the applied voltage, the number of pulses per second generally increases. In addition, the pulse amplitude decreases.

Negative Glow

Upon increasing the voltage further, at a certain voltage the space charge field becomes incapable of reducing the effective electric field intensity in the vicinity of the cathode below the critical value. Hence the discharge becomes continuous causing pulseless current. This mode is known as glow corona (Figure 2.10).

Negative Prebreakdown Streamers

Upon increasing the voltage further, the discharge occupies a major portion of the interelectrode spacing and incomplete streamers are formed. Such streamers are similar to the incomplete streamers already discussed under uniform field breakdown (section 2.5). This corona mode is known as prebreakdown streamer corona. Upon increasing the voltage further, a complete breakdown will take place.

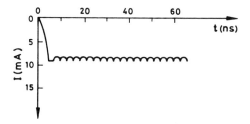

Figure 2.10 Negative glow corona current.

Positive Corona

Onset Streamers

As the voltage is increased till the critical field is established at the anode, electrons are accelerated from the low field region towards the anode and cause ionization in the high field region. At anode, the electrons will be absorbed quickly whereas the positive ions accumulate around the anode (Figure 2.11). The discharge stops when the effective field near the anode drops below the onset conditions. However, the discharge restarts when the positive ions are cleared away from the anode towards the cathode and the field in the vicinity of anode recovers to a higher value. In some special cases, streamers extending tangentially onto the anode can be formed. These are called burst pulse streamers. Such pulses are characterized by slow rise time and small magnitudes.

Positive Glow

As the voltage is increased further, the field at the anode will be high enough to cause discharge even when there is a positive space charge near the anode. In this case a continuous current will flow from glow corona.

Positive Prebreakdown Streamers

Upon raising the voltage further, incomplete streamers are formed resulting in prebreakdown streamer mode of corona. As the voltage is further increased, complete breakdown occurs in the gap.

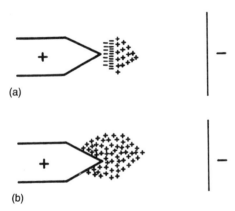

Figure 2.11 Positive corona avalanche: (a) avalanche in progress; (b) avalanche stops.

In power systems, corona occurs on overhead transmission lines and can display all of the modes discussed earlier due to sinusoidal nature of the AC voltage. Such corona increases the attenuation of high voltage surges as they propagate on the line conductors. However, it causes radio interference, audible noise and power losses. Corona also finds many industrial applications, such as in high-speed photocopying machines, in precipitators used for gas pollution control, in Van de Graff HVDC generators and in discharging undesirable charges from airplanes and plastics, etc. Recently pulsed corona is being used for air and water purification and for many other industrial applications.

2.6.3 Breakdown in Nonuniform Field Gaps

In a nonuniform field gap, α and γ are no longer constants. These vary with the field between the two electrodes and hence equations for current growth have to account for such a position related dependence of α and γ. If the field factor, $f = E_{max}/E_{av}$, is less than five, the field is called quasi-uniform and the discharge phenomenon would be similar to that in a uniform field gap. However, if $f > 5$, the field is called nonuniform. In nonuniform field gaps, the corona plays an important role in the final breakdown of the gap and hence breakdown voltage is strongly influenced by the presence/absence of corona as well as by the prevailing corona mode(s).

Once the corona starts, the applied electric field becomes distorted by the space charge field, and hence the breakdown process becomes much more complex. If the highly stressed electrode is positive, the space charge acts as an extension of the anode. On the other hand, if the highly stressed electrode is the cathode, the space charge acts as a shield that decreases the field in its vicinity and thus such a configuration needs a higher voltage for complete breakdown. Consequently, in nonuniform field gaps, breakdown voltage is lower for positive polarity as compared to the negative one. Therefore, the breakdown voltage characteristics for positive polarity direct voltage applied to the sharp point is usually more important for practical applications. Figure 2.12 illustrates schematically the dependence of the breakdown voltage on the gap length and the sphere diameter for sphere-plane air gaps under positive DC voltage. There are three main regions where the transition between them is sphere diameter dependent:

Region I. For short gaps, the field is almost uniform and the breakdown voltage depends mainly on the gap length. There is no corona in this case.

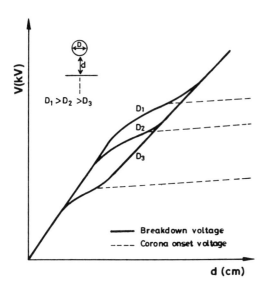

Figure 2.12 Corona and breakdown voltage as a function of gap spacing for sphere to plane geometry.

Region II. For moderate gap lengths, the field shows moderate nonuniformity. No corona occurs in this region also and the breakdown voltage increases with the sphere diameter as well as with the gap length.

Region III. If $d > 2D$, the field is highly nonuniform and the breakdown is preceded by corona. The corona onset voltage depends mainly on the sphere diameter while breakdown voltage is gap length dependent.

2.7 TIME TO BREAKDOWN

The breakdown process develops over a certain period of time like any other natural phenomenon. Generally this time is very short and not noticeable under DC or AC applied voltages. However, under impulse voltages, this time becomes important, since it may be comparable with the impulse voltage front time or its duration. Impulse voltages are generally used to simulate the lightning or switching surge overvoltages that occur in power systems. In most cases, the insulation design of power system components is based mainly on the magnitude of surge overvoltages.

Gas Dielectrics

The time lag between the instant when the applied voltage is of sufficient magnitude to cause breakdown and the actual event of breakdown can be divided into statistical (t_s) and formative (t_f) time lags. The former is the time required for an initiatory electron to appear in the highly stressed region of the gap after the application of an impulse. The formative time lag is the time necessary for the breakdown process to be completed after the initiatory electron becomes available (see Figure 2.13). Thus the impulse breakdown voltage V_i is higher than static or DC breakdown voltage V_s. The total time lag, $t = t_s + t_f$, will depend on the overvoltage value, $\Delta V = V_i - V_s$, and it will decrease as ΔV increases. The statistical nature of time lag leads to a probabalistic variation of the breakdown voltage. Consequently, varying shapes of volt-time characteristics are noticed for different surge voltages. Its engineering significance will be highlighted in the next chapter.

2.8 DISCHARGES UNDER NANOSECOND PULSE VOLTAGES

In the presence of electric field, the number of electrons grow exponentially since $n_x = n_o \exp(\alpha x)$. At the critical avalanche length x_c, the number of electrons at the avalanche head becomes $n_x = n_c$, where $n_c \approx 10^8$ for air. The avalanche space charge field becomes comparable to the applied electrostatic field and conditions for the formation of a streamer are satisfied. Once a streamer forms, breakdown may take place. For the most conservative case, $n_o = 1$, and the critical avalanche length is given as

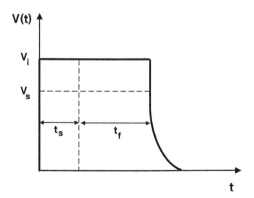

Figure 2.13 Time lag components under a step voltage.

$$X_C = \frac{\ln n_c}{\alpha} \quad (2.47)$$

If $X_C > d$ (gap length), then the primary avalanche is insufficient to transform into a streamer and cause a breakdown. Therefore, subsequent avalanches (2nd, 3rd, etc.) are necessary to cause a breakdown. If $X_C < d$, a primary avalanche can become a streamer provided the avalanche generates enough photons to ionize the gas molecules near the avalanche head. Such photons are produced when excited molecules go to the ground state in about 10^{-9}–10^{-7} seconds [1]. This de-excitation time (T_{dex}) should be more or at least comparable with the electron drift time T_e, which is given as:

$$T_e = \frac{\ln n_c}{\alpha \, \nu_e} \quad (2.48)$$

where ν_e is electron drift velocity in the avalanche. Thus, from equations (2.47) and (2.48) the conditions for streamer formation can be summarized as:

$$X_C = \frac{\ln n_c}{\alpha} < d; \; T_e = \frac{\ln n_c}{\alpha \, \nu_e} > T_{dex} \quad (2.49)$$

These conditions can be achieved if a high voltage pulse of nanosecond duration is applied to the electrode gap. To complete a discharge under such voltage conditions, a large number of electron avalanches must be produced by photoelectrons near the cathode. In multi-avalanche initiation, a space charge current comparable with the maximum circuit current is possible. To reach this condition, an electric field must be applied with a value much higher than that associated with DC, AC and switching or lightning impulse voltages. Using nanosecond high voltage pulses, very intense corona can be produced with high applied electric fields. One of the advantages of nanosecond pulsed discharge is the low energy consumption, since only electrons are accelerated while there is insufficient time for ions to speed up. In such circumstances the gas temperature remains low and the spark is avoided except at extremely high applied field values. Nanosecond pulse discharges are finding increasing applications in recent years. The generation of high voltage pulses of nanosecond duration and their applications will be outlined in chapter 10.

2.9 GAP-TYPE DISCHARGE

A gap-type discharge is a complete electrical breakdown of the insulation between two charged surfaces. It occurs in the form of a spark at microgaps

(<1 mm) where the potential gradient is large enough to initiate a discharge. Either both electrodes of the gap or at least one electrode is capacitively coupled to a voltage source or to ground. Figure 2.14 shows a typical arrangement for gap-type discharge where one of the electrodes is capacitively coupled to the high voltage source. Associated with this discharge is a current pulse characterized by a very high crest value (from a few amperes to several tens of amperes) and extremely short rise time (≈ 1.0 ns). This discharge could be repetitive depending on the gap location. This type of discharge is a broad spectrum electromagnetic noise source (up to hundreds of MHz) which can be radiated or conducted over long distances causing severe interference in nearby computer, communication and control systems [6,7].

The gap-type discharge is not restricted to EHV lines but it could also occur on an improperly designed or constructed distribution system provided that the stress in the gap reaches a certain critical value. The following are some examples of gap-type discharges where the spark may occur between bad contacting metallic parts exposed to high voltages: (1) between caps and pins of an insulator, (2) between cross-arms loosely attached to the wooden distribution towers, (3) at the junctions of insulator strings and the transmission line towers and (4) between the spacers and subconductors of a transmission line's bundle conductors. Bad contacts between metallic parts can be caused by corrosion, dust or dry pollution. Other factors causing miscontact are vibrations due to wind, insufficient mechanical loading and the changes in dimensions due to temperature var-

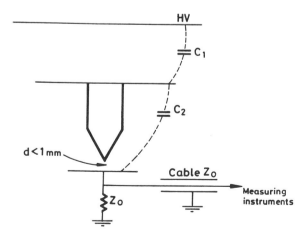

Figure 2.14 Gap discharge arrangement.

iations. A microdischarge may also occur during rain in a microgap between a water drop and the transmission line, or between two water drops close to the transmission line [8,9]. Figure 2.15 displays a typical discharge current waveform resulting from a gap-type discharge.

2.10 CHOICE OF DIELECTRIC GASES

The desirable properties of an insulating gas were summarized earlier in section 1.3.1. SF_6 and air are the most commonly used gases for insulation in high voltage power systems as both of these gases possess most of the desired properties. Although SF_6 has been used satisfactorily and almost exclusively for pressurized gas-insulated high voltage transmission and substation equipment, there have been many investigations into alternative gases. The gas with higher dielectric strength offers the possibility of operating existing designs at higher voltages or reducing the equipment size. Reduction in size means less quantity of gas and lesser amount of sheathing and other insulating materials, etc. The other properties considered for a better gas are its immunity toward particle-initiated breakdown, to surface roughness and area effects [9]. Similarly a gas to be used in circuit breakers as a replacement for SF_6 must possess better arc interrupting properties as well.

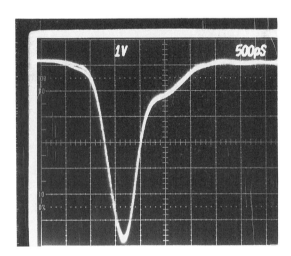

Figure 2.15 Typical gap discharge current.

The dielectric strength of a gas depends on its ability to attach free electrons over a wide energy range. However most dielectric gases capture electrons effectively only in a very restricted low energy range; e.g., SF_6 attaches electrons at energies ≤ 0.4 eV. Beyond this energy range, SF_6 electron attachment cross section (σ_a) decreases rapidly [10], so electrons that possess higher energies escape the attachment. Behavior of σ_a of SF_6 also explains the considerable decrease in the dielectric strength of SF_6 under nonuniform fields, surface roughness, and particle contamination (see chapter 4 for details) [11]. In new gases, a higher σ_a is sought and considerable research has resulted in a variety of new insulating gases. Table 2.4 shows few properties of some dielectric gases. It is clear that insulating gases such as perfluorocarbons that possess higher σ_a exhibit

Table 2.4 Relative Dielectric Strength of Some Gases in Uniform Fields at 1.5 bar ($SF_6 = 1$)

Name	Formula	Boiling point (°C)	Dielectric strength
Hexafluoro-2-butyne	C_4F_6	−25	2.2
Trifluoromethyl sulfur pentafluoride	CF_3SF_5	−20	1.55
Trifluoromethane sulfonyl fluoride	CF_3SO_2F	−22	1.49
Thionyl fluoride	SOF_2	−44	1.42
Trifluoronitromethane	CF_3NO_2	−31	1.34
Perfluoro-n-butane	C_4F_8	2	1.5
Chloropentafluoroethane (F-115)	C_2F_5Cl	−39	1.13
Perfluoro-butane	C_4F_{10}	−2	1.06
Sulfur dioxide	SO_2	−10	1.0
Sulfur hexafluoride	SF_6	−64	1.0
Dichlorodifluoromethane (F-12)	CCl_2F_2	−30	0.99
Octofluoropropane (F-218)	C_3F_8	−37	0.98
Perfluorodimethyl ether	C_2F_6O	−59	0.84
Hexafluoroethane (F-116)	C_2F_6	−78	0.79
Bromotrifluoromethane (F-1381)	$CBrF_3$	−58	0.75
Chlorotrifluoroethylene (CTFE)	C_2F_3Cl	−28	0.69
Nitrous oxide	N_2O	−89	0.5
Carbon tetrafluoride (F-14)	CF_4	−128	0.42
Air	$N_2 + O_2$		0.37
Nitrogen	N_2	−196	0.37
Carbon dioxide	CO_2	−79	0.32

Source: Ref. 9.

higher breakdown voltages. Besides satisfactory dielectric strength and cost, properties like toxicity, low vapor pressure, boiling point, tendency to carbonize and environmental hazards are also important. As shown in Table 2.4, perfluoro-n butane (C_4F_8) possesses a value of dielectric strength which is higher than SF_6 by a factor of 1.5. But its high boiling point (+2.0°C) as compared to −64°C for SF_6 prevents its application in cold climates or in a compressed state. The greatest potential for new practical gaseous insulators therefore lies in the development of synergistic multi-component mixtures, the components of which are selected on the basis of their individual physico-chemical properties [12,13].

REFERENCES

1. E. Naser, *Fundamental of Gaseous Ionization and Plasma Electronics*, Wiley Interscience, New York, 1971.
2. M.S. Naidu and V. Kamaraju, *High Voltage Engineering*, Tata McGraw-Hill Publishing Company Ltd., New Delhi, India, 1982.
3. J.M. Meek and J.D. Craggs (eds.), *Electrical Breakdown in Gases*, John Wiley and Sons, New York, 1978.
4. G.A. Mesyata, Y.I. Bychekov and U.V. Kremnev, Soviet Physics WS Pekhi, Vol. 15, No. 3, pp. 282–297, 1972.
5. W. Janischewskyj and A.A. Al-Arainy, Proceedings of the 1988 U.S.-Japan Seminar, "Electromagnetic Interference in High Advanced Social Systems," Honolulu, Hawaii, August 1–4, 1988.
6. A.A. Al-Arainy, "Laboratory Analysis of Gap Discharges on Power Lines," Ph.D. thesis, University of Toronto, Canada, 1982.
7. W. Janischewskyj and A.A. Al-Arainy, IEEE Trans. on PAS, Vol. 100, No. 2, pp. 539–551, 1981.
8. A.A. Al-Arainy, "The Effects of Rain on Electromagnetic Characteristics of Corona," M.Sc. thesis, University of Toronto, Canada, 1977.
9. L.G. Christophorous, *Gaseous Dielectrics*, Vol. II, Pergamon Press, New York, 1980.
10. L.G. Christophorou, IEE Conf. Publication, No. 165, pp. 1–8, 1978.
11. R.E. Wooten, S.J. Dale and N.J. Zimmerman, Proc. of the Second Int. Symp. on Gaseous Dielectric, Knoxville, Tennessee, 1980.
12. L.G. Christophorou and L.A. Pinnaduwage, IEEE Trans. on Elect. Insul., Vol. 25, No. 1, pp. 55–74, 1990.
13. L.G. Christophorou and S.J. Dale, *Encyclopedia of Physical Science and Technology*, R.A. Meyors (ed.), Vol. 4, pp. 246–262, Academic Press, New York, 1987.

3
Air Insulation

3.1 INTRODUCTION

The basic processes which lead to the electrical breakdown of gases were summarized in chapter 2. Air is the most commonly used gaseous insulation medium in high voltage power networks because it is free, is abundant and becomes self-restoring after a breakdown. Thus, electrical breakdown behavior of air is very important for designers and operators of high voltage equipment. For this reason, electrical breakdown and prebreakdown of air gaps have been thoroughly investigated since the start of this century and a vast amount of literature and data are available on this subject. Based on such information, international recommendations for air clearances have been established and are being used for the design of air insulated, high voltage power lines and other equipment. This chapter provides a brief summary of the breakdown characteristics of air gaps and the most important factors which can influence air insulation characteristics from the power system engineer's point of view.

3.2 AIR INSULATION APPLICATIONS AND MODELING

Air is used as an insulant for outdoor as well as indoor high voltage power networks. For insulation purposes, it is used to provide the phase-to-phase as well as phase-to-ground insulation. In addition, air is also used in chopping, spark and measurement gaps. Due to a wide range of such applications, the electrodes normally used in air insulated components have a great variety. However, most of these cases can be modeled by some simple electrode configurations. Table 3.1 summarizes the most common sections of air insulated power network and their commonly adopted electrode models that are used for the evaluation of these component's/section's dielectric behavior.

Table 3.1 Applications of Air Insulation in Power Network and Laboratory Models Used for Investigations of Their Dielectric Characteristics

Network section	Function	Model used for simulation
Phase to phase insulation	To provide insulation between two phases of an AC transmission line or between opposite poles of a bipolar DC line	Conductor-conductor, i.e., parallel cylinders or rod-plane electrodes to simulate the worst case of a sharp point on one conductor opposite to the other conductor
Phase to tower or phase to ground insulation	To provide insulation between the phase conductor and the grounded tower or the ground itself	Conductor-plane or rod-plane electrodes
Small diameter conductor to grounded object insulation	To provide insulation between a sharp conductor and a flat grounded object in front of it. This type of geometry has the lowest air breakdown strength	Point-plane or rod-plane electrodes
Sphere-sphere gaps	High voltage measurements and high voltage switches in impulse generators	Sphere-sphere electrodes
High voltage protective or measuring gaps	To bypass high voltage surges to ground by a spark discharge. Also used for impulse chopping and for high direct voltage measurements	Rod-rod electrodes

3.3 VOLTAGE STRESSES

High voltage power network components, besides normal operating voltages, are subjected to temporary power frequency overvoltages caused by network faults. In addition, transient voltage surges caused by atmospheric lightning phenomena (lightning surges) and system switching operations (switching surges) overstress the insulation. The dimensioning of the system insulation is dictated by the air gap's breakdown behavior when subjected to standard lightning or standard switching impulses depending upon the rated voltage of the component. For equipment with a line voltage rating of ≤ 300 kV$_{rms}$, lightning impulses are more critical, whereas for higher voltage ratings, switching impulses assume a greater importance. Therefore, for the design of air insulated components, lightning and switching impulse breakdown data are very important.

3.4 IMPULSE BREAKDOWN PROBABILITY

As discussed in chapter 2, the AC or DC breakdown voltage of a uniform field gap depends upon the gas pressure (P) and the gap distance (d), provided the temperature is kept constant. When the field distribution is nonuniform, the breakdown voltage is also influenced by the presence of corona and space charges in the gap, and gaps with asymmetrical field distribution exhibit pronounced polarity effect. Consequently, when positive polarity voltages are applied to the stressed electrode, these result in a lower breakdown voltage value than the corresponding breakdown voltage value for the negative polarity. Thus, positive polarity breakdown data assume greater importance in such cases.

Under impulse voltage applications, breakdown voltage of an air gap exhibits statistical variations due to time lag effects discussed in chapter 2. Consequently, when a given gap is repeatedly subjected to impulses of a fixed amplitude, there exists a certain probability of breakdown. For impulse voltages, the probability function, which shows variation of the breakdown probability with applied impulse voltage magnitude, is very important. From such a function, the following parameters can be determined:

V_{50} = 50% breakdown voltage level, i.e., the impulse amplitude which exhibits 50% probability of breakdown.

V_0 = 0% breakdown voltage level, i.e., the highest withstand voltage, or the maximum impulse amplitude at which no breakdown occurs.

V_{100} = 100% breakdown voltage level, i.e., the lowest impulse amplitude at which a breakdown is always ensured.

σ = standard deviation of the breakdown voltage.

COV = σ/V_{50} = coefficient of variation of the breakdown voltage.

Air gaps mostly display a Gaussian or normal probability distribution, and thus the probability versus voltage graph is a straight line when plotted on a probability graph paper as shown in Figure 3.1. Sometimes the breakdown probability function is not normal due to a change in either the breakdown mechanism or some experimental variables, and a mixed probability distribution function is observed. If the probability function is normal, σ can be calculated from:

$$\sigma = V_{84} - V_{50} = V_{50} - V_{16} \tag{3.1}$$

Moreover, $V_0 \approx V_{50} - 3\sigma$ whereas $V_{100} \approx (V_{50} + 3\sigma)$. Thus, for an air gap's breakdown data, V_{50} values are usually quoted.

3.5 BREAKDOWN VOLTAGE CHARACTERISTICS

Table 3.1 shows that different electrode configurations used for simulating practical air insulated apparatus. For most applications, the gaps employed have nonuniform field distribution and, therefore, breakdown data of such gaps are very important. In addition to the electrode configuration, the other factors which can influence the breakdown behavior of an air gap are the voltage waveform, voltage polarity, air pressure, temperature and humidity as well as presence of atmospheric pollution. The influence of such factors on breakdown characteristics of air gaps is described next.

3.5.1 Basic Electrode Shapes

Since the laboratory measurements are carried out to simulate the actual electrodes used in the high voltage system, the electrode arrangements that are most commonly used for air gap breakdown studies are:

1. Sphere-sphere electrodes which form a symmetrical and a uniform field distribution gap. Sphere-sphere gaps have the highest average breakdown strength. The breakdown voltages for such gaps depend upon sphere diameter, gap length, voltage polarity and voltage waveform and are discussed in chapter 10 in detail.

Air Insulation

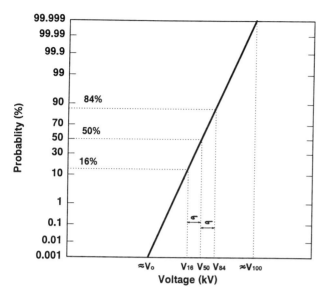

Figure 3.1 Normal breakdown voltage probability distribution.

2. Rod-rod electrodes which form a symmetrical but a nonuniform field distribution in the gap and have several applications. These are discussed in Table 3.1.
3. Rod-plane electrodes which form an asymmetrical and a nonuniform field distribution in the gap. Such electrodes have generally the lowest average breakdown strength.

3.5.2 Breakdown Characteristics of Rod-Plane Gaps

The conservative approach in dimensioning the phase to phase and phase to tower clearances of a transmission line is to use the rod-plane gap's breakdown data. For medium and high voltage lines, the minimum air clearance is determined by AC voltage stress, insulator flashover problems, corona effects and lightning impulse level. However, for extra high voltage lines, positive switching impulses are the most critical. Therefore, breakdown of rod-plane gaps under positive switching impulse voltages have been studied by many investigators. Several researchers have proposed empirical formulas for the relation between breakdown voltage (V_{50}) and gap length (d) for rod-plane geometry under positive switching impulse voltages. Table 3.2 presents several of these formulas and the range of the gap length for which they are applicable. Figure 3.2 shows the relation between V_{50} and gap spacing using the above mentioned formulas. Obviously there is no single formula that can cover the whole range of the tested gap of up to 30 m. For gaps employed in power equipment, most of these formulae give similar results. It is interesting to see some saturation in the breakdown voltage as gap length increases. The scientific community is not sure about the behavior of longer rod-plane gaps, i.e., d > 30 m, unless measurements for such long gaps are carried out. The saturation tendency shown in Figure 3.2 suggests an upper limit on the possible transmission voltage level for overhead UHV lines.

3.5.3 Effect of Electrode Shape

Paris [1] showed that V_{50} of a given air gap geometry stressed by positive switching impulses is proportional to the V_{50} value of a rod-plane geometry of the same gap length. This behavior can be expressed as follows:

$$V_{50}^x = K V_{50}^{r-p} \qquad (3.10)$$

where $V_{50}^x = V_{50}$ value of any air gap configuration x, $V_{50}^{r-p} = V_{50}$ value of a rod-plane air gap of the same gap length and K = gap factor.

Several other researchers investigated this subject further in order to establish the limitations of the gap factor K and the ranges where it can be applied [2–5]. Based on the above references, and the present authors

Table 3.2 Breakdown Voltage Formulae for Rod-Plane Gap under Positive Switching Impulse

V_{50} (kV)	Range of d (m)	Reference	Equation No.
500 ($d^{0.6}$)	$2 \leq d \leq 8$	Paris [1]	(3.2)
3400/(1 + 8/d)	$d \leq 15$	Gallet et al. [6]	(3.3)
450 [1 + 1.33 ln (d − lnd)]	$d \leq 10$	Lemke [7]	(3.4)
$[1.5 \times 10^6 + 3.2 \times 10^5 \, d]^{0.5} - 350$	$d \leq 20$	Waters [8]	(3.5)
$1260r[1 - (r/d)]^{0.5} \tanh^{-1}\sqrt{1 - (r/d)}$	$d \leq 20$	Aleksandrov [9]	(3.6)
1400 + 55 d	$13 \leq d \leq 30$	Pigini et al. [10]	(3.7)
[[(1556 + 50d)/(1 + 3.89/d)] + 78]	$d \geq 4$	Rizk [11]	(3.8)
1080 ln (0.46 d + 1)	$d \leq 25$	Kishizima et al. [12]	(3.9)

d = gap length, r = rod radius (both in meters).

Air Insulation

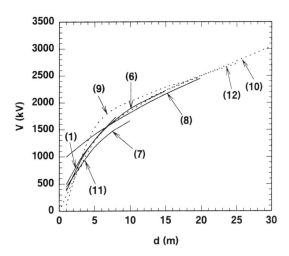

Figure 3.2 V_{50} breakdown voltage as function of gap length using the various formulae mentioned in Table 3.2. Numbers within parentheses indicate reference number.

verifications, Table 3.3 was constructed which gives the values of K for most practical electrode configurations. This table shows that the gap factor increases as the electrode geometry departs from the most divergent case, i.e., rod to plane geometry to a less divergent field. Normally if $K > 1.6$, the withstand voltage for the negative polarity switching impulses becomes less than that for the positive polarity impulses [5]. Therefore, in such a case, the negative polarity breakdown data will assume greater significance and will form the design basis.

3.5.4 Influence of Voltage Waveshape

The magnitude of the sparkover voltage also depends on the shape of the applied voltage waveform since the amount of injected space charge in the gap in the prebreakdown period depends on the time taken by the voltage to reach its crest value. The charges accumulated in the gap during the predischarge process modify the electric field distribution in the gap, and hence influence the sparkover mechanism.

Figures 3.3 and 3.4 show the breakdown voltage as function of gap length for the different voltage waveforms for rod-plane and rod-rod electrodes, respectively. It is clear that for the same gap spacing, the rod-plane gaps have similar or somewhat lower breakdown voltages than rod-rod

Table 3.3 Gap Factor K for a Number of Practical Electrode Configurations

Gap configuration	Gap factor (K)
Rod-plane	1
Conductor-plane	1.12 – 1.25 (depends on d)
Horizontal rod-rod above ground	1.35 – (d/H$_1$ – 0.5)
H$_1$ = Height of rods above ground	
Vertical rod-rod	$1 + 0.6 \dfrac{H}{H + d}$
Conductor-rod	(1.1 to 1.15) exp [0.7 H/(H/H + d)]
Parallel conductors	1.6 – 1.75 (depends on d)
Conductor-rope	1.4
Conductor-cross arm	1.45 (typical value)
Conductor windows	1.25 (typical value)

H = length of grounded rod, d = gap length.

configuration. Moreover, the positive impulse breakdown voltages are generally lower than the corresponding negative polarity values. In some rare cases, especially in the presence of high humidity, an air gap may breakdown at a lower negative voltage than the positive one [11].

The results of Figure 3.3 also show the existence of critical crest time corresponding to the minimum breakdown voltage for positive rod-plane gaps. With an increase in the impulse front time, the breakdown voltage generally decreases up to a certain critical value. When the front time is further increased, the breakdown voltage starts to increase. Thus a U-shaped curve (or a U curve) type of relation between the breakdown voltage and impulse front time is obtained. Typical U curves are shown in Figure 3.5. From air insulation applications point of view, the worst combination, i.e., the combination which produces the lowest breakdown voltage, is the rod to plane electrodes and positive switching impulse wave having a critical front time (t_{cr}). The critical front time (t_{cr}) in μs can be related to the gap spacing and electrode shapes as follow [13]:

$$t_{cr} = [50 - 35(K - 1)]d \qquad (3.11)$$

where K is the gap factor. In the design of EHV and UHV lines, the impulses are assumed to have rise time equal to t_{cr}. The reported research regarding the influence of impulse tail time ($t_{0.5}$) (see chapter 10 for details of $t_{0.5}$) on the breakdown voltage value is scarce. One such study [14] showed that V_{50} and V_{10} increase with wavetail reduction provided t_{cr} is kept more or less constant. In general if $t_{0.5} \gg t_{cr}$, the breakdown voltage becomes less sensitive to the changes in $t_{0.5}$ values.

Air Insulation

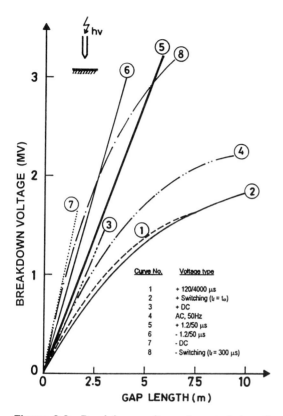

Figure 3.3 Breakdown voltage characteristics of rod to plane gaps.

3.5.5 Influence of Atmospheric Parameters

Atmospheric parameters such as temperature, pressure, humidity, rain (or snow) and wind can influence the breakdown voltage of a gap. It is found that an increase in the absolute temperature (T) which causes an increase in the distance between molecules has the same effect as a decrease in pressure (P) and vice versa. These two variables, i.e., P and T, can be combined in the relative air density (RAD) which is defined as:

$$\text{RAD} = \left(\frac{P}{P_o}\right)\left(\frac{T_o}{T}\right) = \left(\frac{P}{1013}\right)\left(\frac{273 + 20}{273 + T_a}\right) \quad (3.12)$$

where P_o = standard atmospheric pressure at sea level = 1013 mbar, T_o = standard ambient temperature = 293°K (or 20°C), T_a = actual ambient temperature in °C and P = actual pressure in mbar.

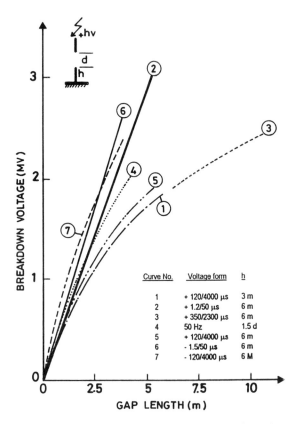

Figure 3.4 Breakdown voltage characteristics of rod to rod gaps.

In general, an increase in RAD causes an increase in the breakdown voltage. The air humidity influences some physical discharge parameters such as the ionization and the attachment coefficients. Consequently, the breakdown voltage values are influenced by the changes in the absolute air humidity (H), expressed as the weight of water molecules present per unit air volume (g/m^3). Generally the breakdown voltage increases with H over the practical range of $2 \leq H \leq 25$ g/m^3. The extent of this increase depends upon the field configuration, gap length, waveform and polarity of the applied voltage, and the value of H. Although the values of humidity correction factor k_h are required by the standards [15,16], there is still some disagreement between researchers about the value of this factor [3,17,18]. According to IEC 60-1 [15], the following correction factors are applicable for breakdown voltages measured at nonstandard atmospheric conditions:

Air Insulation

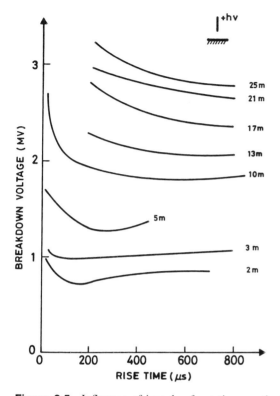

Figure 3.5 Influence of impulse front time on the breakdown voltage of positive rod to plan gaps. (From Ref. 6 © IEEE, 1975.)

1. Air density correction factor (k_1), which is given as:

$$k_1 = (RAD)^m \tag{3.13}$$

where m is an exponent which depends on the electrode geometry, gap length and the applied voltage waveform.

2. Humidity correction factor (k_2). When the absolute humidity, H, is different from the standard humidity of 11 g/m³, then the following humidity correction factor should be applied:

$$k_2 = k^w \tag{3.14}$$

where w is an exponent similar to m and k is given by:

$$k = 1 + A[H/RAD) - 1] \tag{3.15}$$

Here A = 0.01, 0.012 and 0.014 for impulse, AC and DC applied voltages, respectively.

Figure 3.6 shows the values of exponents m and w as a function of g, where g is given by:

$$g = \frac{V_{50}}{500 dk(RAD)} \qquad (3.16)$$

Thus, the actual air gap breakdown voltage V_a is related to the breakdown voltage V_s at standard atmospheric conditions (at T = 20°, P = 1013 mbar and H = 11 g/m³) and the above mentioned correction factors by the following relation:

$$V_a = V_s \cdot k_t = V_s(k_1 \cdot k_2) \qquad (3.17)$$

Table 3.4 shows some typical values of T, P and H in different geographical regions and their influence on the breakdown voltages. From Table 3.4 it can be seen that P, T and H play an important role in the design of air insulation. For example, a certain air gap located in region 3 will breakdown at a voltage which is only ≈65% of the voltage needed for the breakdown of the same gap when located in region 2.

Wind has some influence on corona characteristics but in general has little effect on breakdown voltage values provided the gap distance remains unchanged. However, the wind can cause motion of transmission line conductors and can bring the two phases of a transmission line closer, thereby increasing the possibility of breakdown. Consequently, the sparkover volt-

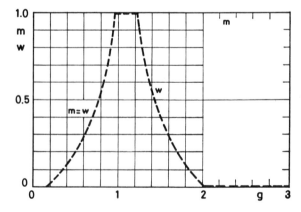

Figure 3.6 The value of exponents m and n in equations (3.13) and (3.14) as per IEC 60-1.

Table 3.4 Examples of Influence of T_a, P and H on Impulse Breakdown Voltages of Rod-Plane Air Gap

$T_a(°C)$	P(mbar)	H(g/m³)	K_1	K_2	$V_a = K_1 K_2 V_s$	Region
20	1013	11	1	1	1.0 V_s	1. Standard conditions
−30	1013	5	1.2	0.93	1.125 V_s	2. Very cold, dry area at sea level
40	850	3	0.795	0.928	0.738 V_s	3. Hot dry area at high altitude
40	1013	25	0.936	1.16	1.08 V_s	4. Hot humid area at sea level
−30	850	5	0.94	0.92	0.865 V_s	5. Cold dry area at high altitude

age can be considerably reduced by the wind. Hence, the influence of swing angle on the switching and lightning impulse breakdown voltages should be included in the transmission line design criteria.

The rain can also affect the breakdown voltage characteristics of open air gaps. It can considerably reduce the breakdown voltages of open air gaps employing large area electrodes. The rain droplets form sharp protrusions over such electrode surfaces, thereby changing the electrode field configuration from a quasi-uniform field gap to close to a nonuniform field point-plane gap, accompanied by an associated decrease of the breakdown voltage.

3.5.6 Influence of Sand/Dust Particles

Airborne particles resulting from sand and dust storms may also influence the insulation behavior of open air gaps. Most of the earlier studies on this regard were confined only to small gaps that are of limited practical value. However, recently authors [19–23] designed and used an environmental chamber to simulate the natural sand and dust storms that are frequent in the desert regions. Various types of voltages and electrode shapes were employed in order to investigate the breakdown characteristics. The breakdown behavior of polluted rod-plane, rod-rod, sphere-plane and sphere-sphere gaps were studied and typical results are summarized here. These findings are applicable to lightning impulses of ≤1000 kV and switching impulses of ≤850 kV with H ≤ 11 g/m³.

Rod-Plane Gaps

The dust and sand particles slightly reduce the breakdown voltages of rod-plane gaps under positive impulse voltages. The highest measured reduction for positive impulses was about 3%. However, the atmospheric pollution has a major effect on the breakdown voltages of such gaps under negative impulses. The effect depends on gap length, cathode radius (r) and the impulse voltage waveform, and can best be summarized by Figure 3.7 for lightning impulses [20]. In this and the subsequent figures of this section $V_p = V_{50}$ value when the gap is contaminated with sand and dust particles, whereas $V_c = V_{50}$ value for the same gap in the absence of any contamination, i.e., clean gap. It is clear from this figure that in small gap lengths, V_{50} decreases by up to 35, whereas in medium gaps V_{50} increases

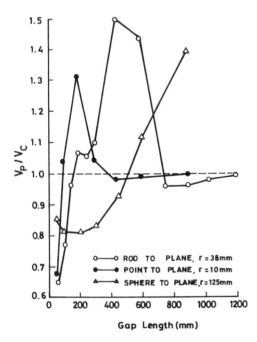

Figure 3.7 The relative value of polluted gap V_{50} (V_p) to clean gap V_{50} (V_c) as function of gap length under negative lightning impulses for asymmetrical geometries studies. (From Ref. 20 © IEEE, 1991.)

Air Insulation   63

by up to 75% due to atmospheric pollution. For larger gaps, atmospheric pollution has minimal influence. The results also show that for most of the gap lengths employed in power lines, the breakdown voltages under negative polarity impulses are either higher or practically similar to the values observed under positive polarity impulses even under extreme particulate pollution. Only in very short contaminated gaps, negative polarity has lower breakdown voltages than the corresponding positive polarity values. The outdoor substation equipment often uses large surface area electrodes to minimize corona effects and/or to grade voltage distribution uniformly. Such large area electrodes result in air gaps with quasi-uniform field distribution. As shown in Figure 3.7 particle contamination can significantly reduce the negative lightning impulse breakdown voltages of such quasi-uniform field gaps. Similar results were also found for switching impulses [21]. Consequently, in polluted environment, the breakdown voltages under negative polarity impulses may become even lower than the corresponding positive polarity values. Therefore, in the selection of proper clearances for station equipment, negative impulses may become more critical than the positive ones, and if the equipment has to operate in an environment where sand and dust storms are frequent, the influence of such contamination must be carefully considered.

Rod-Rod Gaps

An important aspect of the insulation design of overhead lines and substation equipment is to ensure that the flashovers associated with overvoltages are restricted to protective gaps. Rod gaps of various configurations are widely used for this purpose. The most commonly adopted configurations are either square cut or hemispherically terminated rod-rod electrodes.

The studies under lightning and switching impulses show that dust and sand pollution has a considerable effect on the average breakdown voltage gradients of rod-rod gaps. The magnitude of this effect is dependent on polarity, electrode shape and gap length. Figure 3.8 compares the V_{50} values of rod-rod gaps under clean and polluted conditions when subjected to lightning impulses. It is clear from this figure that the pollution effect can be divided into three distinct gap regions: small, medium and large. Small gap region exhibits pollution related reduction in V_{50}. In the medium gap region, the pollution increases the V_{50}, and at larger gaps, pollution has no significant influence on V_{50}. Similar results were found for switching impulses as well [21]. The gap range in which pollution displays severe effects is confined to impulse voltage of ≤ 450 kV. Thus, the systems operating at medium voltages of 33 and 66 kV are most susceptible to pollution related deviations in the protective gap's performance characteristics.

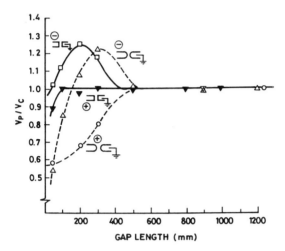

Figure 3.8 The relative value of polluted gap V_{50} (V_p) to clean gap V_{50} (V_c) as function of gap length under lightning impulses for rod-rod gaps. (From Ref. 19 © IEEE, 1991.)

It has also been observed that the square cut rod gaps have more immunity towards the influence of contamination than the hemispherical rods [21].

For hemispherical and square cut rods, the V_{50} values under polluted conditions do not vary by more than $\pm 2\%$ as compared to the clean gap values, provided the ratio d/r is kept ≥ 30 where d = gap length and r = rod radius. This is equally applicable for both polarities and both types of impulse voltages. Similar to V_{50} studies for gaps where d/r is kept ≥ 30, the scatter in the values of breakdown time (T_b) is considerably reduced for polluted gaps while the mean T_b values do not deviate more than $\pm 10\%$ as compared to clean gaps T_b values.

In rod-rod gaps, most of pollution related effects can be attributed to surface adhering dust particles. American standard CD801-1968 (1973) gives V_{50} values for 20 ~ 2400 mm rod-rod gaps when subjected to lightning impulse with an accuracy of $\pm 8\%$. Similarly results of V_{50} from several of the European high voltage laboratories for rod-rod gaps give differences as high as $\pm 10\%$ [24]. Therefore, for longer gaps, changes of up to $\pm 2\%$ caused by sand and dust storms can be considered practically insignificant, and protective rod gaps can be safely designed for sand and dust storm hit areas, based on the clean gap criteria, provided the square cut rod electrodes are selected with $d/r \geq 30$.

Sphere-Sphere Gaps

Sphere gaps are commonly used for measurements of peak values of high voltages with a measurement accuracy of ±3%. The presence of sand/dust pollution in the air gap or on the sphere surface can significantly influence its breakdown behavior. Figure 3.9 displays the (V_p/V_c) ratio as a function of d/D for spheres of two different diameters where D = sphere diameter [23]. This figure shows that for the gaps which are used in high voltage measurements (d/D ≤ 0.5) air pollution causes a reduction in breakdown voltages. Similar results were also obtained for lightning impulses. It is well known that certain sphere gaps should be adequately irradiated to get reproducible breakdown voltages with an accuracy of ±3% [15,25]. Al-Arainy et al. [23] observed that if small diameter spheres are clean but kept "hidden" from the light generated by the impulse generator spark gaps, the breakdown voltages can be up to 100% higher than those given in the standard tables [see Ref. 25]. If the gap receives enough ultraviolet light, the presence of dust pollution does not have any major influence on the breakdown voltages of such gaps. On the other hand, in unirradiated gaps, the sand/dust pollution reduces V_{50} significantly.

3.6 VOLT TIME CURVE AND INSULATION COORDINATION

In order for an insulating media to breakdown the following requirements are needed:

1. The existence of at least one free electron to start the ionization process
2. Enough voltage to cause ionization and avalanche formation
3. Enough time to complete the breakdown process

In nature, free electrons are available from sources such as photoionization and cosmic rays. The number of free electrons in a specific volume is a statistical matter depending on many factors. It is well known that the breakdown time lag depends on the amount of overvoltage. The time lag can be defined as the time elapsed between the moment the voltage reaches the breakdown level and the completion of the breakdown process. Thus, the breakdown voltage value is related to the time taken for breakdown to materialize. This relation is called volt-time characteristic or, in brief, V-t curve. This curve depends mainly on the insulating material, voltage waveform and the electrode shape, in addition to pressure, humidity, etc. The V-t curve for a high voltage device can be established by

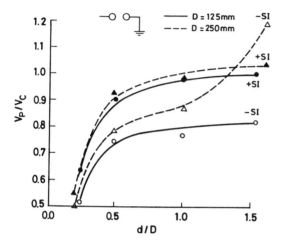

Figure 3.9 The relative value of polluted gap V_{50} (V_p) to clean gap V_{50} (V_c) as function of d/D under switching impulses (D = sphere diameter, d = gap length).

applying a standard impulse of fixed shape but with different peak values and then finding for each application, the breakdown time and the highest voltage to which the gap was stressed up to the breakdown instant. From these data V-t curve can be constructed.

The importance of V-t curve comes when designing an overvoltage protective scheme. Figure 3.10a shows a protective device in parallel with the protected object. Their V-t curve must be in the fashion shown in Figure 3.10b. This means that the protective device, like the rod-rod air gap, will always breakdown before the protected object such as insulator, bushing, transformer, etc. gets damaged by the overvoltage. If the two curves intersect, a protection coordination problem arises and the protective gap only provides a partial overvoltage protection over a certain range of voltage (or time).

In high voltage power networks, rod-rod air gaps or "arcing horns" of various types are used for overvoltage protection of different pieces of equipment such as transformers, bushings, insulators, etc. Since these are usually located outdoors, they are subjected to the variations in atmospheric parameters and pollution levels. In addition, the rod tips may be covered by a film of fine sand and dust particles for extended time periods if the rain is rare like in the desert areas. Several studies have been reported for clean bi-rod air gaps and are summarized by IEEE Committee reports [26,27]. As alternative to bi-rod air gaps, the multiple rod gaps employing (three rod system) are also being used as protective apparatus for medium

Air Insulation 67

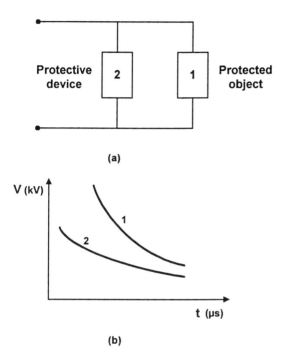

Figure 3.10 Insulation coordination arrangement: (a) protection schemes, (b) V-t curves for good insulation condition.

and high voltage power networks in many parts of the world. These are preferred, since they can be made to sparkover more consistently and in lesser time than single long gap, and comparatively at a lower voltage. They also interrupt power follow current with greater reliability [28].

In selecting gaps for overvoltage protection, the rod end profile, its tip radius, gap length and gap configuration (e.g., single gap or multiple gaps) and horizontal or vertical gaps are the main design parameters. For polluted environment the choice should be of rod end shapes which show minimal influence toward dust pollution. As dust pollution has similar affects for single or multiple gaps as well as for horizontal and vertical gaps [29], the main design parameters are the rod radius, its end profile and gap length. For gap configurations employing rods of bigger diameters and smaller gap lengths, there may be significant differences between the V-t curves under clean and polluted conditions even for smaller breakdown time values as shown in Figure 3.11. It is found that rods with square cut ends and smaller diameters are preferable since these offer immunity toward dust pollution

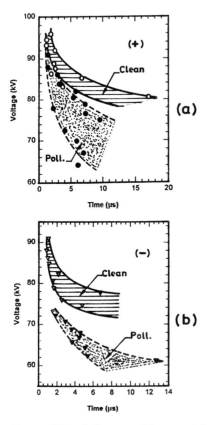

Figure 3.11 Influence of dust particles on V-t characteristics for multiple rod air gaps (gap length = 2 cm + 2 cm).

related influences [29]. Therefore, the rod radius should be as small as possible to minimize the pollution effect on the rod gaps overvoltage protection performance. However, it should not be so small that the gap is always coronating under normal AC voltages.

Table 3.5 shows a typical example of a protective gap. This table includes the recommended gap lengths for single as well as multiple practical rod gaps used on 13.8 kV and 33 kV systems located in semi-arid regions. The rods are assumed to have cylindrical shape with square cut end profiles and rod diameters are not greater than 8 mm. The extreme range of breakdown voltages shown in Table 3.5 includes the range of breakdown probability from 5% to 100%, while considering the temperature variations from 0 to 50°C, pressure variation from 925 to 940 mbar and the influence

Air Insulation 69

Table 3.5 Recommended Gap Lengths for Single and Multiple Rod Gaps Used Across Transformer Bushings Installed in Semi-Arid Region

Nominal system voltage (kV)	BIL (kV)	Recommended single gap (cm)	Extreme range of breakdown voltage (kV)	Recommended multiple rod gap (cm)	Extreme range of breakdown voltage (kV)
13.8	95	6.5	58–83	1.5 + 1.5	64–86
33	170	16	106–145	5.5 + 5.5	110–150

Source: Ref. 28.

of dust particles. Since the above variations will be different for different regions, the protective gap spacing for each regions need to be adequately adjusted.

3.7 PHASE TO PHASE BREAKDOWN CHARACTERISTICS

In the early days of UHV line design, the phase to phase spacing was determined by corona and not by the flashover voltage requirements. The higher transmission voltages have resulted into higher possible switching surges. In addition, the economizing of the transmission line design has reduced the interphase clearances. The air clearance between two phases of a UHV line are in the range of several meters and the breakdown processes in such long gaps have some differences with those corresponding to short or medium gaps.

The discharge associated with positive polarity switching impulses for such long gaps may be divided into three stages: corona formation, leader propagation and the final jump [30]. In most practical geometries, where the field is very divergent, corona inception voltage is much lower than the breakdown voltage. The space charge resulting from corona formation significantly influences the gap's electric field value. Depending on the applied voltage waveshape, the ionization may continue for some time (dark period), where at the end of this period a secondary (burst) corona occurs. At the positive electrode, corona is generally followed by the initiation of a highly ionized channel called leader. The leader's behavior in the gap depends on the geometry of the electrodes and on the shape of the applied voltage. If the voltage is not sufficient, the leader stops propagating and the gap does not breakdown. However, if the voltage is high enough, the leader approaches the grounded electrode where its velocity suddenly increases and with the final jump a complete breakdown of the gap takes place. The critical time to crest (t_{cr}), discussed earlier, is directly linked to the duration of the leader propagation before the final jump [5]. In this regard $(t_{cr})^- < (t_{cr})^+$ since negative leader propagates faster than the positive one. Also t_{cr} increases linearly with gap length and is influenced by the electrode shape, especially the cathode.

In most cases the breakdown probability of long gaps under positive switching impulses can be approximated by normal distribution function. However, in cases where two or more modes of breakdown may occur, abnormal breakdown probability distribution may take place [30]. Time to breakdown T_{BD} also has, in general, normal distribution for long gaps since the leader is dominant at such gaps. Abnormal distribution may occur when $d < 5$ m where the leader corona inception time lag is significant in proportion to T_{BD} [30].

The interphase stress is more complex than between one phase and the ground due the unpredicted time and locations of surges at the two phases. The strength of phase-to-phase insulation is function of the total voltage between the phases and the individual phase-to-ground voltage [3]. The phase-to-phase insulation is a system of two energized electrodes and the ground to which these two different voltages to ground are applied. Phase-to-ground (earth) breakdown voltage is always higher than the phase-to-phase flashover voltage, thus the former is protected by the latter. This is because the spacing between phases and the ground is higher than the interphase gap. In addition the magnitude of overvoltage between phase to phase is 1.4 to 1.8 times higher than the voltage between a phase and the ground [31].

In studying the phase to phase breakdown, the relative values of positive (V^+) and negative (V^-) impulses on the two phases is important. Let α be:

$$\alpha = \frac{|V^-|}{|V^+| + |V^-|} \quad (3.18)$$

Assuming the two impulses are synchronized then the breakdown voltage increases approximately linearly with α [32]. Generally α ranges from 0.3 to 0.7, but for practical purposes, its range can be limited to 0.3 to 0.5 because if $\alpha \geq 0.5$ then $V^- \geq V^+$ and it is not of much interest to the system engineer since $V_{50}^+ \leq V_{50}^-$.

The relative occurrence time of the two impulses in the opposing phases is important since the interphase withstand voltage can be reduced appreciably if the negative surge of V^- preceeds the positive surge of V^+ [31]. This variable time delay is denoted by $\Delta t = t_f^- - t_f^+$ (when the negative surge preceeds the positive surge). The gap insulation is strongly influenced by the space charge created during the negative impulse application. The pre-existing negative space charge field assists the positive leader inception and propagation. For $0.2 \leq \Delta t \leq 300$ ms, the breakdown voltage is lower than when only the positive impulse occurs.

In addition to the single surge or double surges of "standard" shape discussed above, there exist in practice other types of surges such as double or multiple peak surges and irregular surge shapes which result from sudden waveshape changes during the propagation of the surges [31]. The detailed knowledge of the above subjects are important when designing and operating live line maintenance schemes [33].

The bundles in the phases can be arranged in optimal way to strengthen the phase to phase insulation strength. Alexandrov et al. [32] found that the bundle optimization can result in average interphase dielectric strength of 5.4 kV/cm and 4.5 kV/cm for interphase spacings of 1 m and 8 m, respectively.

3.8 ARC DISCHARGE

Spark is defined as a transient plasma channel which bridges two electrodes separated by an insulating gas. If this spark is continuous, then it is called an arc discharge. Arc discharge produces very intensive heat and light and thus it is useful in some industrial applications such as heating, welding and arc discharge lamps. In power systems, the arc discharge is generally undesirable. However, it does take place during circuit breaker opening and during short circuits. There are many types of arc discharges which can be classified according to the method of electronic emission from cathode or according to gas pressure used. In power systems, the arc is generally classified according to the conditions associated with the arc such as:

1. High short circuit current arc
2. Short gap arc, where arc column is effected by the electrodes
3. Long gap arc, where the arc is independent of the electrodes such as in a lightning discharge

When analyzing a network containing an arc an accurate equivalent circuit should be used to represent the arc. One way of representing an arc is shown in Figure 3.12 where arc resistance R, inductance L, and the back voltage E are functions of arc length, arc current and the rate of current change with respect to time [34]. The above parameters are generally found experimentally.

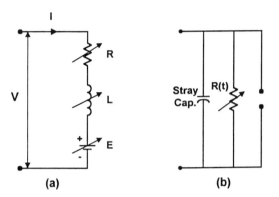

Figure 3.12 Arc equivalent circuit: (a) DC and AC arc during burning period, (b) AC arc during the extinction period. (From Ref. 35.)

Circuit breakers are designed to interrupt the AC current at the moment of current zero. Such an interruption leads to transient recovery voltage TRV across the open breaker contacts. If TRV is high, arc may reignite and the breaker may fail to interrupt the current. Usually the arc quenching process is affected by the heat generation and heat dissipation rates. In an arc, the heat is dissipated mainly by diffusion. The heat dissipation can be increased by arc movement caused by electromagnetic forces. This technique is employed in the interruption of arcs in circuit breakers, gap-type surge arresters and chopping gaps or arcing horns. Another technique which is successfully used in air blast circuit breakers to interrupt arcs is the use of air blast to elongate the arc and increase the arc heat losses.

3.9 UNDESIRABLE EFFECTS OF CORONA

Corona in air has the advantage of attenuating the high voltage surges propagating along the transmission line since the transmission line is usually highly coronating at such high voltages. However, corona on power network has major undesirable effects such as corona loss, audible noise, radio interference and chemical decomposition of air. A brief discussion of some of these effects is presented below.

3.9.1 Corona Loss

In transmission lines, the copper losses (I^2R) at rated current ranges between 20 and 200 kW/km. The fair weather corona loss which is independent of load current is generally designed to be around 10% of the I²R losses. In foul weather, the corona loss increases by at least ten-fold. Beside the weather influence, the corona loss also depends on many other factors such as the difference between the operating voltage (V) and corona onset voltage V_c, i.e., (V-V_c), conductor surface gradient, conductor surface smoothness, system frequency and conductor sizes. Fair weather corona loss can be estimated based on experimental results of similar lines. Corona loss can be neglected if the conductor gradient is low, and the conductors have acceptable audible noise performance. EPRI proposed the following equation to determine foul weather total local corona loss P_L for three phases (in kW/km) for the UHV range [3]:

$$P_L = P_{FW} + \left[\frac{V}{\sqrt{3}} Jr^2 \ln(1 + 10R)\right] \sum_{i=1}^{n} E_i^5 \qquad (3.19)$$

where P_{FW} = total three phase fair weather corona loss (kW/km), V = rms line voltages (kV), J = loss current constant (7.04 × 10⁻¹⁰ for 400 kV lines; 5.35 × 10⁻¹⁰ for 500 and 700 kV lines), r = conductor radius (cm), n = total number of such conductors in the bundle, E_i = maximum gradient for subconductor i and R = rain rate (mm/h).

In some special cases such as very dirty transmission line conductors or transmission line passing through forest having high population of insects, the corona loss may become a significant part of the total power delivered.

3.9.2 Audible Noise

A broad band audible noise (AN), in the human hearing range, is generated during the corona discharge. In addition to the lateral distance from the transmission line, audible noise level depends on the factors mentioned above for the corona loss. Generally audible noise is negligible during fair weather, but during heavy rain it could reach above 60 dB at locations close to the transmission line and it could become very annoying. It is a normal practice to limit the AN at the right of way to around 52 dB. The following empirical formula can be used to calculate the AN level for 3 phase bundled conductor lines during the heavy rain [3]. The AN level is given by dB with 20 μPa sound pressure level as the reference:

$$AN = 20 \log n + 44 \log D_s - 665/E$$
$$+ K_n - 10 \log D_{lm} - 0.02 D_{lm} + A_N + B_n \left[22.9(n-1)\frac{D_s}{D}\right] \qquad (3.20)$$

where n = number of subconductors in the bundle, D_s = subconductor diameter (cm), D = bundle diameter, E = conductor surface gradient (kV/cm), D_{lm} = distance from line to the measurement point (m), K_n = 7.5 for n = 1, K_n = 2.6 for n = 2 and K_n = 0 for n ≥ 3, A_N = 75.2 for n = 1 and 2 and A_n = 67.9 for n ≥ 3, and B_N = 0 for n = 1 and 2 and B_N = 1 for n ≥ 3.

3.9.3 Radio Interference

The current produced by the pulsative corona is responsible for electromagnetic interference (EMI). The current injected in the transmission line conductor due to corona can be approximated by the following analytical expression [30]:

$$i(t) = I_o(e^{-t/\tau_1} - e^{-t/\tau_2}) \quad (3.21)$$

where τ_1 and τ_2 are the front and tail time constants, respectively. The maximum pulse amplitude I_m is about 0.6 I_0. The approximate values of τ_1, τ_2 and I_m for the negative and positive polarity corona pulses are shown in Table 3.6. These current pulses may be represented by a Fourier frequency spectrum with spectral intensity $S(\omega)$ given as:

$$S(\omega) = \int_0^\infty i(t) \, e^{-j\omega t} \, dt \quad (3.22)$$

When the corona pulse is injected into the line, it splits into two equal parts which propagate along the conductor on both sides of the injection point. Each spectral component of such pulses will behave differently depending upon its wavelength. In the frequency range of 0.15 to 30 MHz, the direct electromagnetic radiation from the corona pulses does not contribute much to the EMI which is caused primarily by the propagation along the conductor of various spectral components of the current pulses. When the wavelength of a spectral component is long, a system of two orthogonal fields, one electrical and one magnetic, associated with this spectral current component propagate along the line. This constitutes a guided plane wave with a relatively low attenuation. Therefore, the inter-

Table 3.6 Corona Current Parameters

Corona pulse polarity	τ_1(ns)	τ_2(ns)	I_m(mA)
Negative	6	45	2.7
Positive	30	180	60

ference field is dominated by the aggregation of the effects of all discharges spread over some tens of kilometers on both sides of the measuring location. For spectral components above 30 MHz, the wavelengths are close to, or less than the line clearances and the noise propagation is primarily by radiation.

The quality of radio reception depends upon the signal to noise ratio. To characterize this ratio correctly, it is essential to define noise by a measurable quantity. In general, the instantaneous intensity, S(t), of corona generated noise varies continually and in an erratic manner; but if its average energy for a long time, e.g., one second is constant, then this noise is called stationary random. The RMS value (RN) of the part of noise contained in a narrow frequency band $\Delta\omega$ centered at ω_o (rad/s) is expressed as:

$$\text{RN} = \sqrt{\left(\frac{S^2(\omega)\,\Delta\omega}{2\pi}\right)} \qquad (3.23)$$

Therefore, a radio noise measuring set is basically a selective voltmeter characterized by a passband with bandwidth equal to $(\Delta\omega/2\pi)$ which can be tuned to a center frequency $f_o = (\omega_o/2\pi)$. The radio noise is usually measured in the 0.15-30 MHz frequency range. The measured noise level is proportional to the square root of the bandwidth. Different types of detector weighing circuits such as average, peak, and quasi peak are used for RI measurements.

In radio noise meters complying with international standards [35,36], it is preferred to express the "quasi-peak" value rather than average, peak or true rms value. This type of detector leads to a more realistic measuring device as it represents the psychological effects of the EMI experienced by the listener. The quasi-peak (QP) detector is basically a diode which charges a capacitor placed in parallel with a resistor. When noise is applied to this device, after filtration by the passband of the receiver and suitable amplification, the voltage on the capacitor floats at a value a little lower than the peak value of the noise signal. The charge time constant of the detector is ≈ 1 ms whereas the discharge time constant is either 160 ms or 600 ms as per ANSI or IEC (CISPR) detector specifications, respectively.

Based upon extensive EMI data from a large number of lines, a simple formula was proposed by CIGRE which has a good applicability to many types of lines [35]. The range of parameters for which CIGRE formula was derived are as follows:

Nominal line voltage: 200 – 765 kV_{rms}
Maximum electric gradient: 12 to 20 kV_{rms}/cm
Conductor radius: 1 to 2.5 cm

Number of subconductors per bundle: 1 to 4
Subconductor spacing: 10 to 20 times the conductor diameter

This formula gives a noise figure for the most probable interference level using CISPR detector in dB above 1 μV/m, for dry, aged and moderately dirty line conductors at 2 m above the ground level at a horizontal distance of 15 m from the external conductor, at a measurement frequency of 0.5 MHz. In this method, first the interference level due to each phase is calculated as:

$$NP_1 = 3.5 \, g_{m1} + 12r_1 - 33 \log \left(\frac{D_1}{20} \right) - 30 \quad (3.24)$$

$$NP_2 = 3.5 \, g_{m2} + 12r_2 - 33 \log \left(\frac{D_2}{20} \right) - 30 \quad (3.25)$$

$$NP_3 = 3.5 \, g_{m3} + 12r_3 - 33 \log \left(\frac{D_3}{20} \right) - 30 \quad (3.26)$$

where g_{mi} = maximum conductor gradient at the surface of phase i and D_i = the distance between phase i and the reference measuring point.

Then the summation of these three fields is made in the following way. If one of the fields is at least 3 dB greater than the others, then the other fields are neglected. Otherwise, we use:

$$NP = \frac{NP_a + NP_b}{2} + 1.5 \text{ dB} \quad (3.27)$$

where NP_a and NP_b are the two highest among the 3 values obtained from equations (3.24), (3.25), and (3.26).

For double circuit lines, the interference fields produced by each of the 6 conductors are calculated at the measuring point. The fields produced by the phases corresponding in time are added quadratically and then the summation is made on the three resulting fields.

To obtain the noise level NP at a frequency different from 0.5 MHz, it is sufficient to apply the correction given by equations (3.28) or (3.29) to the value of NP_o, i.e., NP calculated at 0.5 MHz as follows:

Four double circuit and triangular lines:

$$NP = NP_o - (18 \log F + 10 \log^2 F + 4.3) \quad (3.28)$$

For horizontal lines:

$$NP = NP_o - (23 \log F + 12 \log^2 F + 5.8) \quad (3.29)$$

Air Insulation

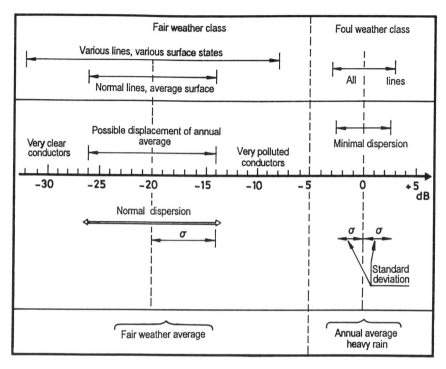

Figure 3.13 Predictions of transmission line EMI for average types of weather. (From Ref. 41 © IEEE, 1973.)

Likewise, the interference level at a lateral distance D other than the reference distance of 15 meters, i.e., D_o, will be obtained by:

$$NP = NP_o - 20 K_1 \log \left(\frac{D}{D_o}\right) \qquad (3.30)$$

where $K_1 = 1.4$ for triangular, vertical and double circuit lines, and $K_1 = 1.6$ to 1.9 for horizontal lines depending upon the voltage level.

The EMI level, for atmospheric conditions other than mean dry fair weather, can be estimated using Figure 3.13. Figure 3.14 shows the measured RI levels from 380 kV lines operating at $E_{max} = 12.1$ kV/cm located in a semi arid dry region [37].

The radio noise problems are not merely restricted to extra and ultra high voltage transmission lines. In many instances distribution lines operating at medium high voltages and much reduced surface gradients can

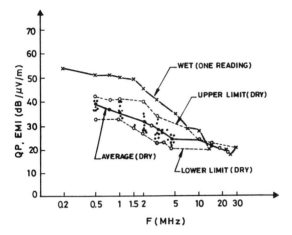

Figure 3.14 Frequency spectra of EMI from 380 kV lines (maximum conductor gradient = 12.1 kV/cm). (From Ref. 37 © IEEE, 1989.)

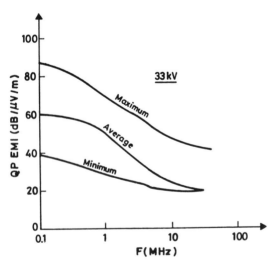

Figure 3.15 All weather frequency spectra for EMI radiated from 33 kV distribution lines (based on over 100 measurements made under different lines). (From Ref. 38 © IEEE, 1989.)

Air Insulation 79

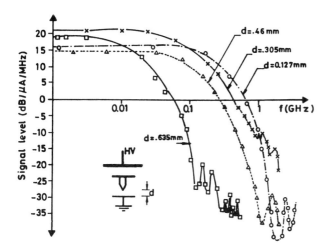

Figure 3.16 Frequency spectra for the gap-type discharge generated from geometry shown in Figure 2.14.

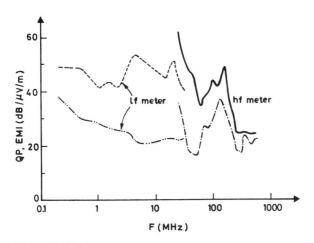

Figure 3.17 Frequency spectra of EMI from a 132 kV line (maximum conductor gradient = 14.4 kV/cm). The measurements were made near a substation (upper curves) and 300 m away from the substation (lower curves). (From Ref. 37 © IEEE, 1989.)

also exhibit significant levels of radio interference. The possible noise sources on such lines are corona from hardware and poorly constructed joints, gap-type discharges and corona on insulations. Since the height of such lines is usually much smaller as compared to EHV and UHV lines, the listener is "closer" to the "noise source," Figure 3.15 (page 78) shows typical frequency spectrum of radio noise level measured on 33 kV distribution lines located in arid dry regions [38].

3.10 TELEVISION INTERFERENCE

The outdoor high voltage network may cause severe television interference (TVI) at nearby locations. TVI is not caused by corona but by the gap-type discharge as discussed in chapter 2. The spectrum of the generated noise is not well defined like RI from corona, but varies significantly depending on geometrical and operating conditions [39]. The level and bandwidth of the gap-type discharge interference depends mainly on voltage level, microgap dimensions and impedance of the circuit external to the gap discharge [39,40]. Figure 3.16 (page 79) shows noise generated from a gap-type discharge model [25] using peak detector [40]. Outdoor measurements show similar results. As an example, Figure 3.17 (page 79) shows EMI measurements using a quasi peak detector for a 132 kV double circuit line located in semi-arid land [37].

REFERENCES

1. L. Paris, IEEE Trans., Vol. PAS-86, No. 8, pp. 936–947, 1967.
2. Y. Aihora, IEEE Trans., Vol. PAS-97, No. 2, pp. 342–347, 1978.
3. *Transmission line Reference Book*, EPRI, California, 1982.
4. K. Schneider and N. Weck, Electra, No. 35, pp. 25–48, 1974.
5. CIGRE Task Force, "Guidelines for the Evaluation of the Dielectric Strength of External Insulation", CIGRE 33.92, (WG-07) 2 IWD, Paris, France, 1992.
6. G. Gallet, G. Leroy, R. Lacey, and I. Kromer, IEEE Trans., Vol. PAS-94, No. 6, pp. 1989–1993, 1975.
7. E. Lemke, Z. Electr. Inform. Energetechnik, Leipzig, Germany, Vol. 3, No. 4, pp. 186–192, 1973.
8. R.T. Waters, "Spark Breakdown in Nonuniform Field", in *Electrical Breakdown of Gases*, by J.M. Meek and J.D. Craggs (eds.), John Wiley & Sons, New York, pp. 510–513, 1978.
9. G.N. Aleksandrov, Zurnal Tecknicheskoi Fiziki, Vol. 9, pp. 744–756, 1969.
10. A. Pigini, G. Rizzi, R. Barmbilla, and E. Garbagnati, "Switching Impulse Strength of Very Large Gaps", 3rd ISH Milan, Paper No. 52.15, 1979.

11. F. Rizk, IEEE Trans. on Power Delivery, Vol. 4, No. 1, pp. 596–606, 1989.
12. I. Kishizimask, Matsumoto and Watanabe, IEEE Trans., Vol. PAS. 103, No. 6, pp. 1211–1216, 1984.
13. L. Thione, Electra, No. 94, 1983.
14. M. Dietrich, J. Wolf, E. Lemke and J. Kurcera, "Influence of the Tail Duration on the Positive Switching Impulse Breakdown of Large Air Gaps", Proc. 4th ISH, Athens, Greece, 1983.
15. IEC-60-1,2,3,4, "High Voltage Testing Techniques", 1989.
16. "IEEE Standards Techniques for High Voltage Testing", IEEE Std. 4-1978.
17. A. Albert, IEEE Trans., Vol. PAS-100, No. 7, pp. 3666–3672, 1981.
18. P. Zacke, IEEE Trans., Vol. PAS-96, No. 2, pp. 701–708, 1977.
19. M.I. Qureshi, A. Al-Arainy and N.H. Malik, IEEE Trans. on Power Delivery, Vol. 6, No. 2, pp. 706–714, 1991.
20. A. Al-Arainy, N.H. Malik and M.I. Qureshi, IEEE Trans. on Elect. Insul., Vol. 27, No. 2, pp. 193–206, 1991.
21. M.I. Qureshi, A. Al-Arainy and N.H. Malik, IEEE Trans. on Power Delivery, Vol. 8, No. 3, pp. 1045–1051, 1992.
22. A.A. Al-Arainy, N.H. Malik and M.I. Qureshi, IEEE Trans. on Dielectrics and Elect. Insul., Vol. 1, No. 2, pp. 305–314, 1994.
23. A.A. Al-Arainy, N.H. Malik and M.I. Qureshi, The Arabian Journal for Science and Engineering, Vol. 20, No. 3, pp. 495–509, 1995.
24. H. Batz, "Comparative Impulse Tests with Impulse Voltage on Rod Gaps", CIGRE Report No. 325, Paris, France, 1962.
25. IEC Publication 52, "Recommendation for High Voltage Measurement by Means of Sphere-Gaps (One Sphere Earthed)," Geneva, Switzerland, 1960.
26. IEEE Committee Report, IEEE Trans., Vol. PAS-86, pp. 1432–1437, 1967.
27. IEEE Working Group Report, IEEE Trans., Vol. PAS-93, pp. 196–205, 1974.
28. Ohio Brass Company, "Hi-Tension News", Vol. 48, No. 9, pp. 2–3, 1979.
29. A.A. Al-Arainy, Journal of King Saud University, Engineering Science, Vol. 9, 1997.
30. Les Renardieres Group, CIGRE Publication No. 53, pp. 31–153, Paris, France, 1977.
31. Special issue on "UHV Air Insulation: Physical and Engineering Research—Part 1", IEE Proceedings, Vol. 133, Part A, No. 7, 1986.
32. G.N. Alexandrov, G.V. Podporkgn, Yu. G. Seleznev and A.D. Sivayev, 4th ISH, Athens, Greece, Paper No. 44.11, 1983.
33. WG No. 7 of Committee 33, *Dielectric Strength of External Insulation Systems Under Live Working*, CIGRE, Paris, 1994.
34. M. Khalifa, (ed.), *High Voltage Engineering: Theory and Practice*, Marcel Dekker, New York, 1990.
35. CIGRE Committee Report, "Interference Produced by Corona Effects of Electric System: Description of Phenomena, Practical Guide for Calculation", CIGRE (Electra), Paris, France, pp. 89–97, 1971.
36. ANSI, "Specifications for Radio Noise and Field Strength Meters 0.015 to 30 megacycles/second", New York, 1963.

37. A. Al-Arainy, N.H. Malik and L.N. Abdulal, IEEE Trans. on Power Delivery, Vol. 4, No. 1, pp. 532–538, 1989.
38. N.H. Malik and A. Al-Arainy, IEEE Trans. on EMC, Vol. 31, No. 3, pp. 273–279, 1989.
39. W. Janischewskyj and A. Al-Arainy, "Statistical Characteristics of Microgap Discharge", U.S.-Japan Seminar on EMI in Highly Advanced Social Systems, Honolulu, August 1988.
40. A. Al-Arainy, "Laboratory Analysis of Gap Discharge on Power Lines", Ph.D. thesis, University of Toronto, Canada, 1982.
41. IEEE Radio Noise Subcommittee Report, IEEE Trans. on PAS, Vol. 92, pp. 1029–1042, 1973.

4
SF_6 Insulation

4.1 INTRODUCTION

It is well known that heavy gases belonging to the seventh group of the periodic table (fluorine, chlorine, etc.) have a considerably higher dielectric strength compared to air under similar experimental conditions. The high breakdown strength depends mainly on their capability of taking up free electrons, thereby forming heavy negative ions. Gases having these properties are called *electronegative*. Of the many available electronegative gases, sulfur hexafluoride, SF_6, has especially gained importance because of its chemical stability as well as its high breakdown strength. It is therefore the dielectric gas of choice. Furthermore, due to its effectiveness in the extinction of arcs, it is used extensively in circuit breakers as well. This chapter provides a comprehensive review about the basic properties, dielectric characteristics and applications of SF_6 insulation.

4.2 BASIC PROPERTIES OF SF_6 GAS

SF_6 is a colorless, odorless, nontoxic, nonflammable and inert gas. Some of its important physical properties are summarized in Table 4.1 [1]. At about 14 bars, SF_6 liquifies at 0°C; whereas at −40°C, the gas liquifies at about 3.5 bars [1]. Thus, in extremely cold regions, SF_6 liquefaction is a potential problem if high gas pressures are used in such regions.

Table 4.1 Physical Properties of SF_6

Molecular weight	146.6
Melting point	−50.8°C
Liquid density (at 25°C/50°C)	1.33/1.98 g/ml
Gas density (at 1 bar and 20°C)	6.2 g/l
Relative density	5.1
Critical density	0.735 g/ml
Critical temperature	45.6°C
Boiling point	−63.0°C
Sublimation temperature	−63.9°C
Critical pressure	35.56 bar
Specific heat (at 30°C)	0.599 J/g
Thermal conductivity	0.1407 W/m°C
Dielectric constant	1–1.07
Vapor pressure (at 20°C)	10.62 bar

At atmospheric pressure, SF_6 is chemically stable up to at least 500°C. At higher temperatures, it can decompose into several species. The gas itself is inert and does not cause corrosion to most common metals used in electrical equipment such as copper, aluminum and steel within the usual temperature range of operation of such equipment. However, SF_6 will undergo some degree of decomposition and oxidation in an electrical discharge, particularly when oxygen and water vapor are present. Some of the byproducts of SF_6 decomposition are quite toxic or corrosive, and there has been increasing concern about the influence that these byproducts can have on system reliability and safety. When SF_6 is dissociated in an electrical discharge, the products of dissociation tend to recombine at a rapid rate to reform SF_6 and thus the dielectric strength of the gas quickly recovers to its original level. It is this characteristic of SF_6 that helps to make it a good arc-interrupting medium.

SF_6 decomposition in electrical discharges (see Table 4.2 for some reactions) produces species such as SF_5, SF_4, S_2F_{10}, SOF_2, SOF_4, SO_2, HF, S_2OF_2, SF_2, S_2F_2, SF_8, SiF_4 and metal fluorides. Some of these products are formed under high power arc discharges [3] while others are produced even under low power corona discharges in the presence of oxygen and water [2]. SF_6 itself is nontoxic and a human can survive indefinitely in a mixture of 20% O_2 and 80% SF_6. However, SF_6 will not support life and can cause suffocation. Being extremely dense, it will accumulate in low-lying areas, requiring care if being exhausted to atmosphere. Moreover, some of the discharge byproducts of SF_6, such as S_2F_{10}, SOF_4, and HF,

Table 4.2 Some Possible Reactions in SF_6

$SF_6 + e$	$\rightarrow$	$SF_4 + 2F + e$
$SF_6 + e$	$\rightarrow$	$SF_2 + 4F + e$
$SF_6 + e$	$\rightarrow$	$SF_5 + F + e$
$SF_5 + SF_5$	$\rightarrow$	S_2F_{10}
$SF_4 + H_2O$	$\rightarrow$	$SOF_2 + 2HF$
$SF_4 + O$	$\rightarrow$	SOF_4
$SOF_2 + H_2O$	$\rightarrow$	$SO_2 + 2HF$

can be very toxic, having working day exposure limits in the range of a few ppm. Therefore, it is prudent to assume that any SF_6 equipment which has been in service for some time may contain some toxic species, thereby requiring extreme care in its handling. It is generally believed that if moisture is absorbed by molecular sieves and oxygen can be prevented from being present in SF_6 equipment, the metal fluorides which constitute the main discharge byproducts are relatively harmless [4]. Thus chemical adsorbents such as sodalime and activated alumina are used in some SF_6 equipment to keep the harmful byproduct concentration within tolerable limits.

In general SF_6 has good heat transfer characteristics. Considering specific heat, thermal conductivity, molecular weight, and viscosity of SF_6, it can be shown that SF_6 has better heat dissipation than that of air. For a cooling gas flow at the sonic velocity, as is typical in circuit breakers, SF_6 has a convective cooling efficiency of about four times that of air. Thus, SF_6 has superior cooling characteristics in an arc environment.

4.3 BREAKDOWN PROCESSES IN SF_6

As discussed in chapter 2, a gas is normally an almost perfect insulator. Like other gases, the primary ionization process in SF_6 is by electron/gas molecule collision and is represented by the Townsend first ionization coefficient α. The most important secondary processes in SF_6 are photoionization of gas, photoemission from electrodes and thermal ionization in case of arcs. In SF_6, negative ions are formed by the direct or dissociative attachment processes. The attachment processes are described by the electron attachment coefficient η.

Both coefficients α and η are strongly dependent on the applied electric field E (in kV/cm) and gas pressure P (in kPa). The net ionization coefficient, $\bar{\alpha} = \alpha - \eta$, can be expressed as [5]:

$$\frac{\bar{\alpha}}{P} = \frac{\alpha - \eta}{P} = K\left[\frac{E}{P} - \left(\frac{E}{P}\right)_{crit}\right] \text{ (cm kPa)}^{-1} \quad (4.1)$$

where K = 27 and $(E/P)_{crit}$ = 877.5 V (cm kPa)$^{-1}$. Here E_{crit} is the critical or limiting field strength. If applied field E > E_{crit} and $\alpha > \eta$, gas ionization will take place leading to discharges. On the other hand, if E < E_{crit}, then $\alpha < \eta$ and ionization and consequently discharges cannot occur. Therefore, from theoretical considerations, electrical breakdown in SF_6 gas should not occur if the applied electric field is less than E_{crit} = $(E/P)_{crit} \cdot$ P. Consequently, SF_6 has a dielectric strength of $\approx$ 90 kV/cm at 1 bar. At a pressure of 3.5 bars, SF_6 will have a dielectric strength of more than 300 kV/cm.

For pressures of technical interest, electrical breakdown in SF_6 occurs as a result of streamer formation. In the streamer mechanism (see chapter 2 for details), it is assumed that the growth of a single electron avalanche becomes unstable before reaching the anode, which results in the formation of streamers. The field in a streamer is $\approx E_{crit}$. In nonuniform field gaps, strong streamers can transform into highly conducting leader channel with electric field of only a few kV/cm, which ultimately causes the collapse of the applied voltage. Mathematically, the streamer breakdown criterion as proposed by Pedersen [6] can be expressed by equation (4.2).

$$\int_0^{x_c} (\alpha - \eta)dx = \ln(N_c) = M \quad (4.2)$$

where N_c is the critical number of ions in an avalanche which has travelled a distance x_c (i.e., the critical avalanche length), in the direction of the applied field when it transforms into a streamer. M is a constant which is considered to have a value between 10 and 20 [7]. The gas volume where an avalanche can grow into a streamer is called the critical volume. Thus, the applied field should be such that equation (4.2) is satisfied for streamer formation and hence for breakdown to occur. A critical analysis shows that equation (4.2) can predict measured breakdown voltages in a uniform field gap with good accuracy in the range of $0.6 \leq Pd \leq 100$ kPa cm [7].

4.4 UNIFORM FIELD BREAKDOWN

To obtain minimal insulation spacing in SF_6 equipment, wherever possible, uniform or slightly non-uniform field gaps with field utilization factor

($\mu_f = E_{av}/E_{max}$) greater than 0.2 are used. Therefore, uniform field breakdown data are very important. Uniform field breakdown characteristics are normally represented by the Paschen curve and the ranges in which Paschen's law is valid. Paschen's curve for SF_6 gas for AC voltages of 50 or 60 Hz [8] and DC applied voltages [5] have been reported and can be approximated by the equation [9];

$$V_b = 1.321 \, (Pd)^{0.915} \qquad (4.3)$$

where the breakdown voltage V_b is in kV and Pd is in kPa cm. At pressures above 200 kPa, deviations from Paschen's law are generally observed [5].

4.5 NONUNIFORM FIELD BREAKDOWN

Whereas in uniform and moderately nonuniform field gaps an electron avalanche usually results in the formation of a streamer which leads to a spark breakdown of SF_6, in highly nonuniform field gaps the situation is somewhat different. In this case, the field near the sharp electrode is much higher than its corresponding value in the rest of the gap. Therefore, the resulting streamers are confined only to the highly stressed region near the sharp electrode causing corona discharges. Thus, the steamer breakdown criterion gives corona onset voltage in highly nonuniform field gaps whereas the breakdown spark takes place at a higher voltage. This type of breakdown is called *corona-stabilized* breakdown. As the gas pressure is increased, the breakdown mode changes and at a certain pressure, the first avalanche can transform into strong enough streamers (and a leader discharge) which can lead to a spark breakdown without any stable corona. This type of breakdown is called *direct* or *corona-free* breakdown. These phenomena lead to nonlinear breakdown voltage-pressure characteristics which are influenced by the radius of the sharp electrode, the gap separation, the voltage polarity as well as the voltage waveform. Figure 4.1 shows breakdown and corona onset voltage characteristics for a rod-plane gap under positive and negative direct voltages. Impulse and AC applied voltages also produce such nonlinear breakdown voltage-pressure characteristics. Similarly, practical coaxial electrodes with metallic contaminating particles also exhibit this type of breakdown behavior. Under certain experimental conditions, the transition from corona-stabilized to corona-free breakdown is very sudden [10]. Moreover, the pressure where such a transition occurs is strongly influenced by the gas composition [11]. Another important factor which can influence the nonuniform field breakdown characteristics is voltage waveform. Thus, the breakdown voltage characteristics of practical systems are influenced by several parameters and an in-

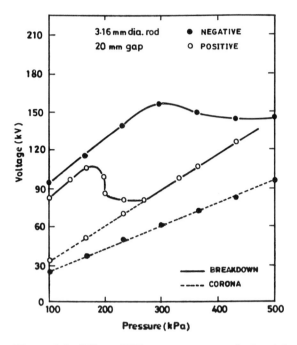

Figure 4.1 Effect of SF_6 gas pressure on the breakdown voltage behavior of rod-plane gap under direct voltage. (From Ref. 10 © IEEE, 1979.)

crease in gas pressures does not always result in an increase in the breakdown voltage. The mechanism of leader formation and breakdown in nonuniform electrode system with stress enhancement is described next [12].

If a small stress enhancement on the conductor results in a small critical volume of gas stressed above ~90 kV/cm-bar and a free electron occurs within that volume, streamers will be generated and fill the critical volume (see Figure 4.2a). Once the volume is filled, discharge activity stops and the positive and the negative ions separate in the field. As the field at the boundary of the critical volume was, by definition, the maximum sustainable in the gas, any ion separation at this boundary results in an increase of the field above the critical value. Thus charge separation results in a breakdown from the critical volume boundary back to the stress enhancement. This breakdown develops into a leader, with a field along its length of only a few kV/cm (as opposed to 90 kV/cm-bar for a streamer). Thus, the leader represents an extension of the stress enhancement into the volume of the gas (Figure 4.2b). The end of the leader creates a new critical

SF$_6$ Insulation

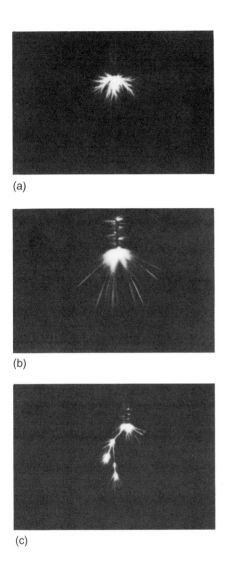

Figure 4.2 Corona discharges in SF$_6$–N$_2$ mixtures: (a) streamer corona; (b) streamer and leader corona; and (c) streamer and stepped leader corona.

volume which fills with streamers, and the process repeats. Through this stepwise breakdown process (Figure 4.2c), a small stress enhancement which creates only a small critical volume can lead to total breakdown of the system [13,14]. Figure 4.2 shows still photographs of streamer, leader and stepped leader corona discharges under positive impulse voltages in SF_6-N_2 gas mixture.

4.6 ESTIMATION OF MINIMUM DISCHARGE VOLTAGES

The estimation of minimum discharge voltages, i.e., corona inception or breakdown voltage, is of considerable importance from engineering point of view. A method capable of predicting discharge voltage with reasonable accuracy can save considerable amount of time and money required in the design and development of new SF_6 equipment and can provide some insight into the insulation behavior of SF_6 gas. Existing methods of discharge voltage calculations can broadly be classified into two categories, namely: those based on streamer breakdown theory and those based on concept of critical field strength. Both methods give reasonably close results to experimental values in most cases and are summarized here.

The streamer criterion, as given in equation (4.2), offers a very promising approach for the solution of engineering problems. This criterion states that the streamer formation results in partial or total breakdown of the insulation depending on the degree of nonuniformity of the electric field in the gap. In a uniform field gap, streamer formation will directly lead to a breakdown while in a nonuniform field it may cause a corona or a complete breakdown. Application of this criterion shows that the minimum discharge voltage V_b (in kV) in SF_6 insulated systems can be expressed as [15]:

$$V_b = 0.8775 \; \mu_f S_f C_f \, Pd \qquad (4.4)$$

where μ_f = the field utilization factor; P = gas pressure in kPa; d = gap length in cm; $S_f < 1$ is the electrode surface roughness factor, which depends upon the conductor surface conditions and gas pressure and can be determined from experience or experiments; and $C_f > 1$ is the curvature factor, which depends upon the shape and size of electrodes as well as gas pressure.

As mentioned earlier, if $E < E_{crit}$ and $\alpha < \eta$, then avalanches cannot form. Hence E_{crit} is the lowest value of stress at which SF_6 gas can breakdown. In the critical field intensity method, the breakdown voltage of SF_6 is related to E_{crit}, gap length, gas pressure and field utilization factor. Using this approach, empirical equations have been proposed to estimate discharge voltages in many configurations [1]. Under impulse voltages, the probability of breakdown is also influenced by the statistics of the initiatory electrons. Such electrons are mostly produced by detachment of negative ions. A method capable of predicting the inhomogenous field breakdown for a variety of electrode configurations is discussed by Wiegart et al. [13,14].

4.7 FACTORS AFFECTING DISCHARGE VOLTAGES

The level of dielectric strength of compressed SF_6 outlined in the foregoing sections may not be fully attainable in reality. Several factors are known to have an effect on the dielectric strength of SF_6 when used in practical systems. The most important of these factors are discussed here.

4.7.1 Contamination

Any fixed or free metallic particle present in SF_6 gas can lower its corona onset and breakdown voltage considerably. The breakdown voltage in contaminated gas can be as low as 10% of the corresponding clean gas value. The breakdown voltage depends on the particle shape, size, material, location, motion, gas pressure and nature of the applied field [16]. Figure 4.3 shows the effect of some of these parameters on breakdown voltage [17]. Free particles present in the SF_6 system can gain charge and move between the electrodes under applied stress. During the motion, the particle can also have corona at its sharp edges [18]. Gas breakdown usually occurs when a metallic particle moves very near to the opposite electrode. The breakdown is triggered as a result of the high pulsed field which appears when a charged particle approaches the opposite electrode, and creates a microdischarge

Insulating particles such as glass and fibers do not have any significant effect on the breakdown voltage. However, dust particles can lower the breakdown voltage by as much as 30%. The effect of dust particles is particularly pronounced under AC voltages. Therefore, cleanliness is very important in SF_6 insulated equipment. Particle traps are usually provided in large SF_6 insulated equipment to capture and immobilize the free conducting particles which can pose a serious risk to the insulation integrity. Another approach which may offer some immunity from free conducting particle-related dielectric strength reductions is the use of insulating coatings on conductors [1,19].

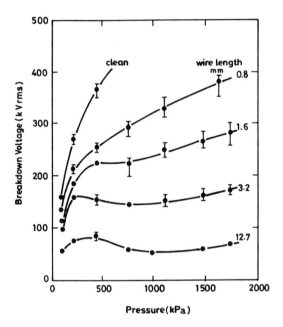

Figure 4.3 Breakdown in SF_6 initiated by free copper wire particles in a 150/250 mm coaxial electrode system. Particle diameter is 0.4 mm. (From Ref. 17 © IEEE, 1973.)

4.7.2 Electrode Factors

The condition, area and material of an electrode surface can also influence dielectric strength of SF_6. When the applied field exceeds ≈100 kV/cm, SF_6 breakdown strength decreases with increasing electrode area. This influence is more pronounced at higher field gradients and also depends on the voltage waveshape and polarity. The negative switching impulse breakdown gradients are less influenced by the electrode area as compared to the power frequency breakdown gradients. In SF_6, the breakdown voltage at higher pressures obeys the extreme value distribution, i.e., the breakdown is governed by the weakest point in the gap. Since in large electrode area systems such points are expected to increase in numbers, lower breakdown gradients are expected for such systems.

Electrode surface roughness can cause a large reduction of breakdown voltages in SF_6 insulated apparatus. Surface roughness can cause localized microscopic regions near rough (or sharp) points where the electric field is much higher than the macroscopic average field in the gap. Depending on gas pressure and applied voltage, such regions of enhanced field can satisfy streamer criterion at lower average stress values, resulting in a large reduction of the breakdown gradient. For a hemispherical protrusion of radius R above a flat plane, Pedersen [20] has shown that the surface roughness will not affect the threshold of breakdown if PR < 0.4 kPa cm. However, as the product PR increases above 0.4 kPa cm, the average breakdown field can become as low as 30% of the value for the clean and smooth electrode system. For practical GIS, the operating pressures are usually in the range of 3–4 bars. If P = 4 bar, R must be less than 10 μm to fully achieve the dielectric strength of SF_6. Such a good surface finish is usually very difficult, if not impossible, to achieve in practical systems.

Although over the range where Paschen's law is valid, the electrode material has no influence on the dielectric strength of SF_6 the electrode material does affect the high pressure breakdown when the applied stress exceeds ~200 kV/cm. Under such conditions, stainless steel electrodes exhibit a higher dielectric strength than copper electrodes. The dependence of breakdown strength on electrode material and electrode area is especially evident in uniform and quasiuniform field electrode systems.

4.7.3 Insulating Spacers

Insulating spacers are used to support the high voltage conductors within the earthed casing in SF_6 insulated equipment and a flashover along the solid/gas interface has a high probability. The spacer flashover voltage is usually lower as compared to the corresponding value for gas gap breakdown. Thus, the spacer efficiency—which is defined as the flashover voltage with spacer present divided by the breakdown voltage for same gas gap but without the spacer—is usually less than one. This efficiency generally decreases with increasing gas pressure and depends upon the design of spacer (shape, material, etc.), its contact with the electrodes, applied voltage waveform and polarity as well as presence of contaminants, e.g., dust, metallic particles and moisture [21]. A less than perfect contact between the spacer and the electrode greatly affects the flashover voltage. The presence of moisture adversely affects the dielectric strength of the spacer/gas interface, which can drop by more than 50% when the moisture is increased, as shown in Figure 4.4 [22]. Generally, the "triple" junction between electrode, spacer and gas is very important and is designed to optimize the field distribution in this region. Presence of particles on the spacer are extremely dangerous for integrity and life expectancy of SF_6 equipment. Once a particle is deposited onto a spacer surface, it is likely to be held on this surface by electrostatic charges. Corona from the particle's tip may cause SF_6 decomposition to highly corrosive species such as hydrofluoric acid (HF), which can attack the spacer surface and cause

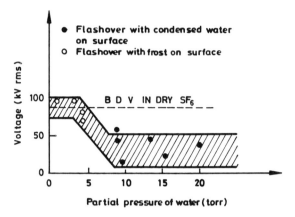

Figure 4.4 Relation between spacer flashover voltage and partial pressure of water. (From Ref. 22 © IEEE, 1971.)

eventual failure. Similar situation arises when corona occurs on any floating component. Under DC stresses, the spacer can get charged which can complicate things further. Therefore, extreme care is required in the design and installation of spacer within SF_6.

4.7.4 Gas Purity

Whereas gas purity might affect the chemical and biological properties of SF_6, small percentages of common gases such as air, N_2 and CO_2 when mixed with SF_6 do not greatly reduce its dielectric strength. Figure 4.5 shows the variation of relative breakdown voltage with SF_6 content in SF_6-N_2 mixtures [23]. The results for SF_6-air, SF_6-CO_2 and SF_6-N_2O mixtures are also similar [23–25]. Thus, whereas a small percentage of air or N_2 contaminants in SF_6 insulated equipment will not adversely affect its dielectric strength, the influence of moisture on insulators and the reaction of moisture and O_2 with SF_6 decomposition byproducts can, in the long run, affect the integrity of SF_6 insulated apparatus and should be avoided.

4.7.5 Gas Temperature

SF_6 insulated equipment may be subjected to extremely hot or cold ambient conditions. In some areas, the outdoor equipment may face temperatures as low as $-50°C$, which can cause liquefaction of the gas at normal pres-

SF$_6$ Insulation

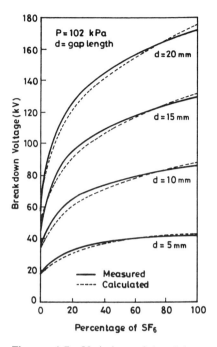

Figure 4.5 Variation of breakdown voltage with SF$_6$ percentage in SF$_6$–N$_2$ mixtures. (From Ref. 23 © IEEE, 1979.)

sures that are commonly employed in such equipment. Therefore, low temperature breakdown behavior of SF$_6$ is of significant importance.

In 20 to 50°C temperatures, SF$_6$ exhibits a constant dielectric strength as long as gas has a constant density. In the temperature range of 50 to 325°C, density (N) reduced critical field intensity of SF$_6$, i.e., $(E/N)_{crit}$ increases by about 10%. This increase has been attributed to an increase in the electron attachment at higher temperatures [24]. In another study, low temperature DC breakdown of SF$_6$ insulated system was investigated for temperatures ranging from −50°C to 24°C. It was found that, although nonuniform breakdown was barely affected, the uniform field breakdown strength of SF$_6$ insulated system was temperature dependent and decreased by about 10% as the temperature was reduced. The decrease appeared at a temperature threshold between −25 and −30°C, and remained constant down to −50°C [26]. Thus, even when constant density is maintained, the operational safety factor of HV equipment using SF$_6$ insulation may be reduced if the equipment temperature drops below −20°C.

4.8 ARC INTERRUPTION IN SF$_6$

The arc interruption behavior of SF$_6$ gas is of considerable practical interest for the use of this gas in circuit breakers. A circuit breaker is required to isolate the faulted part of a network from the rest of the system (Figure 4.6). The breaker is tripped by the protection system when it detects the presence of high fault current I$_f$ flowing through the network. The breaker contacts, which are initially closed, start opening upon the activation of the trip mechanism and an arc is formed between the contacts. The arc current is equal to I$_f$ and depends upon the system configuration, type and location of the fault as well as the fault impedance. The electrical power input to the arc depends upon arc current and arc resistance. Thermal losses occur in the arc by diffusion, conduction, radial and axial convection and radiation processes, and depend upon the gas pressure, temperature and turbulence in the arc region as well as the arc length. Depending upon the balance between the electrical input power and the thermal losses, the arc is extinguished usually at current zero.

Upon current interruption, transient oscillations are generated in the network. These oscillations produce transient recovery voltage (TRV) across the open breaker contacts. For successful interruption of the arc, the dielectric strength of the gas should recover very quickly. Moreover, the residual conductivity should quickly decrease after the current zero to keep the arc extinguished. If these conditions are not met, the arc is re-established and the breaker fails to interrupt the current. For SF$_6$ circuit breakers, two types of failures with different behavior have been noticed [27,28]. Immediately after current zero, if the rate of rise of TRV [i.e., RRRV (rate of rise of recovery voltage)] is greater than a critical value, the decaying arc channel is re-established by ohmic heating caused by residual conductivity. This period, which consists of the first 4 to 8 μs after current zero, is mainly controlled by the arc thermal energy balance and is known as the thermal interruption mode. If the thermal interruption is successful, the TRV can reach to such a high peak value V$_p$, that the breaker sometimes fails through the dielectric breakdown of the open contact gap. This is known as the dielectric failure mode and is usually 50 μs

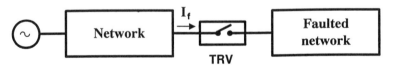

Figure 4.6 Isolation of a faulted part of the network by a circuit breaker.

after current zero. Thus, the behavior of SF_6 and its byproducts in thermal and dielectric recovery phases determines the arc interruption characteristics. From a wide variety of possible fault locations and network conditions, it is found that the terminal fault, i.e., a fault that occurs at the terminal of the circuit breaker, produces the highest stresses for the dielectric failure mode [23]. For the thermal failure mode, the critical fault is the one that occurs on a line some distance (a few km) from the circuit breaker.

The thermal recovery characteristic is usually expressed in the form of a critical boundary separating *fail* and *clear* conditions on a RRRV–dI_f/dt or RRRV–I_f diagram. Typically, the boundary obeys the relationship of the form:

$$\text{RRRV} = A\left(\frac{dI_f}{dt}\right)^{-n} \tag{4.5}$$

where A is a constant and n = 1–4.6 [27]. The thermal recovery performance can be improved by increasing SF_6 pressure (P) in the arc chamber since RRRV is proportional to $(P)^m$ where m = 1–3, and by a suitable design of the interrupter head geometry.

For the dielectric recovery regime, the characteristic is also represented by the critical boundary separating successful *clearance* and *fail* on a V_p–I_f diagram. The dielectric recovery performance may be improved by increasing the number of interrupter units in series. By combining the thermal and dielectric recovery characteristics, the overall limiting curves for the circuit breaker performance are obtained as shown in Figure 4.7, where

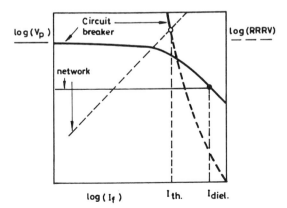

Figure 4.7 Superposition of thermal and dielectric limiting curve diagrams along with network responses. (From Ref. 27.)

the network response under both thermal and dielectric modes is also given. The maximum current which can be interrupted according to thermal mode is I_{th}, whereas the corresponding value under dielectric mode is shown as I_{diel}. Figure 4.8 shows the overall circuit breaker performance characteristics for one and two interrupters in series.

In SF_6 breakers, the arc control is usually achieved by gas blast or electromagnetic methods [28]. In gas blast circuit breakers, the contacts are separated along the axis of a gas flow guiding nozzle so that the arc is subjected to the convective effects of the gas flow. The thermal recovery performance of such breakers depends upon the fault current level as well as gas pressure and gas flow parameters. In electromagnetic circuit breakers, the arc is moved by the action of Lorentz force produced as the fault current flows through a magnetic field which is produced by a coil. The resulting arc motion can be used to control the thermal recovery performance of the arc. Therefore, for improving the thermal and dielectric recovery characteristics, the choice of SF_6 pressure, number of interrupters in series and the methods of arc control and arc movement are important parameters. The other factors which can influence the arc interruption characteristics are nozzle geometry and materials of interrupter contacts, moisture and foreign particles as well as effects of high frequency transients [28].

4.9 GAS INSULATED SWITCHGEAR

SF_6 is used in metal enclosed gas insulated switchgear (GIS) ranging from medium voltages to 800 kV. In GIS, different components of the switchgear are enclosed in metallic housing in adjacent compartments and insulated with compressed SF_6. This arrangement offers the following advantages:

1. GIS are compact as compared to open air substations. Compared to conventional open air station, GIS needs only about 10–15% floor area.
2. They provide total protection from rain, fog, atmospheric pollution, chemicals, etc.
3. They are safe, noise-free, reliable and require minimum maintenance.
4. They are prefabricated and are of modular construction, thereby allowing easier installation and flexible design even under adverse site conditions.

However, GIS requires continuous gas monitoring, gas tight construction and pressure relief devices. Moreover, due to sensitivity of the dielectric

SF$_6$ Insulation

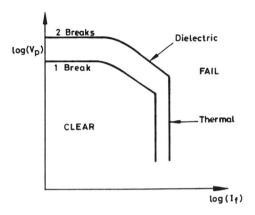

Figure 4.8 Overall circuit breaker performance. (From Ref. 27.)

strength of SF$_6$ gas to several factors discussed in section 4.7, extreme care is required in manufacture, installation and maintenance. GIS can be located indoors or outdoors. Though basic design of both types is the same, the equipment for outdoor GIS requires additional weatherproofing to suit the climatic conditions.

4.9.1 General Design

There are two basic types of designs of GIS. At low-rated voltage of ≤200 kV, all three phases have usually a common enclosure, i.e., 3-in-1 construction; whereas at higher-rated voltages, each phase uses its own separate enclosure, i.e., single-phase construction. If single-phase auto/reclose is a requirement, then for the circuit breaker at least, single-phase construction is preferable even at lower voltages. In a 3-in-1 construction, a single-phase fault within the GIS usually transforms into a phase-to-phase fault and consequently an enclosure burn-through is avoided. Sometimes a single-phase arrangement is adopted for the circuit breaker while a 3-in-1 arrangement is used for the "back parts," i.e., busbars, connectors, etc.

The whole GIS is divided into a number of adjacent gas-tight compartments. Basic components of a GIS are circuit breakers, disconnectors, earthing switches, busbars, connectors and current and voltage transformers, etc. Each component is designed to withstand the expected electrical, thermal and mechanical stresses under normal and fault conditions. Figure 4.9 shows typical feeder bay for a double-busbars GIS.

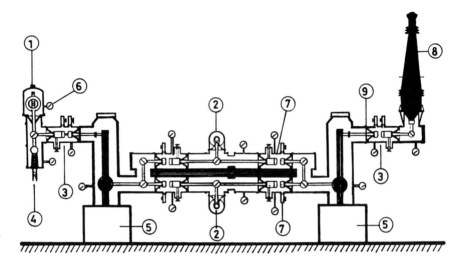

Figure 4.9 Typical feeder bay for double-busbar GIS: 1 = voltage transformer, 2 = busbars, 3 = circuit breaker, 4 = cable box, 5 = hydraulic mechanism, 6 = gas pressure (or density) meter, 7 = disconnector and earthing switches, 8 = SF_6–air bushing and 9 = barrier insulator.

Usually bus conductors are made of aluminum tubes whereas copper is used for switch and circuit breaker contacts. The tube diameter and wall thickness are determined by voltage and current ratings of the equipment. The outer enclosure is made from aluminum or steel, the choice depending upon economic and technical considerations. The insulators are usually made of filled epoxy resins and are of conical, disc or tube designs. They are carefully designed to provide enough creepage length and suitable stress distribution along the surface.

The GIS designs are in modular form and various components can be assembled as desired. SF_6 gas pressure between 2 and 5 bars is used and is determined based on voltage rating, cost, equipment size and reliability considerations. Systems operating at low pressures will have relatively larger size, but will be more defect tolerant and comparatively more reliable than the high pressure systems. Typical service stress levels used for GIS are about 7 to 8 kV_{rms} per mm per MPa of gas pressure. However, the barrier insulators are restricted to stress levels of less than 4 kV_{rms} per mm. Thus, the working stress of GIS is significantly lower than the breakdown strength of SF_6. By this approach, the manufacturers hope to ensure the long-term dielectric integrity of SF_6 equipment, provided normal quality and service procedures have been maintained.

Provision is generally made for a solid grounding of the entire GIS enclosure. In addition, automatic grounding switches are provided for grounding of the cables whenever the isolators are opened.

4.9.2 Circuit Breakers

When a circuit breaker starts operating, the arc is established through a nozzle by the separation of the contacts and/or gas flow and is subjected to a gas blast which abstracts energy from the arc. In the early SF_6 circuit breakers, the gas flow was achieved by the use of a two pressure system, whereby the operation of a blast valve allowed gas to flow from a high pressure reservoir through a nozzle into a low pressure reservoir. After arc extinction, the gas was recycled back into the high pressure reservoir. This system required two separate gas reservoirs with associated seals, a compressor and gas handling plant, and heaters to prevent gas liquefication at low temperatures. Furthermore, in such a system, there was the necessity of synchronizing the blast valve and the contact driving system.

The majority of interrupters today are of much simpler single-pressure "puffer" type design. In this design, during the tripping operation, gas in the arcing chamber is compressed to a high pressure within a piston-and-cylinder arrangement, and then directed across the arc to effect the extinguishing process, finally expanding back to its normal pressure. Depending upon the method by which the gas is directed to the arc the puffer breakers are classified as (see Figure 4.10):

1. Mono-blast, where the gas is forced in a single direction
2. Duo-blast, where the gas is forced in two opposite directions, through two nozzles of equal diameter
3. Partial duo-blast, where the gas is exhausted through two or more unequal diameter nozzles

The interrupters are accommodated in aluminum or steel tanks with the number of interrupters in series depending on the voltage and short circuit current ratings. To ensure equal interrupting duty, voltage grading capacitors are fitted across multibreak units. Puffer type interrupters having 63 kA rating at 520 kV with one break per phase are common now. The SF_6 interrupter can be incorporated in either *live tank* or *dead tank* configuration. The type chosen will depend on economics and/or the type of application. For example, the dead tank construction in which all interrupters are enclosed within an earthed pressure vessel is essential for use in complete metalclad installation, although the same breaker may also be used with terminal bushings in open air type layouts. The circuit breaker tank may be mounted in a horizontal or a vertical configuration.

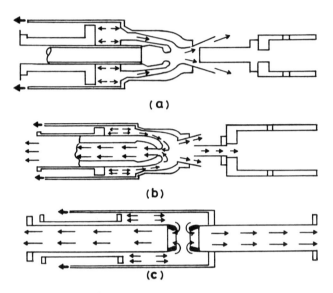

Figure 4.10 Types of gas blast in interrupters: (a) monoblast; (b) partial duo blast; and (c) duo blast.

A high performance puffer system requires a powerful operating mechanism to ensure a high velocity for the contact/cylinder arrangement. Motor-wound spring, pneumatic or hydraulic power units are usually used for this purpose. Arcing contacts are generally fitted with copper-tungsten alloy tips to reduce erosion. Interrupter nozzles are made of polytetrafluoroethylene (PTFE). The breakers are designed for long life and maintenance-free operation. Whenever maintenance is required, the dismantling of the breaker should be carried out in a clean and dry environment. The moisture reacts readily with arc left out byproducts causing formation of hydrofluoric acid, which reduces the insulating capacity of dielectric materials. During arcing, some metallic fluorides are formed and settle on breaker parts as a fine nonconducting dust. This must not be inhaled and must be carefully cleaned off with a dry cloth as soon as the breaker is opened. The dielectric integrity of the equipment is not influenced by these compounds.

4.9.3 Disconnectors

Disconnectors are located on each side of the circuit breaker and are usually the off-load type. They, however, should be able to break small charging currents without generating excessive overvoltages. Usually, disconnectors are designed such that they can be accommodated in busbars or connectors. Their operating mechanisms are hydraulic, hydrostatic or motor operated; the last type being the most common, in which case the motor can be directly mounted on the chamber.

4.9.4 Earthing Switches

Earthing switches are either slow or high speed. Slow speed or off-load earthing switches are used for protection during maintenance and are only operated when the high voltage is off. These are either accommodated within the same chamber as the disconnector, or are housed in a separate compartment. They are normally equipped with a manual operating mechanism, though power drives are also possible. For applications where interlocking with the remote end is not possible, fast earthing switches are required. Such switches can be operated under full voltage and short circuit power and are required to interrupt induced currents from energized parallel lines as well. These switches have motor-wound spring operating mechanisms to produce the required speed of operation.

4.9.5 Voltage and Current Transformers

Voltage transformers can be mounted external to the GIS. In the metalclad construction, electromagnetic type of transformers with secondary windings accommodated in a separate SF_6 insulated chamber and a multi-turn primary connected to the HV bus conductor through a sealed joint are normally used. Alternatively, a capacitive voltage transformer (CVT) can be designed by adding a measuring electrode near the GIS enclosure.

Current transformer can be located near or away from the circuit breaker. The secondary windings can be accommodated over the primary conductors in the circuit breaker tank, in turrets attached to the circuit breaker casing, within standard GIS chambers or externally in modules of metal or glass fiber concentrically mounted over connection chamber [4].

4.9.6 Surge Arresters

These are normally located external to the GIS in open terminal pattern. For cable connected installations and for transformers and reactors directly coupled to GIS, a metal enclosed design is used. Only available for single-phase construction, these comprise SF_6-insulated metal enclosures in which the active components (nonlinear resistors) are installed and connected through a sealed joint to the main conductors. In modern designs, metal oxide nonlinear resistor blocks, without series spark gaps, form the active component resulting in a very compact design and fast operation. A hood and graded electrodes may be required to achieve an acceptable voltage distribution across the active element.

4.9.7 Busbars and Connectors

Busbars and connection chambers of various configurations, such as straights, elbows, tees or angles, are bolted together with O-ring seals. Joints of the conductors have provisions for thermal expansion. Similarly bellows are provided to accommodate for changes in the length of the compartments and variation of dimensions due to manufacturing tolerances. The whole GIS assembly is supported on ground mounted steel frame work.

4.9.8 External Connections

SF_6-air bushings are used to connect a GIS to the overhead lines, transformers and other air insulated components. These consist of a porcelain enclosure through which the HV conductor passes. The bushings are normally divided into two compartments and filled with SF_6 gas at different pressures. A bushing can be bolted directly onto the GIS and can be supported on the wall or roof of an indoor station. Alternatively, the primary conductor is encased with solid insulation and sealed in a porcelain enclosure filled with a nonsetting compound of SF_6 gas.

If a GIS is to be connected to a cable network, the cable connections are brought to the SF_6 filled cable end boxes or terminating chambers where these are bolted to the GIS. The interface seal has to be carefully designed so that SF_6 does not penetrate the oil or vice versa. The failure of such a seal can lead to serious consequences.

4.9.9 Gas System

The whole GIS installation is filled with SF_6 gas at the desired pressure. Since different compartments of a GIS are fitted with gas-tight seals, any gas leak within one compartment does not affect the performance of other compartments. Similarly, arcing byproducts within circuit breaker and disconnector chambers do not enter the main busbars. Furthermore, busbar and connectors, etc., are broken into a number of discrete sections to increase reliability and to minimize effect of internal faults. Each gas-tight section is provided with gas filling and emptying stop valve, a pressure relief device, and a pressure gauge or a gas density meter. The density meters also provide alarm and lockout signals.

4.9.10 Testing

All GIS components are subjected to type and routine tests. Type tests for equipment of ≤245 kV rating require lightning impulse and AC voltage tests. For system voltage ≥300 kV, switching impulse test is an additional requirement. Before shipment to construction site, each transportable assembly is subjected to an AC voltage routine test. During assembly on the shop floor, stringent cleanliness and inspection procedures must be maintained to prevent any contamination. In addition, all seals are tested with a leak detector. Usually an internal arcing test is also done to ensure the proper working of pressure relief devices and to check the possibility of "burn through" of the enclosure wall. Each unit of circuit breaker is subjected to synthetic testing to verify its performance.

At the installation site, tests are performed to check for leaks at every flanged joint; moisture and air content of SF_6 gas in the system; continuity, resistance and voltage drop tests on conductor joints; and functional checks on various GIS components, such as breakers, isolators, earthing switches and control instrumentation. In addition, HV and partial discharge (PD) tests may also be performed. For GIS operating at EHV levels, techniques for PD monitoring are now well established and can detect free and fixed particles, electrode protrusions, floating components, loose nonfloating parts and voids in insulators [30,31]. On line or periodic PD monitoring of the equipment is therefore recommended for GIS equipment operating at EHV levels, since any failure of the equipment can have very serious economic and technical consequences.

4.9.11 High Frequency Transients

The breakdown process in SF_6 gas is very fast and voltage collapse can occur in a few nanoseconds. During the switching operation of a disconnector, prestrike or restrike can occur several times before the switching operation is finally completed. Similarly, during a phase-to-earth fault, breakdown takes place in SF_6 equipment. During such breakdowns, very fast transients with rise time of 4–15 ns are generated which last for several hundred milliseconds. Such transients have very high voltage rate of rise (100 kV/ns) and distance rate of rise (5 kV/cm) [32]. Since GIS is a low loss coaxial medium, there is very little attenuation. Hence various reflections of the transients can take place that could lead to overvoltages in excess of basic impulse level (BIL) of the system. Also when the transients travel to system terminations, they couple to the outside world through SF_6-air bushings, cable sealing ends and current and voltage transformers. Consequently, the grounded enclosure of a 500 kV, SF_6-insulated system near gas to air termination, can rise to tens of kilovolts during switching operations.

Although such switching transients are usually of modest magnitudes (<2.5 per unit), these do cause significant number of failures of all GIS failures in the EHV range due to lack of corona stabilization under fast surges, due to their frequent occurrence and due to a reduced BIL-to-system voltage ratio at the EHV range [32]. Such transients can also cause failure of transformers connected to GIS. Major factors which influence the magnitude and frequency of these very high frequency (VHF) transients in GIS are disconnector contact configuration, SF_6 gas pressure, speed of contacts, capacitive current being switched and trapped charge at the instant of prestrike or restrike [4]. During the disconnector operation, repeated sparking between contacts may lead to phase to earth breakdown also.

Besides the influence of these VHF transients on the performance of GIS, these transients can pose potential hazards to the personnel who come into contact with the GIS enclosure during switching operations as a result of transient ground potential rise. In addition, voltages induced on control and secondary wiring systems in the vicinity of the GIS installation can cause failure of electronics and relays. Several means are used to limit the adverse affects of these transients and achieve electromagnetic compatibility in GIS environment [19,32]. Techniques to measure such VHF transients are described in chapter 11.

4.9.12 Overvoltage Protection

The switching and re-energization of long EHV transmission lines can generate very high switching overvoltages. These overvoltages can be reduced by pre-insertion resistors and by point on wave switching. SF_6 breakers with pre-insertion resistors have complex drive mechanisms, are more expensive than standard breakers and are less reliable. An alternative method of reducing the switching overvoltages is to use metal oxide surge arresters (MOA). These have increased reliability, trouble free service and are more economical. When the HV side of a transformer is directly connected to SF_6 circuit breaker by the GIS busbar and disconnectors, a surge arrester is connected near the transformer for overvoltage protection against VHF and other transients. However, even the fast acting MOA cannot fully cope with VHF transients, which can overstress the transformer winding even in the presence of MOA.

The outdoor, open-terminal substation equipment can also be subjected to pollution problems, lightning strikes and other forms of transient overvoltages. Such equipment can be protected using rod gaps or arcing horns and surge arresters. The indoor GIS equipment is normally equipped with MOA and surge diverters [33].

4.10 COMPRESSED GAS INSULATED CABLES

SF_6 is also used in gas insulated cables. Such cables have the potential of transmitting powers in the range of 1000–5000 MVA as compared to lower ratings of 200–1500 MVA for conventional cables. SF_6 insulated cables have:

1. Lower capacitance, charging current and reactive power
2. Lower resistive and dielectric losses, better heat transfer and thermal performance and hence higher ampacities
3. MVAR losses, which are function of current—and for a certain value of current, total reactive power losses can be made equal to zero thereby operating the cable at a unity power factor
4. The potential of increasing the operating voltage of cable by increasing gas pressure without changing the other design parameters

Such cables are of single phase coaxial design or three phases in a common grounded enclosure. The cross-sectional area of conductor is much bigger in such a cable as compared to conventional cables. The other design features are similar to the busbars of the GIS. Typical gas pressures used are about 3–4 bars. Flexible cable designs also have been developed. Spacer flashover, contaminants, and spacer life are generally the limiting factors in the performance of such cables. For best designs, the ratio of the enclosure diameter to the conductor diameter is kept near 2.72. Typically, for 420 kV system with a BIL of 1425 kV operating at SF_6 gas pressure of 3.5 bar, the minimum possible diameters of outer and inner conductors are approximately 14 and 5 cm, respectively. Generally for a given voltage, the surge impedance of a gas insulated cables is about 60 Ω and therefore surge impedance loading of a compressed gas cable is about 3 to 4 times that of a corresponding overhead line. Cables are manufactured in about 5–10 m segments that are joined in the field, requiring considerable excavation and field work that makes such cables expensive. At present, such cables are used in comparatively shorter segments for bulk power transfer.

4.11 OTHER APPLICATIONS OF SF_6

Besides various applications of SF_6 insulation described in the previous sections, this gas is also used in high voltage gas-insulated capacitors, fast-acting high voltage switches, power transformers, van de Graff generators, high fidelity electrostatic loudspeakers, the semiconductor industry as well as sealed multipan windows.

4.12 SF_6 GAS HANDLING

Since SF_6 condenses to a liquid at room temperature at about 20 bars, it is stored in liquid form out of direct sunlight in a safe place. Its low storage pressure greatly improves handling safety and reduces storage costs relative to other commercial gases. Storage carts are available with compressors, vacuum pumps, oil separators, desiccants, storage vessels with refrigeration to limit the pressure during storage and heaters to help in evaporation of the liquid SF_6. Instruments are available for measuring the moisture content of SF_6, the amount of air in SF_6, gaseous decomposition byproduct concentration, etc. The gas may be withdrawn from the cylinder in either the gaseous or liquid phase. When the cylinder is in upright position, SF_6 will issue out as a gas. When the cylinder is inverted, SF_6 discharges as a liquid.

As mentioned earlier, SF_6 itself is nontoxic but can generate toxic byproducts. Therefore, in some power applications, working with an SF_6-insulated apparatus requires use of a respirator, filter gas mask, and/or protective clothing. Some of the toxic byproducts of SF_6 may be absorbed by a 50–50 mixture of soda lime (NaOH + CaO) and activated alumina, i.e., especially dried Al_2O_3. The preferred granule size is between 8 and 12 mesh. Generally the weight of the absorbent should be equal to about 10% of the weight of the gas. The absorbent should be placed so as to provide maximum contact with the gas which may contain the decomposition products.

Sometimes it is necessary to fill SF_6 in the field. This involves initial evacuation down to less than 1 mbar. Then, the equipment is filled with dry N_2 at 3–4 bar and left in this state for some time. The system is then evacuated again and carefully filled with dry SF_6 gas. After some period, the gas samples are taken to check moisture level which should be kept below $\approx$30 mg of water per kg of SF_6. Higher moisture levels can lead to tracking and failure of insulators.

SF_6 gas leaks from GIS and cables usually can be detected by using tracer gases. The commonly used tracer gases are halogen gases (freon), helium or some radioactive gas. The tracer gas is fed into the system and the leak is located by detecting the presence and measuring the concentration of the tracer gas along the equipment route by a variety of methods. In case of a major leak, all personnel should be evacuated from the area. Good ventilation should be provided in the area and the gas should be removed by pumping it through the absorbent and stored in cylinders.

REFERENCES

1. V. N. Maller and M. S. Naidu, *Advances in High Voltage Insulation and Arc Interruption in SF_6 and Vacuum*, Pergamon Press, New York, 1981.
2. R. J. Van Brunt and J. T. Herron, IEEE Trans. on Elect. Insul., Vol. 25, No. 1, pp. 75–94, 1990.
3. G. D. Griffin, I. Sauers, L. G. Christophorou, C. E. Easterly and P. J. Walsh, IEEE Trans. on Elect. Insul., Vol. 18, No. 5, pp. 551–552, 1983.
4. S. M. Ghufran Ali and W. D. Goodwin, Power Engineering Journal, Vol. 2, No. 1, pp. 17–26, 1988.
5. N. H. Malik and A. H. Qureshi, IEEE Trans. on Elect. Insul., Vol. 13, No. 3, pp. 135–145, 1978.
6. A. Pedersen, IEEE Trans. on PAS, Vol. 89, No. 8, pp. 2043–2048, 1970.
7. N. H. Malik, IEEE Trans. on Elect. Insul., Vol. 16, No. 5, pp. 463–467, 1981.
8. T. W. Dakin, G. Luxa, G. Oppermann, J. Vigreux, G. Wind and H. Winkelnkemper, Electra, No. 32, pp. 61–82, January 1974.
9. R. S. Nema, S. V. Kulkarni and E. Husain, IEEE Trans. on Elect. Insul., Vol. 17, No. 1, pp. 70–75, 1982.
10. N. H. Malik and A. H. Qureshi, IEEE Trans. on Elect. Insul., Vol. 14, No. 6, pp. 327–333, 1979.
11. N. H. Malik and A. H. Qureshi, IEEE Trans. on Elect. Insul., Vol. 14, No. 1, pp. 1–13, 1979.
12. S. A. Boggs, IEEE Elect. Insul. Magazine, Vol. 5, No. 6, pp. 16–21, 1989.
13. N. L. Wiegart, F. Niemeyer, F. Pinnekamp, W. Boeck, J. Kindersberger, M. Morrow, W. Zaengl, M. Zwicky, I. Gallimberti and S. A. Boggs, IEEE Trans. on PWRD, Vol.,3, No. 3, pp. 923–930, 1988.
14. N. L. Wiegart, F. Niemeyer, F. Pinnekamp, W. Boeck, J. Kindersberger, M. Morrow, W. Zaengl, M. Zwicky, I. Gallimberti and S. A. Boggs, IEEE Trans. on PWRD, Vol. 3, No. 3, pp. 931–938, 1988.
15. A. Pedersen, IEEE Trans. on Elect. Insul., Vol. 24, No. 5, pp. 721–739, 1989.
16. J. R. Laghari and A. H. Qureshi, IEEE Trans. on Elect. Insul., Vol. 16, No. 5, 1981, pp. 388–398, 1981.

17. A. H. Cookson and O. Farish, IEEE Trans. on PAS, Vol. 92, pp. 871–876, 1973.
18. A. H. Mufti, A. A. Arafa and N. H. Malik, IEEE Trans. on Dielectrics and Elect. Insul., Vol. 1, No. 3, pp. 509–519, 1994.
19. M. Khalifa (ed.), *High Voltage Engineering: Theory and Practice*, Marcel Dekker, Inc., New York, 1990.
20. A. Pedersen, IEEE Trans. on PAS, Vol. 94, pp. 1749–1754, 1975.
21. J. R. Laghari and A. H. Qureshi, IEEE Trans. on Elect. Insul., Vol. 16, No. 5, 1981, pp. 373–387.
22. T. Ushio, I. Shimura, and S. Tominaga, IEEE Trans. on PAS, Vol. 90, No. 5, pp. 2166–2174, 1971.
23. N. H. Malik and A. H. Qureshi, IEEE Trans. on Elect. Insul., Vol. 14, No. 2, pp. 70–76, 1979.
24. L. G. Christophorou and R. J. Van Brunt, IEEE Trans. on Dielectrics and Elect. Insul., Vol. 2, No. 5, pp. 952–1003, 1995.
25. M. Akbar and N. H. Malik, IEEE Trans. on Elect. Insul., Vol. 20, No. 3, pp. 581–585, 1985.
26. M. F. Frechette, D. Roberge and R. Y. Larocque, IEEE Trans. on Dielectrics and Elect. Insul., Vol. 2, No. 5, pp. 925–951, 1995.
27. K. Ragaller (ed.), *Current-Interruption in High Voltage Networks*, Plenum Press, New York, 1978.
28. H. M. Ryan (ed.), *High Voltage Engineering and Testing*, Peter Peregrinus Ltd., London, England, 1994.
29. A. Bradwell (ed.), *Electrical Insulation*, Peter Peregrinus Ltd., London, England, 1983.
30. J. S. Pearson, O. Farish, B. F. Hampton, M. D. Judd, D. Templeton, B. M. Pryer and I. M. Weleh, IEEE Trans. on Dielectrics and Elect. Insul., Vol. 2, No. 5, pp. 893–905, 1995.
31. R. Baumgartner, B. Fruth, W. Lanz and K. Pettersson, IEEE Elect. Insul. Magazine, Vol. 8, No. 1, pp. 16–27, 1992.
32. S. A. Boggs and H. H. Schramm, IEEE Elect. Insul. Magazine, Vol. 6, No. 1, pp. 12–17, 1990.
33. J. R. Simms, IEE Power Engineering Journal, Vol. 1, pp. 215–222, 1987.

5
Liquid Dielectrics

5.1 INTRODUCTION

Electrical insulating liquids are used abundantly in a variety of components of power system networks, such as transformers, power cables, circuit breakers, power capacitors, bushings and switches. Here they serve various functions either alone or in combination with solid insulating materials, such as to insulate components from each other or the ground, to impart cooling action, to fill up voids in composite dielectrics, to impregnate thin sheet insulation, to control and provide high capacitance in power capacitors, for arc extinction in high voltage switches/circuit breakers and also to act as lubricants in tap changers and circulating pumps.

A wide variety of natural and synthetic oils are being used in the electric power industry. Mineral oil is the cheapest and most commonly used oil. However, selection of a typical dielectric fluid depends on its physico-chemical and electrical properties besides the nature of the service it will perform. This chapter provides details of various important aspects of insulating liquids. These include general classification of liquids, essential characteristics of most commonly used and recently introduced oils, reconditioning of service aged oils beside details on conduction and breakdown mechanisms. The references selected for different topics are chosen from a vast literature to assist both research scholars and engineers in gaining in-depth knowledge of the topic concerned.

5.2 CLASSIFICATION OF INSULATING OILS

Insulating oils can be broadly divided into two categories: organic and inorganic. The most commonly used organic liquid dielectrics for electrical power equipment are mineral oils. Beside the naturally occurring oils, a variety of synthetic organic insulating liquids are also available in the market. Most common of these are silicone oils and chlorodiphenyles. Recently some new synthetic organic oils such as high molecular weight hydrocarbons and tetrachloroethylene have also been introduced. They possess excellent dielectric and thermal properties but are expensive. Among inorganic insulating liquids, highly purified water and its aqueous solution are being used for pulsed power applications, such as capacitors and modulators. At present, water-filled discharge lines for pulsed power are under intensive investigation and development [1]. Similarly, liquefied insulating gases are commonly employed for cryogenic applications. A brief description of most commonly used insulating liquids is given below.

5.2.1 Petroleum Oils

The main elemental constitutes of petroleum are carbon and hydrogen together with trace amount of sulphur, oxygen and some metals. These are joined together to form three commonly occurring structures/compounds: paraffins, naphthenes and aromatics. Examples of some of these are shown in Figure 5.1. These compounds are treated and mixed to form insulating oils to meet the desired physio-chemical and electrical properties. Sometime they are also treated with additives. An additive is defined as a substance not normally present in petroleum but its blending becomes necessary to influence the oxidation, gas absorption and pour point characteristics of oils. Antioxidants are widely used in some countries, whereas gas-absorbing additives are used for high voltage cable oils. Pour point depressants are necessary for waxy oils of paraffinic type.

An oil is considered as naphthenic or paraffinic based, if the content of either of them exceeds the other. Furthermore, an oil is considered weakly aromatic if the presence of aromatics is less than 5% and highly aromatic when these exceed 10%. Most of the mineral oils that are being used today in high voltage apparatuses are low pour point naphthenic based because the paraffinic based oils have the tendency of wax formation when operated at low temperatures.

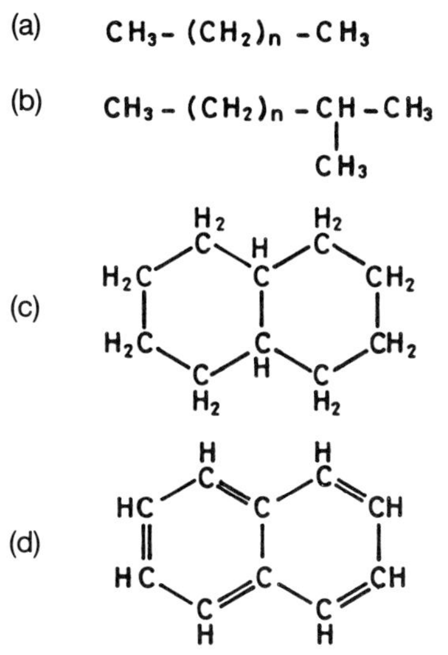

Liquid Dielectrics

(a) $CH_3 - (CH_2)_n - CH_3$

(b) $CH_3 - (CH_2)_n - CH - CH_3$
$\qquad\qquad\qquad\quad\;\; |$
$\qquad\qquad\qquad\;\; CH_3$

(c) [naphthene ring structure]

(d) [aromatic ring structure]

Figure 5.1 Molecular structure of some basic constituents of mineral oils: (a) straight chain paraffin; (b) branched chain paraffin; (c) naphthene; (d) aromatic ring.

5.2.2 Synthetic Hydrocarbons

Among synthetic liquid dielectrics, polyolefins are the dielectrics of choice for applications in power cables and some other electrical systems. Over 55% of synthetic materials produced worldwide today are polyolefins [2]. Most commonly used olefins are poly-butylene and alkyl-aromatic hydrocarbons (e.g., alkyl-benzene). Their composition is simpler than mineral oils, but the general characteristics are very similar.

Poly-butylenes

These are produced by the polymerization of short chain length hydrocarbons. Figure 5.2 illustrates their structure. The carbon chain extends in the range C_9 to C_{25}. Their properties are given in Table 5.1. They possess low pour point and high viscosity index, i.e., small change in viscosity with temperature. Their gas absorption characteristics and dielectric properties are slightly better than petroleum oils.

Alkyl-benzenes

These are available in a range of mixtures composed of compounds having alkyl chains up to 25 carbon atoms attached to a benzene ring. Table 5.2

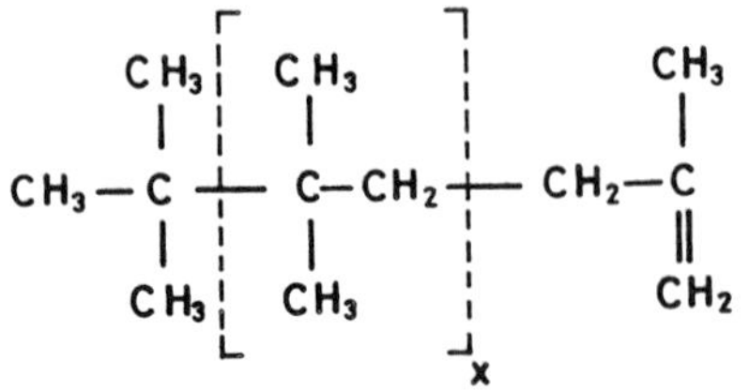

$$CH_3-\underset{\underset{CH_3}{|}}{\overset{\overset{CH_3}{|}}{C}}\left[\underset{\underset{CH_3}{|}}{\overset{\overset{CH_3}{|}}{C}}-CH_2\right]_x CH_2-\underset{\underset{CH_2}{\|}}{C}-CH_3$$

Figure 5.2 Structure of branched chain poly-iso-butylene.

shows their properties. They are unstable in presence of O_2 and are therefore not suitable with breathing transformers. However, they possesses high gas absorption characteristics as shown in Figure 5.3 and also lack in reaction with copper. They are therefore employed extensively in EHV and UHV oil filled cables either alone or mixed with mineral oil.

5.2.3 Chlorinated Hydrocarbons

Two aromatic hydrocarbons, benzene and diphenyle, are chlorinated to produce chlorinated aromatic compounds called askarels or simply polychlorinated biphenyl (PCB). They posses high fire point and excellent electrical properties. In recent years their use has been banned throughout the world, because once they are discharged in the environment they exhibit a strong resistance to biodegradation. They accumulate in biological organisms and end via the food chain in human body, thus posing a serious health hazard.

5.2.4 Silicone Oils

Silicone oils represent an alternative to PCBs but they are rather expensive. Their molecular structure's main chain consists of silicone and oxygen and

Table 5.1 Chemical and Electrical Properties of Poly-butylenes

Characteristic	IEC	ASTM
Breakdown voltage, kV (2.5 mm)	40	>35
Dielectric dissipation factor at 90°C	≤0.0005	0.0003
Resistivity Ωm at 90°C	1.5×10^{12}	$>10^{10}$
Relative permittivity at 90°C	2.2	2.2
Neutralization number mg KOH/g	≤0.03	<0.04
Water content (ppm)	—	40

Table 5.2 Essential Properties of Alkyl-Aromatic Hydrocarbons

Gassing characteristics (hydrogen atmosphere) ml min^{-1} at 80°C	<70
Breakdown voltage, kV (2.5 mm)	>60
Impulse breakdown voltage, kV (25 mm)	
Positive	92
Negative	312
Dielectric dissipation factor at 90°C	0.0004
Resistivity, Ω-m at 90°C	10^{12}
Relative permittivity at 90°C	2.15–2.5

organic groups constitute the side chain. The chemical configuration of polymethylesiloxane is shown in Figure 5.4. Chain lengths of up to 800 siloxane units and relative molecular weights up to 60,000 can be found.

Table 5.3 compares some important properties of commonly used mineral and silicone oils. The dissipation factor of silicone oils is independent

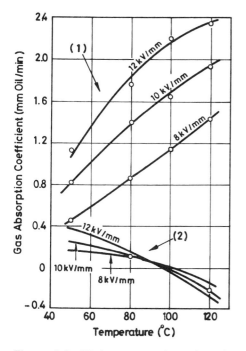

Figure 5.3 Hydrogen gas absorption of (1) alkylbenzene and (2) cable mineral oil. (From Ref. 12.)

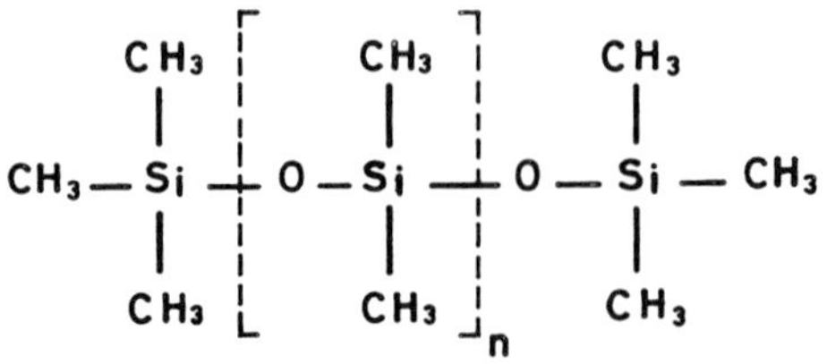

$$CH_3-\underset{\underset{CH_3}{|}}{\overset{\overset{CH_3}{|}}{Si}}-\left[O-\underset{\underset{CH_3}{|}}{\overset{\overset{CH_3}{|}}{Si}}-\right]_n O-\underset{\underset{CH_3}{|}}{\overset{\overset{CH_3}{|}}{Si}}-CH_3$$

Figure 5.4 Molecule of silicone oil; n is in the range of 10 to 1000.

of frequency and temperature. Their high long-term thermal stability at 150°C is particularly notable. Silicone oils are resistant to most chemicals. They are oxidation resistant, even at higher temperature. On thermal dissociation in an arc, nonconducting silicone dioxide (quartz) is formed from the main chain. The higher admissible working temperature compared with mineral oil allows volume-saving designs. Silicone oil is an acceptable substitute for PCBs in transformers despite its slightly inferior nonflammable properties. Its major drawbacks are its poor gas absorption and large water content absorption. However, it is compatible with mineral oil and PCBs, which make it well suited for retrofilling contaminated transformers.

Table 5.3 Properties of Commonly Used Insulating Oils

Property	Unit	Mineral oil	Silicone oil
Breakdown field strength	kV/mm	28	10
Volume resistivity (ρ)	Ω-m	$10^{11} \sim 10^{13}$	10^{13}
Dielectric constant ε_r	—	2.2	2.8
Dissipation factor at 25°C (1 MHz) tan δ	—	0.001	0.0002
Density	g/cm³	0.91	0.96
Thermal conductivity	W/K°m	0.14	0.16
Specific heat (C_p)	cal/g/°C	0.53	0.36
Thermal stability limit	°C	90	150
Flash point	°C	145	>300
Neutralization number (acidity)	mg KOH/g	<0.03	—
Pour point	°C	−40	−55
Dielectric impulse breakdown, negative needle to sphere (25.4 mm gap)	kV	145	—
Water content	ppm	25	50

Source: Refs. 3–5 and 12.

Liquid Dielectrics 117

5.2.5 Esters

Natural ester such as castor oil has been used as a capacitor impregnant for many years, but currently two types of synthetic esters are being used: organic esters and phosphate esters. Others are under investigation.

Organic esters have high boiling points in relation to their viscosity and, therefore, have high fire points. They have a good viscosity-temperature relationship, and respond to inhibitor treatment to produce products of excellent stability. The permittivity of selected esters is higher than that of mineral and the silicone oils, but is lower than that of the askarels. They are used extensively in capacitors.

The phosphate esters have a better fire resistance than silicone oil, but not as good as the askarels. They have poor viscosity-temperature characteristics, relatively poor electrical properties, but a higher permittivity than mineral and silicone oils. They also generally have a poor hydrolytic stability. Due to their high boiling point and low flammability, they are therefore potential candidates both for transformers that are to be installed in hazardous areas as well as in switchgear, where arc suppression in a fluid system is of major importance.

5.2.6 Some Recently Introduced Oils

At present there is considerable short supply of naphthethic-based crude oils. Moreover, in a continuing effort to replace PCBs and continue the search for better quality oils, some new oils have been introduced in recent years. These are being marketed under different commercial names, such as high temperature or high molecular weight hydrocarbon oil, tetrachloroethylene, perfluoropolyether and many others [6].

High Temperature or High Molecular Weight Hydrocarbon Oils

These are alternative to PCB fluids and are called HTHs. They have good electrical insulating properties and adequate heat transfer properties. These are chemically similar to regular mineral transformer oils, but they possess higher boiling points and higher fire points. However, they have higher viscosity which reduces heat transfer capability. General properties of HTHs are given in Table 5.4.

Tetrachloroethylene (C_2Cl_4)

This is also a nonflammable insulating fluid. It can be used in mixtures with mineral oil. It has very low viscosity and therefore gives excellent heat transfer properties. Its mixtures with mineral oils also remain inflam-

Table 5.4 Some Typical Properties of HTH and C_2Cl_4 Fluids

Property	HTH	C_2Cl_4
Flash point, °C	285	none
Fire point, °C	312	none
Expansion coefficient cc/cc/°C at 25°C	0.0008	0.00102
Pour point, °C	−30	−22
Viscosity, cSt		
100°C	16	0.36
50°C	85	0.42
25°C	350	0.55
Specific gravity, g/cm³	0.877	1.620
Dielectric strength, ASTM D-877, kV	43	43
Impulse breakdown, kV		
Negative polarity	118	—
Positive polarity	85	—
Dielectric constant, ε_r	2.38	2.365
Dissipation factor (%)		
100°C	0.4	—
50°C	0.4	0.05
25°C	$<10^{-3}$	—

Source: Ref. 6 © IEEE, 1992.

mable. However some toxic effects of this insulating fluid are also reported [5].

Perfluoropolyether

This has recently been introduced in the European market with the trade name Galden HT40 as replacement for PCBs. It is nonflammable oil, as its boiling point exceeds 400°C. It possess low vapor pressure, thus it can be used as a good heat transfer medium. Its molecular structure is given in Figure 5.5. Since it has no C-H bonds and C-F bonds are much stronger (single bond energy 4.6 eV), bond scission does not take place at normal stress. Therefore no gaseous products are formed. Another interesting feature of HT-40 is the lack of moisture sensitivity, as it has no affinity for water and it does not absorb or mix with water.

If any fluorine evolves during service it does not react with water to form any acid (unlike PCBs which produce highly corrosive HCl). Because of their higher density, water accumulated in transformer will float on top

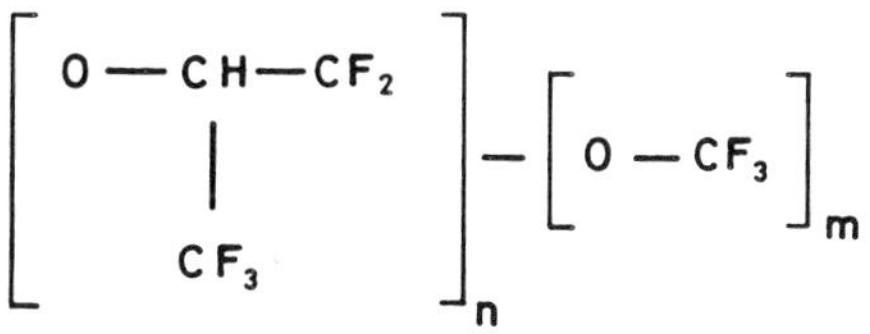

$$\left[\begin{array}{c} O-CH-CF_2 \\ | \\ CF_3 \end{array} \right]_n - \left[O-CF_3 \right]_m$$

Figure 5.5 Molecular structure of perfluoropolyether.

of the oil surface rather than at the bottom, as is the case in conventional petroleum mineral oils. However, its viscosity and cost are high.

5.3 ESSENTIAL CHARACTERISTICS OF INSULATING OILS

Table 5.5 outlines a list of the most important characteristics of insulating oils along with their typical accepted values. However, for the purpose of design and operation, the following are very essential and are therefore dealt with in some detail:

Table 5.5 Properties of Mineral Insulating Oils Considered Prior to, During and After Their Long Use

Characteristic	ASTM test method	IEC publication	Typical values
Dielectric strength	D-877	156	≥ 30 kV (1.0 mm)
	D-1816		≥ 28 kV (1.02 mm)
Dissipation factor	D-924	247	0.1–0.5% (90°C)
Impulse strength	D-3300	897	145 kV
Dissolved-gas content	D-831	567	
Density, specific gravity	D-1298	296	825 ~ 890 kg/m^3
Viscosity	D-445	296-A	3 ~ 16 cSt (+40°C)
Pour point	D-97	296-A	-30°C
Neutralization number	D-974	296	≤ 0.5 mg KOH/g
Antioxidant content	D-2668	666	$\leq 0.3\%$
Water content	D-1533	733	<80 ppm
Gassing characteristic	D-2300 (AB)	628(A,B)	-35 to $+35$ mm^3/min
Interfacial tension	D-971	296	40 ~ 60 mN/m
Resistivity	D-1169	247	3 ~ 10 GΩm

1. Thermal transfer characteristics
2. Chemical stability against electrical stresses
3. Dielectric properties

5.3.1 Thermal Transfer Characteristics

In a liquid filled system (transformer, cable, circuit breaker, etc.), heat is transferred mainly by convection. Under natural atmospheric cooling conditions convection (N) is given as [7]:

$$N = f \left[\frac{K^3 \, AC}{\nu} \right]^n \qquad (5.1)$$

where K = thermal conductivity, A = coefficient of expansion, C = specific heat per unit volume, ν = kinematic viscosity, and n = 0.25 ~ 0.33. It is clear that heat transfer is strongly dependent on K and to a lesser degree on A and C, whereas it inversely varies as viscosity. Generally in liquids, A and C do not vary much, but ν varies greatly. Therefore, the main factors that control heat transfer are K and ν. Obviously, an increasing value is preferable for systems likely to operate continuously at a high temperature. On the other hand, a low value of K and high viscosity can lead to localized overheating or even electrical "burn out." Figure 5.6 shows the variation of viscosity with temperature. It is an important feature

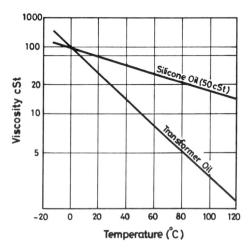

Figure 5.6 Effect of temperature on viscosity in two commonly used insulating oils.

of the liquid that as temperature increases its viscosity falls off. This greatly enhances the cooling effect.

Silicone oils do not exhibit this property appreciably, therefore this viscostatic property coupled with poor heat capacity and falling thermal conductivity can pose severe overheating problems at high temperature in systems based on such insulants. Thermal conductivity and specific heat values are therefore much more useful for evaluating temperature gradients and possible hot-spot formation in places where the oil is trapped. Inside oil-impregnated paper insulation, for instance, a temperature gradient of 15°C has been reported across a layer of only 0.4 mm of oil-impregnated paper covering a copper conductor [8].

5.3.2 Chemical Stability Under Electrical Stress

In service, insulating liquids are subjected to thermal and electrical stresses in the presence of materials like O_2, water, fibers and decomposition products of solid insulation, as well as oil soluble constituents of impregnating varnishes and resins. These, either singly or in combination, promote degradation of the liquid with the result that soluble solid and gaseous products are formed, which can result in corrosion, impairment of heat transfer, deterioration of electrical properties, increased dielectric losses, discharges and arcing. In the absence of any remedial action, this cycle continues and produces an ever-worsening liquid and equipment condition.

Degradation due to electrical and severe thermal stresses is indicative of an equipment design or fault conditions, while the oxidation constitutes the most important "in-service" deteriorating property of insulating liquids in equipment subject to breathing. The rate of oxidation is increased by the presence of materials acting as catalysts. Copper is one of the most active of such materials [9]. Improper impregnation, however, results in enhanced catalysis due to attack on the conductors by varnish acids. It also produces heavy sludge deposits when incompletely cured impregnants dissolve in the liquid filling and precipitate after polymerizing or curing.

To retard oxidation, oils are inhibited with additives called antioxidants. A widely used antioxidant additive is DBPC (2,6 di-tert-butyl peresol) and, to a lesser extent, DBP (2,6 di-tert-butyl phenol). Antioxidants do not prevent copper dissolution in oil (Figure 5.7) but as long as they are present (induction period) they prevent the formation of free-radical peroxides, especially from the saturated hydrocarbon fraction, and the subsequent chain oxidation reactions. During this period they themselves are slowly oxidized and deactivated [8] and, when they have been almost completely consumed (end of the induction period), the above protection is no longer available. Accelerated formation of peroxides is then observed, fol-

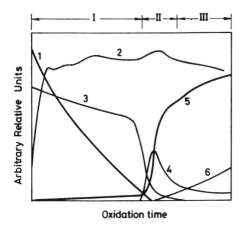

Figure 5.7 Oxidation of insulating oil versus time: I, induction period; II, acceleration period; III, saturation period; (1) antioxidant content, (2) copper content, (3) interfacial tension, (4) peroxide content, (5) acidity, (6) sludge content. (Reprinted with permission from Ref. 1 © ASTM.)

lowed by acids and polar oxidation compounds and, finally, by oil-soluble and -insoluble sludges. The relative saturation in the final stage of oxidation results from the buildup of inhibiting phenolic forms of the aromatics. The electrical properties (tan δ and conductivity) are closely related to the observed oxidation steps and the mechanism applies to paraffinic as well as naphthenic oils [11].

The electrical characteristics such as electric strength, impulse strength, DC volume resistivity, dielectric dissipation factor and relative permittivity are generally of much importance. The maintenance of electrical properties at an acceptable level ensures satisfactory equipment performance, reduces ohmic losses and limits discharge inception within the liquid. The level of test results is affected by the chemical constitution of the liquid, but with the exception of impulse strength and permittivity, is predominantly due to the presence of conducting contaminants such as fibers, water, particulate matter (dirt) dipolar and ionic or dissociated compounds.

Liquids can vary widely in electric strength. Power frequency breakdown depends considerably on the level of moisture and contaminating particles. Similarly these factors also effect the dielectric loss and resistivity of insulating oil. However, with filtration and removal of moisture these can be considerably improved.

5.4 STREAMING ELECTRIFICATION

Transformer failures due to streaming electrification have been reported worldwide to occur within few years after commissioning [13,14]. It has, therefore, received considerable attention since 1980. Charge separation at interfaces between a moving fluid and solid insulation boundary can give rise to the generation of substantial electric fields. Either alone or in combination with the already existing charges in the fluid produced due to aging of insulation and/or due to energization of the equipment can thus lead to insulation failure.

Figure 5.8 illustrates the formation of a double layer at solid interface (e.g., transformer pressboard, paper) due to ionic charges resulting from dissociable impurities in the liquid. Enforced motion of the fluid strips away the ions from this double layer and transports them downstream. The resulting charge separation between the bound charges on the solid surface and convected opposite-polarity charge can generate dangerous potentials. Streaming electrification therefore depends mainly on the ionic species that get adsorbed at the solid surfaces. In transformers, the -OH groups associated with the pressboard/paper cellulose are the likely sites for the attraction of ionic species from the bulk of oil. Their accumulative effect increases with aging and polymerization of cellulose, since as a result the leeching of ionic species is enhanced. Increase in operating temperature

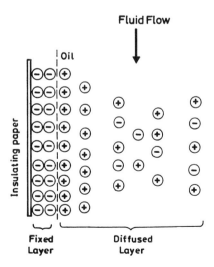

Figure 5.8 Electrical double layer and ion distribution at the insulated paper/oil interface.

and presence of moisture also play a detrimental part to increase it further [14,15]. Beside these factors, the velocity of the fluid is yet another controlling parameter, since the shearing of charges at the solid/liquid interface is intimately related to the rate of flow of the fluid. The transition of liquid flow into turbulence itself becomes another source of charge generation. In transformer, the turbulence in liquid flow can occur in ducts, windings or circulating pumps.

Currently a great deal of effort is being made, both in industry as well as in several research centers, to understand these problems and to identify the practical solutions. The initial industrial response so far has been to reduce design velocities and curtail the use of pumps. The most innovative approach has been adopted in Japan, where the use of charge suppressant additives is widespread. Materials such as alkylbenzene and 1,2,3-benzotriazole (BTA) are successfully being used to suppress the effects of streaming electrification [14,17].

5.5 RECONDITIONING OF INSULATING OILS

Insulating liquids can be kept in service as long as their properties have not deteriorated beyond specified levels, as described earlier (for details see Table 5.5). For instance, their breakdown voltage strength should not decrease below 30 kV. The limits for the resistivity of the liquid are 3–10 $G\Omega \cdot m$, and for the dissipation factor $0.001 \sim 0.005$ at 90°C. In more highly stressed equipment such as cables, the limit for the dissipation factor is set as low as 0.001 [12]. Similarly the level of acidity neutralization factor of insulating liquids could be measured as the amount of potassium hydroxide sufficient to neutralize the acids in 1 g of the liquid and its acceptable level is ≤0.5 mg KOH/g. The acceptable level of water content in power equipment is 15–80 mg/kg. The level is set even lower (0.1 mg/kg) for EHV cables, as their working electrical stresses are very high. Similarly, diagnosis of the lifetime of transformer oil based on furfural dissolved in oil has recently received much attention [8,18]. (For more details see chapter 8, section 8.8.6.) The characteristics of the quality of oil being used should be checked periodically. The frequency of testing depends upon the atmospheric/ambient conditions, power rating, nature loading, construction of the equipment and manufacturer's recommendations.

As the level set for any of the foregoing characteristics is approached, measures for reconditioning the liquid should be taken. There is portable equipment now available for reconditioning insulating oils while in service. A method well known and in use is that of filtering and vacuum drying.

Liquid Dielectrics

While under a vacuum of about 1 kPa, the liquid is heated to about 30–60°C, which is above the boiling point of water at such a reduced pressure. With large oil surfaces exposed to the vacuum, it becomes freed from its moisture content, dissolved gases, as well as acids and particulate matter. The process could include replenishing inhibitors and other additives in the liquid, as shown in Figure 5.9. Reconditioned oils usually have about half the life of new ones [19].

5.6 ELECTRIC CONDUCTION IN INSULATING LIQUIDS

In an effort to understand the breakdown mechanism better, the study of conduction in insulating liquids is being carried out since the period they are serving as insulators or dielectrics in electrical equipment. Many books and reviews published during the last 50 years reflects this aspect [20–27].

5.6.1 Polar and Nonpolar Liquids

Insulating liquids, in general, can be characterized as polar or nonpolar. The dielectrics that have permanent dipoles are known as polar. In such dielectrics, the asymmetry in their molecular structure leads to a permanent displacement of positive and negative charge centers even in the absence

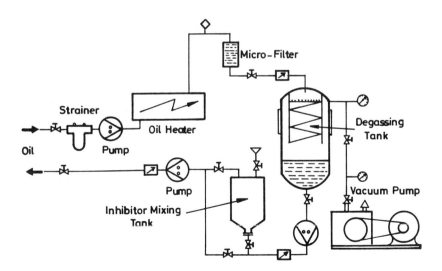

Figure 5.9 Components of a mobile oil reconditioning unit. (From Ref. 19.)

of an external field. Typical examples of polar liquid dielectrics are water, propylene carbonate and nitrobenzene. However they dissolve and dissociate impurities so efficiently that they are difficult to keep in purified state so that they exhibit high levels of resistivity. However, highly pure polar liquids are attractive by virtue of their large permittivities, e.g., pure water ($\varepsilon_r = 80$) and propylene carbonate ($\varepsilon_r = 69$).

The dielectrics that do not form dipoles appreciably in the absence of external electric field are known as nonpolar. The dipoles formed in such dielectrics, on the application of electric field, are not permanent and their atoms return back to their original state once the electric field is removed. Most of the insulating liquids used in power system applications are of nonpolar nature.

When subjected to electrical stress both polar as well as nonpolar liquids are characterized by a universal conduction current characteristic. Fig-

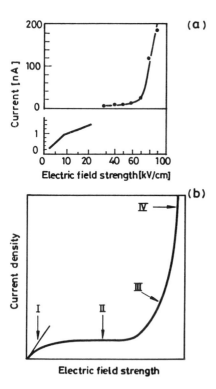

Figure 5.10 Conduction current as a function of electric field strength: (a) transformer oil, test gap = 6 mm (from Ref. 1 © ASTM), and (b) idealized schematic representation of different regions.

ure 5.10a depicts current-field variation in transformer oil, whereas Figure 5.10b shows the general trend in liquids in which three distinct regions are observed. Regions I and II appear under electric stresses that do not exceed around 2×10^6 Vm^{-1} and therefore are considered to represent conduction at low fields; whereas region III appears when these stresses exceed 10^7 Vm^{-1} and represents the case of conduction under high field. In this situation the conduction current increases rapidly with slight increase in electric field strength. A further increase in the electric field strength leads finally to breakdown (region IV).

In liquids, unlike in gases where natural radiation plays important role, the major role is played by the dissociation of ionic impurities and solid impurity particles. The host liquid may dissociate the impurity molecules into ion pairs. These positive and negative ions, which escape recombination, drift to the electrodes and thus give rise to a flow of current in the ohmic region. The ionic drift velocity and therefore the current increases with increasing field, as illustrated in Figure 5.10. At intermediate fields, the charge carriers are produced entirely by electro-chemical processes at the interface between liquids and metal. A so-called double layer is formed at the electrode [1]. Its significance is that positive ions preferentially, but negative ions also, are created at this junction. At certain field strength the ions will move to the electrodes at a rate faster than their generation rate in the bulk of the liquid. The current then depends on their resultant effect. This effect causes growth of space charge at electrodes [20] and thus leads to the situation where the current tends toward saturation, as shown in region II.

5.6.2 Conduction Current at High Fields

When the field applied to insulating liquids is sufficiently high, the current tends to increase sharply for comparatively small increments of field. A number of factors govern this current. Following are the most prominent among them:

1. When the concentration of undissociated impurity molecules is sufficiently high, the increase in electric field will lead to an enhancement of the dissociation process. As a result, more free ion pairs are produced and conductivity increases.

2. Reduction or oxidation of impurity molecules at the liquid/metal interface may yield excess negative or positive ions. Such electrochemical reactions are controlled by the double layer potential which in turn is determined by the density and the nature of absorbed ions.

3. Solid impurity particles are also considered to contribute extensively toward high field conduction. Since these particles may also be of

semiconducting or insulating nature, prolonged action of electric stress at the electrodes will either lead their removal from interelectrode gap or neutralization at the electrodes, thus reducing the conduction current. The temporal dependence of current after the application of stress at a fixed value is found to decay in a manner which can be expressed as [21,28,29]:

$$i(t) = i_o t^{-n} \quad (5.2)$$

where n decreases with time. In commercial liquids, additional conduction current will be caused by the impurities beside particles, like water droplets, acids, resins and cellulose fibers.

4. High field can lead to emission of electrons at cathode, whereas at anode, field ionization can be initiated. According to the model proposed for nonpolar liquids by Schmidt et al. [25], the field strengths required for these processes to materialize are of the order of $1.5 \sim 2.0 \times 10^9$ Vm^{-1}. Recently, Denate et al. [30] demonstrated the occurrence of field emission current in the form of short-duration pulses in cyclo-hexane. Similarly, when the electrode tip is positive, the positive charge carrier formation due to field ionization has been observed both in aromatic liquids as well as in silicone oils [30,31]. Due to the injected space charge under high fields, the convection currents are established which augment the ion drift velocities and hence the conduction. If, at a given condition the injected charge has a density q, then it will give rise to a coulomb's force of qE. Due to action of this force on the bulk of liquid, hydrodynamic instability will be caused. It will develop convective motion in the liquid which is found to be always directed toward the opposite electrode [32]. This interelectrode turbulent motion of liquid is known as electrohydrodynamic (EHD) motion and the mobility of charge carriers in this situation is given as [27,33]:

$$\mu_{ehd} = \sqrt{\varepsilon_o \varepsilon_r / \rho} \quad (5.3)$$

where ρ is the density of the liquid. The transport of ions in this situation is enhanced by a factor $\sqrt{M}$, where M is the ratio of μ_{ehd} to the true mobility of the ions. For some polar and nonpolar liquids, M can vary in the range of $2 \sim 100$ [27].

5. Under high fields, conduction currents are usually accompanied with randomly occurring pulses or bursts of currents. In the past they have been reported to be either due to particles or ionization of microscopic gas bubbles in the bulk of the liquid [20,22]. However, with the aid of elegant ultrafast optical techniques, it has been demonstrated recently that these pulses are associated with bubble generation process [34] and the development and propagation of discharges called *trees* [35].

However, the dynamic nature of the process remained obscure due to the random nature of these events and the inherent delay in the detection system. These difficulties were overcome by using more viscous liquids, since the occurrence of prebreakdown pulses at a point cathode under DC are observed to be strongly influenced by insulating liquid viscosity [36]. Their onset is always initiated at a critical threshold voltage level which is a function of liquid structure and cathode tip radius. Their repetition rate increases in an exponential manner with the increase in voltage. At elevated stress levels, besides the single pulse activity, a regular burst activity also occurs. Each burst is composed of a well-structured set of pulses which increase in amplitude, initially in a linear fashion with time, before becoming irregular. Increasing the viscosity results in an increase in the duration of the burst and the time interval between the pulses within a burst of current. Using the onset of each pulse or burst to trigger the capture of shadowgraph always produces an associated discharge. Its size is found to increase with burst duration. Figure 5.11 demonstrates bursts of current pulses and associated shadowgraphs in silicone oil [37]. These studies also support the view that current pulses encompassed in a burst of conduction current are the consequence of periodic spark discharge inside a vaporized cavity which is formed by the initial current pulse through the process of Joule heating. At elevated stresses these discharges expand into well-defined tree-like structures which span around 2/3 of the gap before the breakdown ensued [36]. Under point anode, steady current is observed but only with occasional random pulses of small amplitude. However, at elevated stresses a sudden rise in current is associated with filamentary discharge, which once initiated always leads to complete breakdown [35,36]. Individual branches of these discharges are filamentary, unlike the discharge patterns observed under cathode which possess a thicker structure and therefore resemble the bush-type and tree-type discharge patterns observed under short-duration high voltage pulses.

5.7 BREAKDOWN IN INSULATING LIQUIDS

Efforts to understand breakdown mechanisms in a variety of liquid insulants have been continuing for many decades. The pertinent features of these studies have been summarized in a number of reviews [1,20,22,24,38]. However, the picture is not impressive as compared to the state of knowledge achieved for gases and solids. This is because the molecular structure of liquids is not simple and not so regular as compared with solids and gases. For instance, transformer oil alone contains well over 100 chemical compounds, and the fact that liquids tend to be contaminated with various impurities is a serious problem for fundamental studies. Moreover, the transition from liquid to gas phase, which takes place during

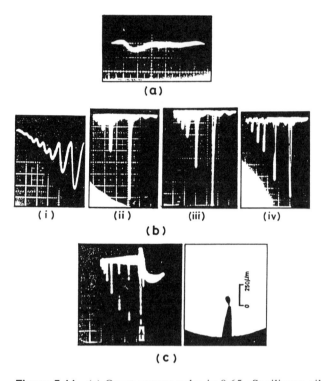

Figure 5.11 (a) Onset current pulse in 0.65 cSt silicone oil. Horizontal axis = 20 ns/div, vertical axis = 4 mA/div; 0.03 pC charge. (b) Regular current burst and the effect of oil viscosity upon pulse repetition rate (i) 0.65 cS oil 40 ns/div, (ii) 10 cS oil, 200 ns/div, (iii) 100 cS oil, 400 ns/div, (iv) 1000 cS oil, 4 ms/div. (c) Conduction current burst (time scale = 400 ns/div) with shadowgraph of associated discharge in 100 cS oil. Arrow indicates instant of light flash. (From Ref. 36.)

the development of breakdown, still further complicates the phenomena and hence their interpretations. Several promising hypotheses of breakdown based on particles and bubble effects were advanced in the late 1960s, but it has been necessary to modify and sometimes even reject them with the emergence of new experimental evidence.

5.7.1 Study of the Breakdown Process Without Optical Techniques

Based on these studies, the breakdown strength of liquids is influenced by various factors such as experimental procedure, electrode material and surface state, electrode geometry, presence of chemical impurities, presence of physical impurities, molecular structure of oil, temperature and pressure [24,38]. From such studies, breakdown theories like electronic theory, suspended particle theory, cavitation theory and bubble theory were postulated.

Electronic Breakdown Theory

This theory was extended at the early stages of investigations and used the concept of electron avalanche operative in gases. Field emitted electrons from a cathode were assumed to collide with the atoms of liquid molecules. If enough energy is transferred during such collisions, some electrons would be knocked off their atoms and drift with the original electrons toward the anode. Thus electron avalanches similar to those in gas discharge develop in the liquid and finally lead to breakdown [39–41]. This model appeared quite reasonable, but since 1960 it has been rejected on the following grounds:

1. The mean free path of electrons in liquids is very short (of the order of 10^{-6} cm) to enable them to acquire ionization potential of 10 eV necessary for liquid molecules. Moreover, there was no direct experimental evidence for an α-process.
2. It failed to explain pressure dependence of the breakdown process. At pressures of 25 atm breakdown strength increases by 50%, whereas at this pressure the mean free path of the electron is hardly altered [43].

Suspended Particle Breakdown Theory

Suspended particles are always an integral part of liquids. In spite of rigorous cleaning techniques imparted both on liquids as well as test cells, submicron sized particles cannot be removed from the system. The relative permittivity of these particles, ε_2, is higher than that of oil, ε_1. If we assume them to be spherical of radius r, and if the applied field is E, then the particles experience a force F such that:

$$F = \frac{1}{2} r^3 \frac{(\varepsilon_2 - \varepsilon_1)}{2\varepsilon_1 + \varepsilon_2} \text{ grad } E^2 \tag{5.4}$$

This force is directed toward areas of maximum stress. Particles will thus align on the high stressed electrode and start forming a bridge, which could lead to gap breakdown. Similarly if particulate matter is fiber it will get polarized due to the presence of moisture on its surface and move along converging fields. Assuming hemispherical tips with radius r, the charge q at either end of fiber would be [19]:

$$\pm q = \pi r^2(\varepsilon_2 - \varepsilon_1)\varepsilon_0 E \tag{5.5}$$

where ε_0 is the permittivity of free space. If the field in the gap is non-uniform and $\varepsilon_2 \gg \varepsilon_1$, the driving force will be

$$F = r^3 \varepsilon_0 \text{ grad } E \tag{5.6}$$

When a fiber reaches either electrode, its outward tip would act as extension of the electrode and cause field intensification and thus attract more fibers, thereby forming a bridge in the gap. This can lead to breakdown via joule heating of the bridge and its surrounding liquid.

Although this theory did explain the strength of liquids containing large amounts of particles, it is unlikely to be extended to pure liquids. Moreover, particles have been seen on several instances to bridge the gap, while discharge occurs in a different region and still at higher voltages [43]. This means breakdown involves some other mechanism. Nevertheless, particles may be instrumental as an aid in the process of breakdown.

Cavitation Theory of Breakdown

This theory was proposed by Krausucki [44], and is based on the concept that whenever a particle comes in the high field region, the presence of enhanced field on its surface will generate an electromechanical pressure (P_m) tending to lift the liquid off the particle surface against the opposing hydrostatic pressure (P_h) and pressure due to surface tension (P_{st}). This action will develop a region of zero pressure, thus forming a vacuous cavity. Electron bombardment of the walls of cavity will sustain its growth, which will eventually lead to breakdown. The critical condition for zero pressure generation and vacuous cavity formation is given as:

$$P_m = P_h + P_{st} \tag{5.7}$$

If m is the field intensification factor at the surface of particle, then $P_m = 1/2\ \varepsilon(mE)^2$ and $P_{st} = 2\psi/r$; where ψ = surface tension and r = particle radius. Putting these values in equation (5.7) leads to the expression:

$$\frac{\varepsilon}{2}(mE)^2 = P_h + \frac{2\psi}{r} \tag{5.8}$$

Krasucki selected m = 4.2, therefore breakdown strength in this case becomes:

$$E_b = 0.337 \left[\frac{1}{\varepsilon}\left(P_h + \frac{2\psi}{r}\right)\right]^{0.5} \text{Vm}^{-1} \qquad (5.9)$$

Using particle radii of 100 Å and 250 Å (particle size usually produced in spark breakdown), Krasucki showed that the pressure variation of the dielectric strength measured by Kao and Higham [42] is contained within the theoretical estimates as shown in Figure 5.12. However this theory fails to explain the effects of electrode material, electrode separation and dissolved gases.

Bubble Theory of Breakdown

According to this theory [45], a low density vapor bubble is generated in the liquid by the injection of large leakage currents at the cathode protrusions. By this process local vaporization can occur in a few milliseconds. If H is the heat needed to vaporize a liquid from ambient temperature T_a, then for a unit mass of liquid:

$$H = C_p(T_b - T_a) + L \qquad (5.10)$$

where C_p = specific heat at constant pressure, L = latent heat of vaporization, T_b = boiling temperature, and T_a = ambient temperature. Near

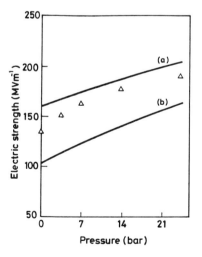

Figure 5.12 Variation of breakdown strength of n-hexane with hydrostatic pressure (a) r = 100 A°; (b) r = 250 A°; (Δ) experiments of Kao and Higham (Ref. 42).

breakdown the emission current from the cathode is space charge limited and is given as:

$$I = AV^x \qquad (5.11)$$

where x is in the range of 1.5 ~ 3 and A is a constant. It follows that the local energy input during the applied high voltage pulse duration τ_w can be expressed as:

$$H = A'E^x\tau_w \qquad (5.12)$$

where A' is constant. Combining equations (5.10) and (5.12) yields the relationship between E and H as:

$$A'E^x\tau_w = C_p(T_b - T_a) + L \qquad (5.13)$$

This is the thermal breakdown criterion and exhibits a marked pressure and temperature dependence since T_b increases with pressure. It also explains the effect of molecular structure of the liquids on breakdown. Figure 5.13 exhibits the fidelity of equation (5.13) when tested with respect to the experimental data of Kao and Higham [42]. However, the main objection to this model has been the simple heat transfer treatment based on the steady state equation for a phenomena which needs to be described by transient heat flow dynamics.

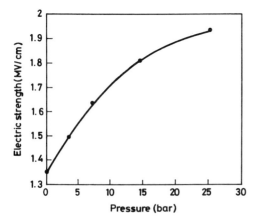

Figure 5.13 Pressure dependence of breakdown strength of n-hexane. Solid line is due to equation (5.13) with x = 1.5, whereas data points are from Ref. 42.

5.7.2 Study of the Breakdown Process Using Optical Techniques

It is clear that no single concept so far proposed can explain all experimental observations in a unified manner. However, with the advent of fast electro-optical techniques, our understanding of breakdown in liquids has been advanced tremendously. With these techniques, once a voltage pulse is applied any perturbations occurring in the electrode gap can be easily visualized under magnification by taking a photograph of each fast-occurring event. Verification of the bubble concept was first put to test using ultra high speed photography by Hakim and Higham [46] and later by several others [47–53]. They all confirmed that near breakdown, streamers emerge from the high voltage needle electrode that resemble in structure like bush or tree as shown in Figure 5.14. If the field at the originating electrode is critical than they grow out in the liquid toward the opposite electrode. Actual breakdown is preceded by the formation of secondary streamers which grow much faster than the primary ones.

To study the phenomena in more detail and also to see photographically the effects of various experimental parameters on the breakdown cri-

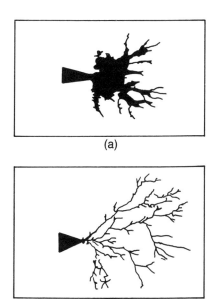

Figure 5.14 (a) Cathode initiated bush type and (b) anode initiated filamentary discharge in n-hexane.

teria (such as chemical additives, particulate additives, viscosity and rate of rise of the applied voltages), liquids varying from simple hydrocarbons and silicone oil to more complex technical fluids were used. Various sophisticated electroptical devices were employed. However, the most popular diagnostic methods presently being used are the shadowgraph and Schlieren techniques of photography. Salient features of these are explained next in brief [36].

Schlieren and Shadowgraph Optical Systems

The main elements of these two systems are shown in Figure 5.15. The light source at A can be a spark in a gas or a nanosecond light pulse laser along with a collimating lens, providing an intense parallel light beam. In case of the Schlieren system, the probing light beam passes through the liquid between the electrodes at B before being focused on the knife edge at C. The knife edge can be moved with a micrometer so as to cut off all or part of the light that would pass C. Any light that might skirt around the knife edge would be focused on to a still or moving film at D. If the knife edge is critically adjusted so that it just cuts off all the light, then on the occurrence of a discharge in the gap there is an associated temperature and density change in those parts of the liquid affected by the discharge. The change in the refractive index of effected part will cause the beam to be deflected. It may then pass beneath the knife edge and be focused at D. The effect is as if the low density region emits light, whereas in reality the light is derived from the illuminating beam and refracted by the discharge. Therefore, this technique is more useful for studies of problems in which a precise density gradient is to be probed, since the angle of refraction of the ray passing through the object is directly linked to the density gradient.

In the case of the shadowgraph system of photography, the knife edge is not used and the probing light rays are either absorbed, reflected or refracted out of the low density region (the discharge) and therefore do not reach the image plane. The region in which this phenomena occurs is thus represented by its shadow in the image plane at D. Comparatively, shadowgraphy is the diagnostic method which is easy to set up and gives an economical and qualitative description of shape, size and different density regions and of strong gradients occurring in the liquid under the application of high voltage.

To capture an event associated with the applied voltage pulse in the electrode gap, synchronization of voltage pulse output at B, the light pulse at A and camera at D is essential. Alternatively, if events associated with current pulse are to be observed, then the synchronization of this pulse

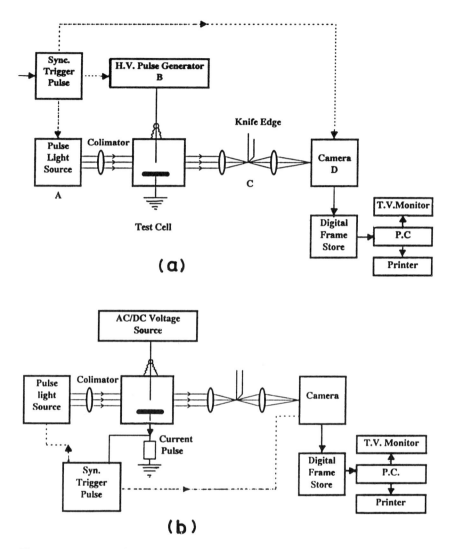

Figure 5.15 Essential elements of schlieren/shadowgraph system with synchronization of (a) applied voltage pulse and (b) induced current pulse.

with light source and camera is essential. Pictures taken at D could be single-frame or multiframe photography. For single-frame operation, use is made of regular cameras; however, for the purpose of multiframing, ultra high speed image converter cameras (ICCs) are being used with or without image intensifiers [1]. By multiframe photography, usually 10–12 frames are obtained with a speed which varies over $2 \times 10^7 \sim 10^5$ frames per second and with exposure time ranging from 100 ns to 10 ms per frame [30]. More recently, digital frame stores are also being used, where the events captured by a CCD camera are converted into a digital form and can be viewed on a TV monitor, and saved or processed by the computer and printed at will [36,54,55].

Experimental Results

All of the results compiled so far confirm the role of the electrode/liquid interface, the discharges (steamers) produced in the prebreakdown regime, and their fast propagation in the interelectrode gap which leads to final breakdown.

Earlier workers were mostly confined to visualization of the failure mechanisms in liquid under unrealistic overvoltages. However, recent investigations [36,54,56] deal in detail with the initial stage of discharge growth and its relation to the molecular/liquid structure and other associated characteristics. Based on these and many other findings [49–56], models of cathode and anode initiated breakdown mechanisms have been forwarded.

Breakdown Processes at Cathode

With the increase in voltage to a critical threshold level, either a thin or spherical image of low density region appears at cathode which expands with time. Experiments carried out in low viscosity fluids and both under DC fields [57] as well as short duration voltage pulses [36,58] have shown that this low density region consists of a vaporized bubble which expands and collapses following Raleighs theory [59]. According to this theory, the time taken for a cavity to collapse (τ) limited by inertia and driven by the ambient pressure is given as:

$$\tau = 0.915 \, R_m \left[\frac{\rho}{p_a} \right]^{0.5} \tag{5.14}$$

where R_m = cavity maximum radius, p_a = ambient pressure and ρ = liquid density. If the cavity is formed from an initial impulse of energy (i.e., in a very short time), then the growth and collapse will be symmetrical in time.

Figure 5.16a illustrates the growth and collapse of cavity produced in 0.65 cSt silicone fluid following the application of a single 632 ns high voltage pulse. The cavity takes several microseconds to grow to R_m before it starts to collapse in size with almost at similar rate to that of its expansion. The cavity size rebounds after reducing down to 25 μm and the

Liquid Dielectrics 139

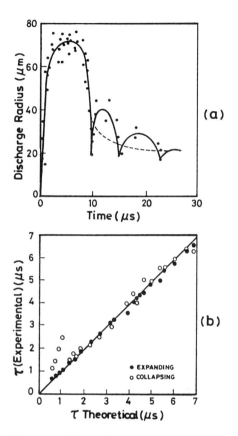

Figure 5.16 (a) Cavity radius versus time in 0.65 cSt silicone oil, under the application of a 632 ns duration 15 kV pulse. (b) Experimental collapse and expansion time of cavity versus theoretical time according to equation (5.14). (From Refs. 36 and 58.)

process is repeated in several cycles of rapidly decreasing amplitudes, until the cavity comes to an equilibrium state with surrounding liquid. Figure 5.16b compares the variation of theoretical value of τ estimated from R_m and based on equation (5.14) to the experimentally observed values of both the contraction as well as expansion of the cavity. It is clear that the experimental values conform in an excellent manner with the theoretical values and strongly exhibit the inertia limited expansion and contraction of a bubble.

Figure 5.17 depicts growth of cavities with time for 0.65, 10 and 100 centistoke (cSt) viscosity silicone oil. These images cover the initial growth

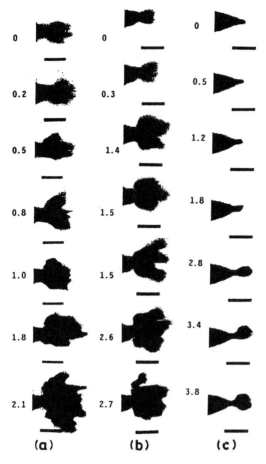

Figure 5.17 Shadowgraphs illustrating the growth of the cavity with time, and the development of wave-like instability in silicone oil, at 7.5 kV (50 μm scale bar on each picture). Delay times are marked in μs. Oil viscosity (a) 0.65 cSt; (b) 10 cSt; and (c) 100 cSt. (From Refs. 36 and 54 © IEEE, 1991.)

of cavity, the formation of the instabilities and then propagation of these instabilities deep into the gap. With the increase in voltage or with the increase in pulse duration more instabilities appear and they elongate into filaments. Thus the cavity develops into a bush-type structure, and these figures are referred in literature as primary streamer elements. If the voltage magnitude is small, these streamers detach from the cathode and start fragmenting into bubbles. However, if voltage is significant, these streamer elements propagate with even larger velocity toward the anode. At breakdown voltage level, when primary streamer elements cross 1/3 to 3/4 of the electrode gap, then secondary streamers emanate from primary ones and finally impinge at the anode, causing breakdown. The secondary streamers possess thinner filaments and their growth rate approaches sonic speed and is almost an order of magnitude higher than that of primary elements. These structures have been found to be conductive and are also found to be suppressed with the increase in ambient pressure. Light pulses are also reported to occur in synchronism with the current pulses that appear as the initial cavity begins to expand. The following description represents a generally accepted mechanism for initiation and propagation of negative streamers.

Energy for cavity generation and expansion comes from a current pulse from the cathode point. Microbubbles (<10 μm in diameter) are created by rapid localized injection of current pulse from the point cathode, and this energy is converted into heat, which causes evaporation of liquid and provides driving force. Beyond kinetic limits, the bubble expands by the dual action of internal vapor pressure and the force generated by the electrostatic field. Inside the cavity the pressure will decrease as the structure expands. As the pressure drops, the mean free path of electrons being injected from the cathode increases and they are thus able to gain energy from the field and produce impact ionization and/or breakage of C-C and C-H bonds of oil. Eventually the Paschen's minimum limit will be reached and a sudden discharge will occur. The external circuit will recognize it as a current pulse. This way the internal field in the cavity controls charge injection while external field, largely determined by the shape of cavity, appears to control its growth.

From this cavity surface appear the streamer filaments. Watson [60] has treated this problem as the Raleigh–Taylor instability. According to this theory if one medium expands within another medium of different density then hydrodynamic instabilities appear at the interface. In the present case, the field-driven interface with laterally mobile charges corresponds to an EHD instability. The amplitude of instabilities grows as the field at the outward tip of the elements increases. Fields in excess of 10^8 V/cm have been calculated to exist in front of these elements. This is how the primary branches of streamers grow at rates that are of the order of 1 km/s or less. The secondary streamers emanate from primary filaments and grow at a rate in excess of 10 km/s. At present, little is known about the cause of these secondary streamers except that presence of impurities may be involved [56].

Breakdown Processes at Anode

Generally the positive streamer growth rates are consistently higher than those for the negative streamers. The structure of the two types are also different. The positive streamers are mostly filamentary as compared to thicker and bushier patterns of negative streamers as shown earlier in Figure 5.14. Initiation voltages are always higher with positive polarity than with negative polarity. Once initiated, provided that the voltage is maintained, all but the smallest positive discharges inevitably lead to breakdown.

There are three stages of growth for streamers that originate at the point anode. The first is the bushy, subsonic structure which occurs mostly near the onset voltage levels or when the insulants are subjected to elevated atmospheric pressure. The second is a thin filamentary structure which propagates near sonic velocity. The third mode is an order of magnitude faster than the second [56,61].

Light emission in the form of pulses also occurs with the growth of positive streamers. Light emission is similar for anode and cathode streamers, but the growth of the anode streamer is much smoother than for the cathode streamer.

Various models have been proposed to explain the propagation of anode-initiated streamers. However each is supported by a limited set of experimental results, and none of them gives an adequate explanation for all the experimental phenomena reported so far. According to the model of Devins et al. [38], the positive steamers grow by a field ionization process. Thermalized electrons present in the liquid (due to the liquid's contact with the anode) are focused near the anode surface, where favorable sites exist for electron transmission that are characterized by a large enhancement factor and a low work function. Such a site then enhances the local field and causes field-induced ionization of the liquid column existing between the focused charge and the anode spot. The conductive streamer can be represented as an extension of the electrode in liquid. The tip of the streamer replaces the surface spot on anode and the field ionization process can repeat itself in any direction in which focused charges can accumulate near the tip. Overall, the streamer grows steadily but in short spurts.

Field ionization of liquid at the head of the streamer filament has also been put forward by Chadband [62]. According to this model, propagation of the streamer tip depends upon the local tip field. The local field is taken as the vector sum of the applied (Laplacian) field and the Coulomb field arising from the positive charge at the streamer tip. Growth occurs through the action of the local tip field on the liquid ahead of the tip. Negative charge is released and falls into the tip, leaving behind a net positive charge which becomes an extension of the tip. This process causes both heating and dissociation of liquid, making the filament visible optically.

Watson [63] considers that positive streamers propagate as columns of hot liquid that convert to high-pressure vapor channels in volatile fluids, whereas Feleci [64], on the basis of energetic considerations, concludes that steamer filaments probably contain an ionized gas phase. Lesaint [65], through meticulous experimental technique, showed that these filaments expand and collapse like cavitation bubbles as they are composed of vapor whose pressure varies with time and space. This vapor is created mainly from energy dissipated at the fast-moving filament tip.

Postbreakdown Events

When a streamer bridges the gap, an arc begins to develop. This development is characterized by a rapid increase in current, a rapid decrease in the voltage and the emission of visible light. During this stage the resistance of the electrode gap approaches the wave resistance of the connecting conductors [1]. It has been found that once the breakdown is initiated in n-hexane, the current rises to values of more than 100 amperes in 4 ns [66]. The establishment of the arc deposits significant energy in a small volume in the fluid and thus raises that volume to a high temperature. Upon arc initiation, an intense emission of light occurs and a shock wave is produced which propagates away from the arc at the sonic velocity. Liquid vaporizes quickly to make a gas column surrounding the arc. This column is quickly transformed into a bubble which expands in time due to the heat energy. The bubble then collapses in a few seconds by disintegrating into smaller bubbles that eventually lead to the liquid surface or are finally absorbed by the liquid.

REFERENCES

1. R. Bartnikas (ed.), *Engineering Dielectrics Vol. II: Electrical Insulating Liquids*, ASTM, Philadelphia, 1994.
2. B. Seymour (ed.), *Handbook of Polyolefins*, Marcel Dekker, Inc., New York, 1993.
3. A.C.M. Wilson, *Insulating Liquids*, Peter Peregrinus Ltd., London, England, 1980.
4. D. Kind and H. Karner, *High Voltage Insulation Technology*, Fried, Vieweg and Sons, Braunschweig, Germany, 1985.
5. W.T. Shugg, *Handbook of Electrical and Electronic-Insulating Materials*, IEEE Press, New York, 1995.

6. R.C.M. Pompilli and C. Mazetti, Electrical Insulation Magazine, Vol. 8, No. 1, pp. 28–32, 1992.
7. A. Bradwell, *Electrical Insulation*, Peter Peregrinus Ltd., London, England, 1983.
8. M. Daval, J. Aubin, Y. Gigurre, G. Pare and Y. Langhame, IEEE Trans. on Elect. Insul. Vol. EI-17, No. 5, pp. 414–422, 1982.
9. T. Tanaka, T. Okamoto, K. Nakanishi and T. Miyamote, IEEE Trans. on Elect. Insul. Vol. 28, No. 5, pp. 826–844, 1993.
10. D.W. Murray, Proceedings of World Petroleum Congress, Vol. 11, No. 4, pp. 447–457, 1983.
11. C. Lamarre, J.P. Crine and M. Duval, IEEE Trans. on Elect. Insulation, Vol. EI-22, No. 1, pp. 57–62, 1987.
12. T. Tanaka and A. Greenwood, *Advanced Power Cable Technology*, CRC Press, Boca Raton, Florida, 1983.
13. T. Tanaka, T. Okamoto, K. Nakanishi and T. Miyamoto, IEEE Trans. on Elect. Insul. Vol. 28, No. 5, pp. 826–844, 1993.
14. J.K. Nelson, IEEE Elect. Insul. Magazine, Vol. 10, No. 3, pp. 16–28, 1994.
15. I.A. Metwally, IEEE Trans. on Dielectrics and Elect. Insul., Vol. 3, No. 2, pp. 307–315, 1996.
16. M. Ieda, Y. Yanari, T. Miyamoto and M. Higaki, EPRI Workshop on Static Electrification, EPRI, San Jose, California, 1992.
17. M. Ieda, K. Goto, H. Okubo, T. Miyamoto, H. Tsukioka and Y. Kohno, IEEE Trans. on Elect. Insul., Vol. 23, No. 1, pp. 153–157, 1988.
18. IEC-Draft 1198, "Methods for Determination of 2-Furfural and Related Compounds", IEC, 1991.
19. M. Khalifa, *High Voltage Engineering: Theory and Practice*, Marcel Dekker, Inc., New York, 1990.
20. A.A. Zaky and R. Hawley, *Conduction and Breakdown in Mineral Oils*, Peter Peregrinus Ltd., London, England, 1973.
21. I. Adamczewski, *Ionization, Conductivity and Breakdown in Dielectric Liquids*, Taylor and Francis Ltd., London, England, 1969.
22. T.J. Gallagher, *Simple Dielectric Liquids*, Calendren Press, Oxford, England, 1975.
23. A. Gemant, *Ions in Hydrocarbons*, Interscience Publishers, New York, 1962.
24. A.H. Sharbaugh, J.C. Devins and S.J. Rzad, IEEE Trans. on Elect. Insul., Vol. EI-13, pp. 249–276, 1978.
25. W.F. Schmidt, IEEE Trans. on Elect. Insul., Vol. EI-19, pp. 389–418, 1984.
26. K.C. Kao, IEEE Trans. on Elect. Insul., Vol. EI-11, pp. 121–128, 1976.
27. N. Felici, IEEE Trans. on Elect. Insul., Vol. EI-20, No. 2, pp. 233–238, 1985.
28. T. Matuszewski, Acta Physica Polonica, Vol. A48, pp. 861–869, 1975.
29. A.A. El-Sulaiman, A.S. Ahmad and M.I. Qureshi, IEEE Trans. on Elect. Insul., Vol. EI-16, No. 5, pp. 453–457, 1981.
30. A. Denat, J.P. Gosse, IEEE Trans. on Elect. Insul., Vol. 23, No. 4, pp. 545–554, 1988.

31. J.G. Rabe and W.F. Schmidt, Journal of electrostatics, Vol. 7, pp. 253–266, 1979.
32. T. Takashima, R. Hanaoka, R. Ishibashi and A. Ohtsubo, IEEE Trans. on Elect. Insul., Vol. 23, No. 4, pp. 645–658, 1988.
33. P. Atten, IEEE Trans. on Dielectric and Elect. Insul., Vol. 3, No. 1, pp. 1–17, 1996.
34. M.I. Qureshi and W.G. Chadband, IEEE Trans. on Elect. Insul., Vol. 23, No. 4, pp. 715–722, 1988.
35. O. Lessaint and R. Tobazeon, IEEE Trans. on Elect. Insul., Vol. 23, pp. 941–954, 1988.
36. M.I. Qureshi, Ph.D. thesis, University of Salford, U.K., 1992.
37. M.I. Qureshi and W.G. Chadband, Proc. of 10th ICDL, Grenoble, France, IEEE Conf. Recd. 90CH 2812-6, 1990.
38. J.C. Devin, S.J. Rzad and R.J. Schwab, Journal of Applied Physics, Vol. 52, pp. 4531–4545, 1981.
39. A. von Hippel, Journal of Applied Physics, Vol. 8, pp. 815–832, 1937.
40. T.J. Lewis, Journal of Applied Phys. Vol. 27, pp. 645–650, 1956.
41. D.W. Swan, Proceedings of the Physical Society, Vol. 78, pp. 423–432, 1961.
42. K.C. Kao and T.B. Higham, Journal of the Electrochemical Society, Vol. 108, pp. 522–528, 1961.
43. A. Ruffini, D.A. Hoch and J.P. Reynders, Proc. 6th International Symposium on High Voltage Engineering, New Orleans, Louisiana, Paper No. 13.14, 1989.
44. Z. Krausucki, Proceedings of Royal Society London, Vol. A-294, pp. 393–404, 1966.
45. P.K. Watson and A.H. Sharbaugh, Journal of Electrochemical Society, Vol. 107, pp. 106–112, 1960.
46. S.S. Hakim and J.B. Higham, Nature, Vol. 189, pp. 966–996, 1961.
47. W.G. Chadband and T.G. Wright, British Journal of Appl. Physics, Vol. 16, pp. 305–313, 1965.
48. W. Hauschild, Ph.D. thesis, Technical University of Dresden, Germany, 1969.
49. W.G. Chadband and J.H. Calderwood, Journal of Electrostatics, Vol. 17, pp. 75–91, 1979.
50. E.O. Foster and P.P. Wong, IEEE Trans. on Elect. Insul., Vol. EI-12, pp. 183–187, 1977.
51. H. Yamashita and H. Amano, IEEE Trans. on Elect. Insul., Vol. EI-20, pp. 247–255, 1985.
52. R.E. Hebner, Annual Report CEIDP, pp. 26–34, 1983.
53. O. Lessaint and P. Gourney, IEEE Trans. on Elect. Insul., Vol. 1, No. 4, pp. 702–708, 1994.
54. S.M. Arghi, M.I. Qureshi and W.G. Chadband, IEEE Trans. on Elect. Insul., Vol. 26, No. 4, pp. 663–672, 1991.
55. M.I. Qureshi and W.G. Chadband, Proc. of the International Conference on Partial Discharge, IEE Publ. No. 378, pp. 9–10, 1993.
56. E.O. Foster, Journal of Physics D. Appl. Phys., pp. 1506–1514, 1990.

57. R. Kattan, A. Denat and N. Bonifaci, IEEE Trans. on Elect. Insul., Vol. 26, No. 4, pp. 656–662, 1991.
58. M.I. Qureshi, W.G. Chadband and P.K. Watson, Proc. of 6th IEE Conf. on Dielectric Materials and Measuring Apparatus, pp. 89–92, Manchester, England, 1992.
59. L. Rayleigh, Phil. Magazine, Vol. 34, pp. 94–98, 1917.
60. P.K. Watson, W.G. Chadband and M.S. Araghi, IEEE Trans. on Elect. Insul., Vol. 26, No. 4, pp. 543–559, 1991.
61. G.J. Fitzpatrick, Proc. of CEIDP, 1985, IEEE Conf. Record 85CH216J-9, pp. 26–32, 1985.
62. W.G. Chadband, IEEE Trans. on Elect. Insul., Vol. 23, pp. 697–706, 1988.
63. P.K. Watson, T. Sufian, W.G. Chadband and H. Yamashita, Proc. of 11th IEEE Intl. Conf. on Conduction and Breakdown in Dielectric Liquids, Baden, Switzerland, pp. 234–238, 1993.
64. N. Felici, IEEE Trans. on Elect. Insul. Vol. 23, pp. 497–503, 1988.
65. O. Lesaint and P. Gourney, Journal of Physics D: Appl. Phys., Vol. 27, pp. 2111–2116, 1994.
66. J. Fuhr and W.F. Schmidt, Journal of Applied Physics, Vol. 59, pp. 3702–3708, 1986.

6
Solid Dielectrics

6.1 INTRODUCTION

Solid insulating materials are used extensively in all types of electrical power networks, devices and substation equipment. The equipment size and operating limitations are dictated by the type and amount of material required for insulation. In the early days of the electrical industry, engineers had to adapt varnishes, natural resins and petroleum residues as saturents and coatings for tapes used to wrap coils and cables. Today, with the advent of polymers, as well as the availability of a variety of glasses and ceramics, an almost unlimited number of insulating materials is available, and the problem becomes one of selection rather than adaption.

Solid dielectrics have high breakdown strength as compared to liquids and gases. A good dielectric should, besides having high dielectric strength, have low dielectric loss and high mechanical strength and stiffness, be free of gaseous inclusions and moisture, and be resistant to thermal and chemical deterioration. It should also be insensitive to ambient condition at its site of application. Ozone resistance, impermeability, hygroscopic resilience, low water absorption, radiation stability, among others, are the additional requirements. Efficient utilization and improvement in the performance of such materials, therefore, calls for not only a detailed knowledge of their essential properties but also the mechanisms which cause their degradation and failure. This chapter illustrates both of these essential prospects of solid dielectrics.

6.2 SOLID INSULATING MATERIALS

A larger proportion of commercial insulating materials that are used in electrical power apparatus are solids. With the advent of new insulating materials in the past three decades, many earlier conventional materials that were commonly used in apparatuses like electrical machines, cables and capacitors are being discarded now. The most common method of classifying solids is by using their chemical composition, such as organic, inorganic and synthetic polymers. Some examples are outlined in Table 6.1. From their application's point of view, they can be classified as: (1) thermoplastic compounds, (2) thermosetting compounds, and (3) embedding and jacketing compounds.

Organic materials are derived from either vegetable or animal matter. They possess good insulating properties. However, they deteriorate rapidly if their operating temperature exceeds 100°C. They are mostly employed after treatment with varnishes or impregnation in oil. Examples of such insulators include paper and pressboard that are commonly used in oil-filled equipment such as cables, capacitors, panel boards and transformers.

Inorganic materials are distinctly different from organic substances. As a rule, they do not show any appreciable fall in either mechanical or electrical quality at 100°C and may retain their properties up to a working temperature of ~250°C. Because of their compact physical structure, they do not absorb oil or varnish, with the exception of the fibrous asbestos material. Inorganic solids are difficult to fabricate but they are very good dielectrics. The most important members of this group are glasses and ceramics. Glass serves as a material for power-line insulators whereas porcelain, in the form of a bushing, was the first ceramic material to be used by the electrical industry [1].

Table 6.1 Classification of Some Commercial Solid Dielectrics

		Synthetic polymers	
Organic	Inorganic	Thermoplastic	Thermosetting
Amber	Ceramics	Perspex	Epoxy resins
Paper	Glass	Polyethylene	Phenolics
Pressboard	Mica	Polypropylene	Melamine
Rubber	Fiber glass	Polystyrene	Urea formaldehyde
Wood	Enamel	Polyvinyl chloride	Crosslinked polyethylene
Resins		Polyamid	Elastomers
		Polycarbonate	

Synthetic polymers include all types of polymeric materials that have been produced by various industrial processes. Polymers are generally divided into two groups: thermoplastic and thermosetting. The former have low melting temperatures, in the range of 100–120°C. However, they are flexible and can be molded and extruded at temperatures below their melting points. These properties make thermoplastics extremely desirable as insulants for high-voltage cables. Thermosetting polymers are heat-curing solids. On heating they acquire substantial mechanical strength and hardness. Epoxies, introduced in 1947, are predominant in embedding compounds. In combination with phenolics they make useful thermosetting compounds. Polycarbonate was introduced in 1950. It has excellent dielectric properties and is also corona resistant. Polyethersulfone resins were introduced in 1973. They possess a temperature index rating of 180°C, which is the highest for any thermoplastic. Specialty resins introduced since 1980 include polyacrylate and polyetherether ketone (PEEK), which can be applied up to a temperature of 200°C and are also highly resistant to hydrolysis [2]. Some important types of insulating materials that are commonly used as high voltage insulants along with potential areas of their applications are described next.

6.2.1 Dielectric Paper and Boards

Dielectric paper and boards are produced from a variety of materials, including wood, cotton, organic fibers, glass, ceramics and mica. The distinction between paper and board is not specific but paper is generally <0.8 mm thick, whereas boards are >0.8 mm thick. For more than 6-mm thickness, boards are laminated with adhesive to get the desired thickness. The boards are also referred to as pressboard, transformer board or fuller board.

The paper normally employed for insulation purposes is a special variety known as Kraft paper. The thickness and density of papers vary depending on their application. Low-density paper (0.8 g/cm^3) is preferred in high-frequency capacitors and cables, while medium density paper is used in power capacitors. High-density papers are preferred for energy storage capacitors and for the insulation of DC machines.

Paper is hygroscopic, therefore, it has to be dried and impregnated with mineral oil, synthetic oils or vegetable oils. The relative dielectric constant of impregnated paper depends upon the permittivity of cellulose (the base material) and the permittivity of the impregnant and the density of the paper. Paper is also used in the form of hardboard and pressboard. Hardboard is produced by compression of paper with epoxy or phenolic resins. It is used as supporting material and insulating barrier. On the other hand, soft paper or pressboard is used with impregnation in transformers and bushings.

6.2.2 Mica and Its Products

Most of the mica used for insulation purpose is a naturally occurring inorganic substance. It occurs in the form of crystalline mineral silicates of alumina and potash. Mica can be split into very thin flat laminae. It has a unique combination of electrical properties, such as high dielectric strength, low dielectric losses, resistance to high temperatures and good mechanical strength. Due to these properties, it is used in many electrical apparatus where high temperatures are experienced. The grading of muscovite mica, the most widely used for insulation purposes, is covered by ASTM standard D351 [3].

Mica is also grown electrothermally with a composition similar to that of natural mica. Its splittings are used extensively for glass-bonded mica insulation. Mica is built into sheet form by bonding together with a suitable resin or varnish. Depending on the type of application, mica can be mixed with a required type of resin to meet the operating temperature requirements. Micanite is another form of mica which is used extensively for insulation purposes. Mica splittings and mica powder are used as fillers in insulating materials, such as glass and phenolic resins. The use of mica as a filler results in improved dielectric strength, reduced dielectric loss and improved heat resistance and hardness of the material. Table 6.2 gives electrical properties of mica used for electrical insulation purposes.

Table 6.2 Electrical Properties of Mica

Property	Natural mica	Synthetic mica
Dielectric strength (at 30°C)	~1000 kV/mm	~1000 kV/mm
Dielectric constant (tan δ) (1 kHz–3 GHz)	6.5–8.7	6.5
Loss tangent		
50 Hz	0.03	—
1 MHz	0.001	0.0002
Surface resistivity (at 60% humidity)	10^{12}–10^{14} $\Omega \cdot m$	—
Volume resistivity (constant up to 200°C)	10^{15}–10^{17} $\Omega \cdot m$	10^{15}–10^{17} $\Omega \cdot m$
Maximum operating temperature	540°C	980°C

Glass-bonded mica is produced by bonding finely grounded (natural or synthetic) mica with low temperature melting electrical grade glass, to form a ceramic that is both machinable as well as moldable. Its important properties are its high service temperature (650°C) and high mechanical strength, and that it is nonflammable, nonarcing, nontracking or outgassing and impervious to moisture. It is being used extensively in circuit breakers, switchgears, arc barriers and bushings.

6.2.3 Glass

Glass is a thermoplastic inorganic material comprising a complex systems of oxides (SiO_2). Glass is defined as a liquid which has cooled to a rigid solid without crystallization. At temperatures below the glass transition temperature, glass is rigid and displays properties of the crystalline state. At the temperatures above glass transition temperature, glass is plastic and viscous. Glasses of interest in electrical insulation are primarily silicate based. The so-called E-glass is used for producing fiber glass which is used for reinforcing plastic materials to obtain high mechanical strength. The dielectric constant of glass varies from 3.7 to 10, whereas the density varies from 2.2 to 6 g/cm³. At room temperature, the volume resistivity of glass varies from 10^{14} to 10^{22} ohm-m. The dielectric loss factor of glass varies from 0.004 to 0.02 depending on the frequency. The losses are highest at lowest frequencies. The dielectric strength of glass varies from 3 to 5 MV/cm, which decreases with increase in temperature, reaching half the value at 100°C. At present the most common application of glass on power networks is in the form of fiber glass, which is used (1) in bandaging core packets of transformers, (2) as resin-impregnated fiberglass cores for composite insulators, (3) as resin-impregnated fiberglass mats and insulating plates, and (4) as fiberglass reinforced plastics in the form of tapes in electrical machines.

Glass in the form of paper is also used for insulation purpose. It is composed of glass microfibers. Its outstanding feature is its thermal stability up to 538°C. Other attributes include high thermal conductivity, low moisture adsorption and good chemical resistance [2].

6.2.4 Ceramic Insulating Materials

Ceramics are inorganic materials produced by consolidating minerals into monolithic bodies by high-temperature heat treatment. Ceramics can be divided into two groups depending on the dielectric constant. Low-permittivity ceramics ($\varepsilon_r < 12$) are used as insulators, while the highpremittivity ceramics ($\varepsilon_r > 12$) are used in capacitors and transducers. Table 6.3 gives selected dielectric properties of some ceramics commonly used for electrical insulation purposes.

Porcelain and steatite are gas tight, corrosion proof, chemically inert to all alkalies and most of acids and therefore resistant to contamination. Steatite possesses higher mechanical strength and low ε_r and is also suitable for making electronic components.

Alumina (Al_2O_3) has replaced quartz because of higher mechanical strength, good insulating properties and good thermal conductivity. Thermal conductivity of plastics range between 0.15 and 0.3 W/mK, in porcelains and glass it ranges between 1.2 and 1.7 W/mK whereas for alumina it stands around 35 W/mK [4]. Its dissipation factor of 0.0002 at 1 MHz is the lowest among all ceramics, although other dielectric properties are not outstanding. Overall, alumina is one of the best ceramic insulations available. It is used to fabricate high-current vacuum interrupters, and for a variety of electrical/ceramic components. Its powder is being used to

Table 6.3 Dielectric Properties of Electrical Insulating Ceramics

Ceramic	Dielectric strength V/mil (ASTM D 149) (kV/mm)	Dielectric constant 1 MHz (ASTM D 150)	Dissipation factor 1 MHz (ASTM D 150)
Alumina (99.9% Al_2O_3)	340 (13.4)	10.1	0.0002
Aluminum silicate	150 (6.0)	4.1	0.0027
Beryllia (99% BeO)	350 (13.8)	6.4	0.0001
Boron nitride	950 (37.4)	4.2	0.00034
Cordierite ($2MgO \cdot 2Al_2O_3 \cdot 5SiO_2$)	200 (7.8)	4.8	0.0050
Magnesia (MgO)	—	5.4	< 0.0003
Porcelain ($4K_2O \cdot Al_2O_3 \cdot 3SiO_2$)	—	8.5	0.005
Quartz (SiO_2)	—	3.8	0.0038
Sapphire	—	9.3–11.5	0.0003–0.00086
Silica (fused)	—	3.2	0.0045
Steatite	—	5.5–7.2	0.001
Zircon ($ZrSiO_4$)	—	5.0	0.0023
Magnesium metatitanate ($MgTiO_3$)	—	16	0.0002
Strontium zirconate ($SrZrO_3$)	—	38	0.0003
Titanium oxide (TiO_3)	—	90	0.0005
Calcium titanate ($CaTiO_3$)	—	150	0.0003
Strontium titanate ($SrTiO_3$)	—	200	0.0005
Barium titanate ($BaTiO_3$)	—	1500	0.015

produce various composite insulators and sheets. Similarly, feldspar coarse grain porcelain insulators have been replaced by fine-grain alumina-porcelain. At present, alumina based porcelains are widely being used for suspension-type insulators, station post insulators, and so on, which demand a high mechanical strength.

6.2.5 Polymers

Polymers consist of long-chain macromolecules with repeating monomer (or mer) units. A polymer is usually named by putting the prefix *poly-* in front of the name of the monomer from which it is derived. For example, the monomer ethylene is the repeated monomer in polyethylene:

$$\begin{array}{c} H \quad H \\ | \quad | \\ C = C \\ | \quad | \\ H \quad H \end{array} \longrightarrow \left[\begin{array}{c} H \quad H \\ | \quad | \\ C - C \\ | \quad | \\ H \quad H \end{array} \right]_n \qquad (6.1)$$

ethylene   polyethylene

A selection of polymers which are commonly used in electrical insulation, together with their monomers is given in Table 6.4. Different molecular units are attached to the ends of the chains (e.g., CH_3 in polyethylene). However, since n (the degree of polymerization) is very large in this case (typically in the range 10^3–10^5), the end units do not usually influence the physical properties of the polymer.

Table 6.4 includes some important polymers which are based on (—C—C—) linkage along the length of the polymer "backbone." These are known as *homopolymers*. The other form is *hetrochain* polymers, in which carbon atoms in the backbone have been replaced by other elements and are placed into categories depending on their characteristic chemical linkages. Simple polymer chains may form branches off the main chain; this is commonly found in polyethylene as shown in Figure 6.1 (on page 155). Such branches can occur every 30–100 monomer units along the backbone and result in side branches which can be short (e.g., up to several monomer units long) or which can be long (e.g., as long as the main chain). Branching can be produced or inhibited to a large extent by altering the polymerization conditions. Branching reduces the potential for regular molecular packing and so lowers the density; for instance, in the case of polyethylene (PE), it produces what is commonly called low-density pol-

Table 6.4 Chemical Structure of Commonly Used Polymers

Generic structure		Name (abbreviation)
-C(X)(X)-C(X)(X)-	X = H X = F	polyethylene (PE) polytetrafluoroethylene (PTFE)
-C(H)(X)-C(H)(H)-	X = CH$_3$ X = Cl X = C$_6$H$_5$ X = OCOCH$_3$	polypropylene (PP) poly(vinyl chloride) (PVC) polystyrene (PS) poly(vinyl acetate) (PVA)
-C(H)(H)-C(X)(X)-	X = Cl X = F X = CH$_3$	poly(vinylidine chloride) (PVDC) poly(vinylidine fluoride) (PVDF) polyisobutylene (butyl rubber)
-C(H)(H)-C(X)(Y)-	X = CH$_3$ Y = COOCH$_3$	poly(methyl methacrylate) (PMMA)
-CH$_2$-C(H)=C(X)-CH$_2$-	X = H X = CH$_3$	polybutadiene (BR) polyisoprene (natural rubber)
⌬-C(CH$_3$)(CH$_3$)-⌬-O-C(=O)-O-		polycarbonate (PC)
⌬-O-⌬-O-⌬-C(=O)-		poly(ether ether ketone) (PEEK)
-(CH$_2$)$_n$-N(H)-C(=O)-(CH$_2$)$_m$-C(=O)-N(H)-		m = 4, n = 6, polyamide 6.6 (PA6.6, nylon 6.6)

yethylene (LDPE). LDPE is mechanically inferior to high-density polyethylene (HDPE), its nonbranched counterpart, but nonetheless it has excellent dielectric properties and is commonly used in high voltage power cables. Beside the branches, polymers also have "cross-links" in which the polymer chains are joined by short, long or even polymeric molecules which effectively form connecting branches. A cross-linked polymer is therefore, in principle, essentially one gigantic molecule and so above its melting point (if it is semicrystalline) or above the so-called glass transition temperature (if it is amorphous) it becomes rubber-like rather than a liquid.

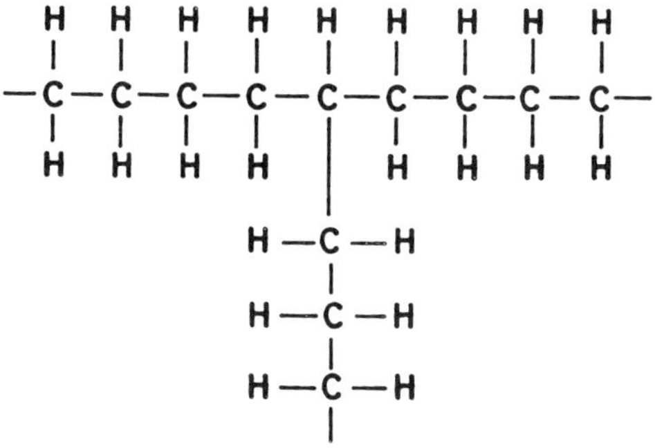

```
 H   H   H   H   H   H   H   H   H
 |   |   |   |   |   |   |   |   |
-C — C — C — C — C — C — C — C — C—
 |   |   |   |   |   |   |   |   |
 H   H   H   H   |   H   H   H   H
                 H — C — H
                     |
                 H — C — H
                     |
                 H — C — H
                     |
```

Figure 6.1 Side branches in the molecular structure of polyethylene.

Cross-linking Techniques

Cross-linking is usually achieved in three different ways that are known as *curing techniques*: catalyst curing, curing with chemical hardener, and radiation curing.

1. In the case of *catalyst curing*, a catalyst (also called an initiator) is mixed into the polymer. After this mixture has been molded or cast into the required geometry, it is subjected to heat and/or pressure to initiate a crosslinking reaction. Such polymeric products are called *thermosets*, since their shape, once set, becomes irreversible. Polyethylene cables are usually crosslinked by incorporating 1–2% peroxide (e.g., dicumyle peroxide) which does not react when the polymer is extruded but reacts later when the cable is either (1) heated by super-heated steam, (2) heated by high-pressure nitrogen gas maintained at an elevated temperature in a long tube, or (3) kept in silane solution. In the latter case, the cables, once formed, are placed in a "sauna" for cross-linking to take place. Polyethylene cross-links by means of polymer-free radicals produced via the initiation process. Since these are usually generated in low concentration, the product is normally only lightly cross-linked. For compounding in line during the extrusion process, there are advantages to using a liquid instead of a powder. Di-tert-butyl peroxide, which decomposes more slowly, is also frequently adopted.

2. *Curing with hardener.* In this case a chemical (hardener) is mixed with the base polymer for its cross-linking. Electrical grade epoxies, and elastomers are the examples of this type of curing technique.

3. *Curing with radiation* is also used to promote cross-linking. However, this process is used only on thin sheets. As the equipment required

in this case is quite expensive, this technique finds only limited applications [2].

Polymers which are not cross-linked can be remolded to other shapes and hence are called *thermoplastics*. Polyethylene, polycarbonate and acetal copolymers are examples of thermoplastics commonly employed in electrical applications. Table 6.5 summarizes salient dielectric properties of some thermoplastic materials that are commonly used for power system applications [2].

Polyethylene

Polyethylene (PE) is a thermoplastic semicrystalline polymer which is widely used in the cable industry. It is produced by polymerization of ethylene C_2H_4 as shown in equation (6.1). The long chain molecules of PE are not chemically bonded to each other, therefore, it is hard at room temperature but shows viscous flow at high temperatures, since these molecules start sliding on each other. It is an amorphous solid with maximum crystallinity of 95%. Higher crystallinity increases its tensile strength, rigidity, chemical resistance and opacity while it reduces its permeability to liquids and gases. Its properties may be modified by additives and fillers,

Table 6.5 Selected Properties of Thermoplastics Used for Electrical Insulation

Polymer	Dielectric strength[a] Volts/mil (kV/mm)	Dielectric constant		Dissipation factor		Maximum service temperature °C	Arc resistance (seconds)
		60 Hz	10^6 Hz	60 Hz	10^6 Hz		
Polyphenylene sulphide	380 (15.0)	3.1	3.2	0.0003	0.0007	205	34
Polyether sulphone	400 (15.7)	3.5	3.5	0.001	0.004	180	70
Poly carbonate	380 (15.0)	3.2	3.0	0.009	0.01	130	120
Acetal homo polymer	380 (15.0)	3.7	3.7	—	0.0048	90	220
Acetal copolymer	380 (15.0)	3.7	3.7	0.001	0.006	105	240 (burns)
Acrylic (PMMA)	500 (19.7)	3.7	2.2	0.05	0.3	95	No tracking

[a] 3.175 mm (125 mil) thick sheets tested as per ASTM-D149.
Source: Ref. 2.

which are usually mixed with PE granules during the extrusion process. Antioxidants are important to prevent its degradation during manufacture and in service. These include amines, hindered phenols and phosphites. Cable-grade PE also includes metal deactivators, since PE reacts with copper conductor and screens. It is inhibited by adding polymerized 1,2-dihydro-2,2,4-trimethylquinoline. Degradation by ultraviolet light is inhibited to a remarkable degree by incorporating <3% of carbon black. Inflammability is reduced significantly by additives such as antimony oxide, aluminum trihydrate and halogenated compounds. Depending upon the reaction process, PE can be produced either as LDPE or HDPE. The crystallinity of LDPE is typically 45–55% while that of HDPE is 70–80%.

As a raw material, PE is available in the form of granules which can be extruded with conductors to form cylindrical cable insulation. HDPE has greater breakdown strength and higher dielectric constant than LDPE. For thin films, breakdown field strengths of more than 200 kV/mm are achieved.

PE does not contain any polar groups and therefore has a low dielectric constant and a very low dissipation factor. Its volume resistivity is extraordinarily high (10^{14} Ωm), and thus stationary space charges can occur in the material which can produce undesirable field distortions. PE can be employed in the temperature range of $-50°C$ to $75°C$. It is combustible, and its chemical resistance is good except to chlorine, sulfur, nitric acid and phosphoric acid. Under the influence of oxygen, its surface becomes brittle.

Whereas LDPE and HDPE are being produced by reaction processes introduced several decades ago, a new family of reaction processes is presently being used to produce linear low-density polyethylene (LLDPE) and medium-density polyethylene (MDPE). As shown in Figure 6.2, LLDPE and HDPE (also MDPE, not shown) molecules are generally linear in structure. The linear resins exhibit short-chain, rather than long-chain, branching. This fundamental difference in molecular structure accounts for many of the major differences in mechanical and dielectric properties as compared to LDPE [5]. LLDPE has become the most prominently used plastic in underground power cable jacketing because of its excellent mechanical properties, installation temperature range, vapor transmission resistance and environmental stress cracking resistance (ESCR). The other two linear polyethylenes, HDPE and MDPE, also have excellent properties but are stiffer because of their higher densities.

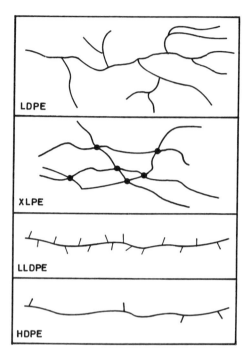

Figure 6.2 Structure of LDPE, XLPE, LLDPE, and HDPE.

Cross-linked Polyethylene

Cross-linked polyethylene (XLPE) is obtained by cross-linking PE as described earlier. Cross-linking renders PE infusible and suitable for service temperatures up to 125°C, thus permitting cables to carry higher current densities. It also for a short time withstands fault temperatures of around 250°C. Its resistance to cold flow and abrasion are superior to conventional PE, while its dielectric properties are comparable. Another advantage of cross-linking is that it makes possible higher filler loading without significant loss of physical properties. This is why XLPE cables have almost completely outclassed compound filled cables in the medium voltage range. However, the main difficulties of PE and XLPE insulation are their sensitivity to partial discharge and the associated question of life time. Cavities of 1–30 μm are unavoidable during manufacture and are potential source of commonly observed partial discharge activity. Table 6.6 gives comparative properties of commonly used cable insulating materials [2,36].

Table 6.6 Comparative Properties of Some Cable Insulating Materials

Property	PVC			Polyethene			Impregnated kraft paper (0.125 mm thick)
	Flexible	Semirigid	Irradiated	Low density	High density	Cross-linked	
Specific gravity	1.37	1.31	1.34	0.92	0.95	0.92	0.65
Elongation (%)	300	300	150–200	600	400	500	2.6
Maximum operating temperature (°C)	60–105	80	105	80	90	125	—
Flame resistance (comparative)	Good	Good	Good	Poor	Poor	Poor–Good	—
Dielectric constant (1 MHz)	6.2 max	4.3	2.7	2.28	2.34	2.3	3.3–3.9
Dissipation factor (1 MHz) 20°C	—	0.1 max	—	0.0005	0.0007	0.0003	0.0026–0.003 (0.14 at 80°C)
Volume resistivity ($\Omega \cdot m$)	5×10^{11}	2×10^{12}	2×10^{10}	10^{14}	10^{14}	10^{14}	10^{13}
Dielectric strength V/mil (kV/mm)	—	—	—	550 (21.6)	500 (19.7)	550 (21.6)	1884 (74)

Source: Refs. 2 and 36.

Polyvinylchloride

Polyvinylchloride (PVC) has been in use since 1930 as cable insulation and recently as jacketing material. It is produced by the polymerization of vinyl chloride. There are several commercial polymerization processes, all carried out in pressure vessels. PVC typically contains 56.8% chlorine which can be increased to 67%. It's chemical structure is as follows:

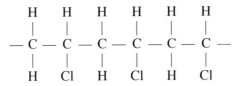

PVC resins alone are hard and brittle up to 82°C. To be made useful for cable applications, they are compounded with plasticizers and additives. A typical electrical grade formulation contains 50% PVC, 25–35% plasticizer and the rest other additives. Trimellitats are favored for cable applications up to 105°C. To achieve balance in properties, more than two plasticizers are generally added. Following is the list of commonly used additives and their functions [2]:

- *Heat stabilizers*, frequently organometallic compounds, retard thermal degradation, prolonging useful life.
- *Impact modifiers* improve impact resistance. These include chlorinated polyethylene, acrylonitrile-butadiene-styrene (ABS), methyl methacrylate-buta-diene-styrene and ethylene-vinyl acetate (EVA) polymers.
- *Fillers*, such as calcium carbonate, reduce compound cost and raise heat deflection temperature. Calcined clay improves dielectric properties.
- *Pigments* impart color and opacity and improve weatherability.
- *Flame retardants*, such as antimony oxide, significantly increase the oxygen index (OI) of PVC compounds.

PVC is also irradiated to improve its properties such as mechanical strength, abrasion, cut through resistance and temperature resistance (which is increased up to 105°C in case of cable grades).

PVC has good resistance to aliphatic hydrocarbons, oils, fats, waxes, alcohols, concentrated inorganic acids and bases, dilute organic acids, salts and most solvents. PVC is attacked by aromatic and halogenated hydrocarbons, phenols, ketones, esters, ethers and concentrated organic acids.

PVC insulation possesses only average electrical and dielectric properties. Its breakdown strength is <30 kV/mm, whereas dissipation factor at room temperature, which depends upon the filler grade, varies in the range 0.002–0.1. This poor characteristic of PVC restricts its application in medium-voltage cable insulation to only around 10 kV. However, it is very widely used for low-voltage cables operating up to 1.0 kV. Beside that, its irradiated version is also commonly used for providing inner and outer protective sheathing of high voltage cables.

Elastomers

Elastomers are polymeric materials which exhibit elastic properties similar to rubber. The most commonly used elastomers on power system network are silicone rubber, ethylene propylene rubber (EPR) and ethylene propylene dien monomer (EPDM).

Elastomers are generally prepared by vulcanization with sulfur, sulfur compounds, or other polymerizing agents. The compounding of elastomers and blends of elastomers are very complex, which often involves around 10–15 ingredients. Depending on the elastomer and curing system, these ingredients may include [2]:

- The elastomer(s), up to 50% by weight.
- Sulfur or sulfur compounds as crosslinking agents, $\leq 2\%$.
- A peroxide curing agent, such as dicumyle peroxide, $\leq 3\%$.
- A rubber accelerator, such as dipentamethylene thiuram hexasulfide, $\leq 2\%$.
- A curing promoter such as N,N'-m-phenylenediamaleide, to improve scratch and insulation resistance, $\leq 1\%$.
- A metal oxide acid acceptor and vulcanizing agent, such as sublimed litharge (PbO) or dibasic lead phthalate, up to 5%.
- A hydrogenated wood resin to activate compounds containing litharge, $\leq 1\%$.
- A stabilizer to improve heat resistance, such as nickel dibuthldithiocarbonate, $\leq 2\%$.
- Carbon black to improve weathering properties, up to 25%.
- A lubricant, such as paraffin or petrolatum, up to 12%.
- A hindered phenol antioxidant, $\leq 3\%$.
- A filler, such as calcined clay or kaolin, up to 50%.
- A plasticizer, such as phthalate ester, chlorinated paraffin or alkyl substituted trimellitate.

Once prepared, the compound is subject to curing. There are three types of common curing techniques: (1) curing using vulcanizing agent such as sulphur or its compounds, (2) lead sheath curing and (3) radiation curing.

Silicone rubber is a high-temperature insulation applied to specialty type of cables and high voltage insulators. It is prepared from dichlorosilane, which is crosslinked or vulcanized, by the action of heat in the presence of a vulcanizing agent (a peroxide). It is therefore called high-temperature vulcanized (HTV) silicone rubber. Its physical properties are enhanced by compounding with fillers such as silica and diatomaceous earths. Silicon rubber insulation can safely operate in a temperature range of $-55°C$ to $200°C$. It has good resistance to ozone, corona and weathering. It also exhibits, good resistance to alcohols, dilute acids, alkalis, salts and almost all types of oils and waxes. Silicone rubbers are, however, attacked by halogenated hydrocarbons, aromatic solvents, concentrated acids and steam.

Besides cable insulation, HTV silicone rubber is being used for the manufacture of outdoor high voltage insulators and also in the form of additional extended sheds on ceramic insulators to enhance their dielectric integrity and their performance under polluted atmosphere [9]. Fiberglass reinforced rod insulators with silicone rubber sheds are accepted as standard overhead line insulators with a good experience for voltage ratings up to 765 kVac and ±500 kVdc [6,7]. Table 6.7 summarizes important properties of HTV silicone rubber.

The other type of silicone rubber that has gained popularity for its applications as outdoor high voltage insulators is room temperature vulcanized (RTV) silicone rubber. RTV is being used in the form of coatings on ceramic insulators. Commercial RTV coating consists of a poly-di-methylesiloxane (PDMS) polymer, plus two fillers such as fumed sil-

Table 6.7 Comparative Properties of Silicone Rubber and EPDM

Property	HTV silicone rubber	EPDM
Dielectric strength (kV/mm)	20.0	19.7–31.5
Dielectric constant (1 MHz)	3.0–3.6	2.5–3.5
Dissipation factor (1 MHz)	0.005	0.007
Volume resistivity ($\Omega \cdot m$)	10^{13}	10^{14}
Specific gravity	1.15–1.55	0.85
Elongation, % (ASTM-D412)	200	200
Comparative abrasion resistance	Fair	Good
Water resistance	Good	Excellent
Maximum operating temp.	200°C	177°C
Flame resistance	Poor	Poor
Resistance to ozone	—	Excellent

Source: Ref. 2.

ica and aluminum trihydrate (ATH), beside a colorant pigment and a crosslinking agent [8]. This coating may also include PDMS fluid, additional fillers, a condensation catalyst and an adhesion promoter for its improved bonding to ceramic surfaces. The driving motivation for their increased acceptance and use on high voltage insulators is their unique characteristic of hydrophobicity, which provides a high surface resistance, even in the presence of moisture and contamination, thereby suppressing the leakage current to low values as shown in Figure 6.3 [8–10]. Uncontrolled leakage current promotes intense dry band arcing, which on polymers, can ultimately lead to material degradation in the form of tracking and corrosion and/or flashover even at operating voltages [11]. Unlike most polymeric insulating materials, silicone elastomers are able to maintain their low surface energy (contact angle >90°). This property causes water to "bead up" rather than form a continuous film of moisture. This is the critical property of silicone elastomer which causes suppression of leakage current. Moreover unlike EPRs, silicone elastomers are immune to the sun's ultraviolet rays. Today the trend of the utilities worldwide is toward using silicone rubber for the sheds of all types of overhead line insulators [12].

Ethylene-propylene rubber (EPR) is primarily an extruded dielectric used in medium and high voltage power cables. Its extensive use is based on superior electrical properties. It has attractive features like wet electrical stability, flexibility, water tree and corona resistance.

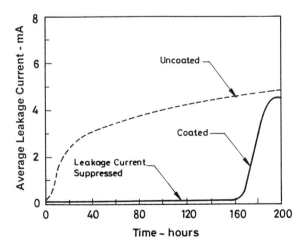

Figure 6.3 Suppression of leakage current for a RTV coated insulator tested in a salt-fog chamber. (From Ref. 8 © IEEE, 1995.)

For the manufacture of cable insulation, ethylene and propylene are mixed with fillers such as clay, talc, whiting, silica and alumina. Other ingredients typically compounded with EPR are metal oxides, plasticizers, antioxidants and curatives. Curing (crosslinking) of this compound is carried out in the presence of peroxides. One of the curing methods—steam, dry nitrogen and pressured salt solution—is selected. However, at present, a debate continues in the industry as to whether microvoids formed during steam curing result in a more rapid degradation of the insulation. Salt solution is preferred in the case of large diameter cables. For better cable performance, choice of filler and its content are of key importance [10].

Today, three types of EPRs are in common use for overhead electrical insulation; namely, ethylene-propylene monomer (EPM), ethylene-propylene diene monomer (EPDM), and a copolymer of ethylene-propylene and silicone (ESP). All three types are highly filled with alumina as well as other types of fillers as described above. A critical filler level is considered important for an optimum end result. Table 6.7 compares the key properties of EPDM and silicone rubber. Early EPRs suffered from tracking and poor resistance to ultraviolet radiation. However, formulations of present-day EPRs are somewhat immune to tracking and exhibit good resistance to ultraviolet radiation, showing only surface ablation. These are being employed successfully for distribution and transmission class insulators of up to 765 kV. However, their long-term performance under polluted environment has been unsatisfactory. The ESP insulators are still relatively new and will require a few more years in use for their potential weaknesses to be discovered [12].

Epoxy Resins

Epoxy resins are a family of thermoset polymers in which two components are mixed to eventually form a glassy product at room temperature which has reasonable electrical insulating properties and is also highly impermeable to water. Epoxies are polymers in which the end groups contain the three-membered epoxide ring as shown in Figure 6.4a, whereas Figure 6.4b depicts the common configuration of the diepoxide structure. Many different curing agents are used to bring about crosslinking of this resin. The action of the curing agent or "hardener" is to open and join into the epoxide rings. Diamine compounds (H_2N—R'—NH_2) are commonly used which provide four sites for attachment as illustrated in Figure 6.4c.

Because of the high chemical reactivity and potentially large number of epoxide rings, crosslinking in cured epoxies can be very high and an extensive network of connections with high mechanical rigidity is pro-

Solid Dielectrics

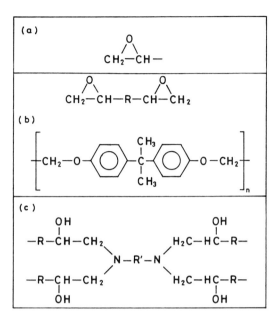

Figure 6.4 (a) Epoxide ring and (b) diapoxide structure, where R is commonly bisphenol-A and n is typically about 10; (c) structure of epoxy resin compound.

duced. Once the compound is cured, the material becomes epoxy polymer [13].

To improve the physical and mechanical properties of the end product and also to control the cost, the epoxy resins are loaded with fiberglass, fumed silica and other inorganic particulate fillers. Cast resin polymers are compounds that are formulated by mixing resin with, hardener, filler, plasticizer, and coloring pigments. These fillers may constitute 50% or more of the compound weight. Table 6.8 summarizes essential properties of some commonly used glass-filled epoxy compounds. The highest recommended temperature for continuous use of nonreinforced epoxies is usually 130°C, but some of the glass-filled compounds can be used up to 250°C. Water absorption of epoxy polymers is generally low, ranging from 0.05 to 0.5%. The higher values apply to epoxy polymers that are reinforced filled. Their relative permittivity lies between 3.5 and 5.0 which is increased to between 4 and 8 for glass-filled epoxies as shown in Table 6.8.

The dissipation factor of epoxy increases only slightly with temperature, as shown in Figure 6.5. However, an initial hump generally appears

Table 6.8 Properties of Thermosetting Molding Compounds

Compound	Dielectric strength[a] Volts/mil (kV/mm)	Dielectric constant		Dissipation factor		Maximum service temperature (°C)	Arc resistance (seconds)
		60 Hz	10^6 Hz	60 Hz	10^6 Hz		
Glass-filled allyle	400 (15.7)	4.2	3.5	0.004	0.01	260	140
Glass-filled phenolic	380 (15.0)	6.0	5.0	—	0.02	232	180
Glass filled epoxy (electrical grade)	390 (15.4)	5.0	4.6	0.01	0.01	204	187
Glass-filled melamine	340 (13.4)	8.0	6.2	—	0.02	204	180+
Glass-filled alkyd/polyester	375 (14.8)	5.3	4.6	0.1	0.02	204	180+

[a]3.175 mm (125 mil) thick sheets tested as per ASTM-D149.
Source: Ref. 2.

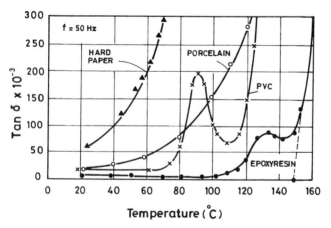

Figure 6.5 Variation of dissipation factor with temperature of solid dielectrics at 50 Hz AC voltage. (From Ref. 14.)

around 130°C. This is attributed to the dipole orientation. Once the polymer is warmed up to allow ionic conduction, the movement of the ions is enough to contribute to heating, and the runaway thermal heating shown by the increasing tan δ values at ~150°C [14,15]. Figure 6.5 also illustrates a much lower dissipation factor of epoxy resin as compared to PVC, porcelain and paper.

Epoxy resins are extremely useful for their high mechanical strength, good adhesion to materials and metal (except polythene), and resistance to moisture. They are being used extensively in high voltage switchgear and electrical machine insulation as well as for the "potting" of electrical components. They are also finding applications in low and high voltage transformer winding encapsulation, as well as spacers for SF_6 GIS installations. Glass-fiber-reinforced epoxy rods for overhead line polymeric insulators are another field of their application.

Bisphenal-A-based epoxy resins (highly filled with quartz) were used in the first generation of polymeric outdoor insulators. During long service, however, cracks were reported in insulators, and were found to be due to low tracking resistance and poor ultraviolet radiation resistance of bisphenal-A. At present it has been replaced by cycloaliphatic-type epoxy resin which contains a hydrated aluminum filler [7]. Today, a large variety of these insulators are in use at voltage levels of ≤69 kV. Their long-term performance in normal atmosphere has been quite successful. However, in polluted atmospheric conditions, their performance has been far from satisfactory. Therefore, their present use is mostly limited to apparatus bushings and bus bar insulators [12].

6.3 DIELECTRIC LOSS IN SOLID INSULATING MATERIALS

It was shown earlier that relative permittivity (ε_r), dissipation factor (tan δ), and hence the loss factor or the loss index (ε_r tan δ) are the characteristics of an insulating material pertaining to its molecular and atomic structure, which play a significant role in its power loss. Before considering this dielectric loss, some basic concepts will be reviewed here.

6.3.1. Electric Polarization

Let us consider a parallel plate capacitor having surface charge density q on its opposite plates of area A and separation d, then the resultant electric field E_o in vacuum is:

$$E_o = 4\pi q \qquad (6.2)$$

If we insert a dielectric material of permittivity ε between the plates, it will cause a decrease in the electric field intensity between the plates having a fixed surface charge density. This is because the dielectric material tends to neutralize the charges at the electrode surface due to the formation of dipole chains of the molecules, as shown in Figure 6.6. This phenomenon is called dielectric polarization (P). The charge density in this case becomes $E_o/4\pi\varepsilon$, where $\varepsilon = \varepsilon_o\varepsilon_r$. Hence

$$P = \frac{E_o}{4\pi} - \frac{E_o}{4\pi\varepsilon}$$

$$P = (\varepsilon - 1)\frac{E}{4\pi} \tag{6.3}$$

These equations provide a direct relationship between P, ε and E, and emphasize that polarization is increased with permittivity and applied stress.

6.3.2 Polarizability of Molecules

In the absence of applied electric field, the net charge on molecules of a dielectric material is zero. But on the application of external field, the electrons are slightly displaced with respect to the nuclei; induced dipole moments result and cause the so-called electronic polarization of materials. For example, if in a neutral H_2 atom the charge of an electron and that of a proton are displaced distance d, then it results as a net-induced dipole moment, $\mu = ed$, where e is the charge of electron. The resultant dipole moment due to electrons in an insulating material can therefore be expressed as:

$$\mu = \alpha_e E \tag{6.4}$$

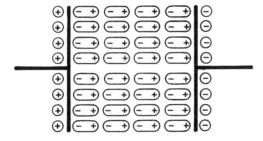

Figure 6.6 Electric polarization in a dielectric.

where α_e is the *electron polarizability* of atoms, which depends on E. For all materials, the value of α_e is constant with frequencies up to the ultraviolet spectrum (~2.5 MHz).

When atoms of different types form molecules, they will normally not share their electrons symmetrically, as the electron clouds will be displaced eccentrically toward the stronger binding atoms. Thus atoms acquire charges of opposite polarity, and an external field acting on these net charges will tend to change the equilibrium positions of the atoms themselves. This displacement of charged atoms or groups of atoms with respect to each other results into *atomic polarization* of materials. In this case, "atomic or ionic polarizability" is given as α_a. The value of α_a is generally constant, with frequencies up to the infrared spectrum. For nonpolar materials, the polarizability is mainly determined by α_e, as α_a generally does not exceed 0.1 α_e [15].

Polar substances, on the other hand, possess permanent dipole moments in their molecules even in the absence of an applied field. In this case, the centroids of positive and negative charge distributions are permanently separated by fixed distances, thus resulting in permanent dipoles. Such dipole moments experience a torque in an applied field that tends to orient them in the direction of the field. This is called *orientation (or dipole) polarization*, P_d. This is of particular significance, since most polymeric insulating materials have permanent dipoles in their molecular structure.

When a field is applied suddenly across a dielectric, it will attain finite polarization in a very short time provided it is only due to α_e or α_a. However, in the case of slower movement of orientation dipoles, the polarization will attain saturation only with some time lag (also called relaxation time, τ_p) due to slower migration of ions. When the applied field is alternating, the dipoles or the charges must change their direction every half cycle. If the frequency is very low, or when the duration of half cycle is longer than τ_p, the polarization will attain its maximum value. But at higher frequencies, the duration of half cycle could be shorter than τ_p, therefore in this case the dipoles cannot follow the change in the field intensity. This results into reduction in polarization, which will ultimately diminish to zero at a very high frequency. Figure 6.7 illustrates variation of polarization (and hence ε) as a function of frequency [15]. The different relaxation times result in frequency limits beyond which the respective mechanisms no longer exist, because the corresponding dipole movement does not occur. This is why the ε must also decrease. At each transition (sudden drop in P_d) of the dielectric constant, the dissipation factor (tan δ) has its maximum value. But only the transition region controlled by P_d, which represents the frequency range in which dipole orientation vanishes, is of interest for

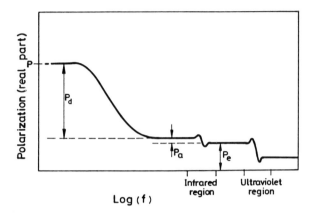

Figure 6.7 Schematic plot of polarization as a function of frequency. (From Ref. 15.)

technical insulation systems. In this transition domain of frequency, which is also known as the *dispersion domain*, the relative permittivity ε_r depends strongly on eigenfrequency ($\omega_o = 1/\tau_p$), which is a characteristic property of a material and varies from one dielectric material to another. For example, for hard pressed board it is 10 kHz, whereas for pure water it is ~100 MHz. It is also a function of temperature.

6.3.3 Losses Under High Voltages

In practical insulating materials, the losses are not confined to polarization processes. The presence of ionic impurities and voids in the bulk of insulating material as well as the presence of trapped space charge may result as ionic flow of current under the application of high electric stress. It will increase with the increase in electrical stress as well as in temperature. In turn, the dissipation factor tan $\delta = \sigma/\omega\varepsilon$ will also follow suit, since the AC conductivity σ follows this trend while ε remains constant with electric field. Similarly in the vicinity of breakdown stresses, increased losses could also result due to electronic emission from electrodes, which can arise either from the Schottky effect [16] or from the "tunneling effect," where electrons would be tunneled into the bulk of the dielectric material through thin (<100Å) dielectric oxide layers that cover the metallic electrodes [15].

In ceramics and glasses, the electronic conduction (even at low temperatures) is mainly governed by the "electron hopping effect" through transition metal oxides. This refers to the oxides in which either all or a portion of the d-shell electrons have been given off by the transitional metal atom to oxygen atom [17].

The above briefly outlines the theory of dielectric loss and the various processes that control it. For more in-depth treatment, the reader is referred to Bartnikas and Eichhorn [15] and O'Dwyer [18]. It is clear that dielectric loss depends, in principle, on the molecular structure of the insulating material, its homogeneity and the included impurities. Efficient utilization of the dielectric in electrical power apparatus, therefore, requires the knowledge of dielectric loss behavior under specific voltage, frequency and temperature conditions. It is much more helpful—and expedient—to compare the loss behavior in different generic materials such as polymers, glasses and ceramics.

In practical polymers, the losses are mainly dominated by dipole orientation or ionic conduction due to contaminants. Exception to this general rule may be observed only in polymers which possess an exceptionally high degree of purity, which leads to lower conductivities and dielectric constants. In the electrical power industry, the most commonly used polymers are XLPE and EPR. Figure 6.8 compares the 60-Hz dielectric constants of commercially available XLPE, filled XLPE and EPR over a practical temperature range [19]. It is clear that ε_r of XLPE is considerably lower than that of filled XLPE over this temperature range. This shows that the addition of "fillers" in polymers leads to "interfacial polarization," which will lead to higher dielectric constant and dielectric losses. If these fillers contain polar species, it will further enhance the loss. The decrease

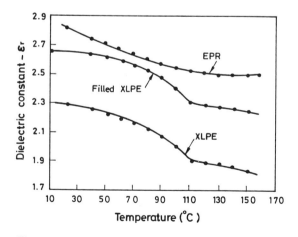

Figure 6.8 Dielectric constant as a function of temperature at 60 Hz for XLPE and EPR. (From Ref. 19.)

of ε_r with temperature for all three materials results directly from a decrease in the density with temperature and hence the number of molecules per unit volume that contribute to orientation polarizability. The sudden drop in ε_r for both XLPE samples at 110°C is due to the vicinity of melting transition temperature.

Figure 6.9 illustrates tan δ values as a function of temperature for the same materials. The losses in EPR and filled XLPE are higher than in XLPE. The increase in tan δ in the case of XLPE beyond 70°C with a maximum at ~120°C is due to the α-type relaxation process that is attributed to the vibrational and reorientation molecular motions within the crystalline regions [20]. For filled XLPE and EPR, the α-type process is also operative but it is rather dispersed and its peak occurs at a considerably higher temperature. It is interesting to note that in extruded XLPE cable material, no distinct α-peak is observed [19]. If moisture is absorbed by the polymers, it leads to the formation of a double phase resulting in interfacial polarization loss. Likewise, any ionic impurities contained in water will give rise to a higher overall conductivity and thus a higher loss.

Whereas the dominant loss portion in polymers is due to orientational polarization, the dielectric loss in glasses and ceramics is mainly electronic and ionic in nature. Electrical grade glasses consist, to a large extent, of

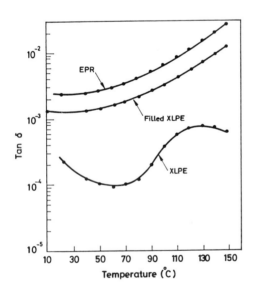

Figure 6.9 Tan δ as a function of temperature for XLPE and EPR at 60 Hz. (From Ref. 19.)

SiO_2, B_2O_3 or phosphoric anhydride (P_2O_5) structures. These are sufficiently open to permit ionic diffusion and migration. The conduction losses in this case mainly result from sodium (alkali) impurity ions. Ceramics are composed of various materials that are formed permanently into durable, hard dielectrics by either firing or sintering processes. Various clays are also used as fillers. These are the source of ionizable impurities that give rise to significant losses. The main charge carriers responsible for generating dielectric losses may be either electrons or ions, or both types of species may contribute. However, because of the complex structure of ceramic materials, it has not always been possible to delineate clearly which charge carrier process may be responsible for the observed dielectric loss [15].

6.4 BREAKDOWN IN SOLID INSULATION

Matter in the solid state consists of atoms held together in a condensed phase. The most striking feature of the majority of solids is that their atoms (or atom groups) are arranged with a high degree of order in some regular repetitious pattern in three dimensions; such solids are called crystalline. Solids whose atoms are arranged in an irregular and more or less random fashion are called noncrystalline or amorphous. Since a large proportion of commercial insulating systems are solids, studies of the breakdown of solid dielectrics are therefore of extreme importance in insulation studies.

The application of a strong electric field to a solid dielectric material can result in motion of free charge carriers, charge injection from the electrodes, charge multiplication, space charge formation and dissipation of energy in the material. Depending on the circumstances, many of these may occur alone or in combination, thus finally leading to the electrical failure of the material, which is also called its breakdown. The theoretical aspects of different mechanisms advanced to explain breakdown have been reviewed by O'Dwyer [18], while breakdown of polymeric insulation has been reviewed recently by Ieda et al. [21].

When breakdown occurs, solids get permanently damaged while gases fully recover their dielectric strength, and liquids partly do, after the applied field is removed. The mechanism of breakdown is a complex phenomena in the case of solids, and depends on the time of application of voltage, as shown in Figure 6.10. It is unlikely that any single mechanism can be responsible for the many diverse phenomena generally associated with the breakdown of solids. In the following sections the principles of various breakdown mechanisms are described in brief. These are:

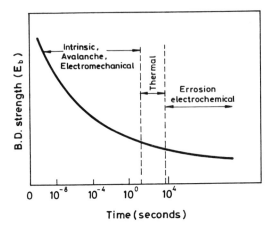

Figure 6.10 Variation of breakdown strength and the operating breakdown mechanism with time of stressing.

1. Electronic breakdown mechanisms
2. Thermal breakdown mechanisms
3. Electromechanical breakdown mechanisms

6.4.1 Electronic Breakdown Mechanisms

In such mechanisms, the field causes either the number of electrons or their energy to reach unstable magnitudes such that they rise catastrophically, causing, at least locally, the destruction of the lattice of the solid material. Electronic breakdown is essentially of two types: intrinsic breakdown and avalanche breakdown.

Intrinsic Breakdown

If the material under test is pure and homogeneous, the temperature and environment conditions are carefully controlled, and the sample is so stressed that there are no external discharges, then under voltages applied for short time the electric strength increases to an upper limit which is called the *intrinsic dielectric strength*. The intrinsic strength is a property of the material and its internal temperature only.

In pure homogenous dielectric materials the conduction and the valence bands are separated by a large energy gap, and at room temperature the electrons cannot acquire sufficient thermal energy to make transitions from valence to the conduction band. The conductivity in perfect dielectrics should therefore be zero. In practice, however, all crystals contain some imperfections in their structures due to missing atoms, and more frequently due to the presence of impurities. The impurity atoms may act as traps for free electrons in energy levels that lie just below the conduction band.

At low temperatures the trap levels will be mostly filled with electrons caught there as the crystal was cooled down during its manufacturing. At room temperature some of the trapped electrons will be excited thermally into the conduction band, because of the small energy gap between the trapping levels and the conduction level. An amorphous crystal will therefore have some free conduction electrons. Such electrons would be accelerated in applied field and, provided the density of such carriers is low, energy would be lost to the lattice of the crystalline material by phonon interaction. In steady-state conditions, the electron temperature will be nearly equal to the lattice temperature. Breakdown for this model is deemed to take place when average ratio of energy gained from the field is greater than that lost in collisions with the host material. This condition is called von Hippel low field breakdown criterion [22].

Unlike in crystalline solids, in amorphous or impure solids the trapped impurity electrons below the conduction band have a higher concentration. In this case, Frohlich [23,24] postulated that electron-electron interactions may take place between the conduction band and trapped electrons so that they form part of the same energy distribution that is defined by the electron temperature. The electron system will gain energy from the field through the acquisition of kinetic energy by the free electrons and lose it by lattice scattering primarily via the far more numerous trapped electrons. For this energy loss to take place, electron temperature must be greater than the lattice temperature. The increase in field will increase electron energy more rapidly than the electrons can transfer it to the lattice, so the electron temperature will exceed the lattice temperature. The effect of the increased electron temperature will be a rise in the number of trapped electrons reaching the conduction band. This increases the material's conduction and, as the electron temperature continues to increase, a complete breakdown is eventually reached. This breakdown mechanism is normally referred to as *electronic thermal breakdown*.

Avalanche Breakdown

In its simplest form, avalanche breakdown may be visualized in a similar way as the Townsend α process in gases. An electron released at the cathode will gain energy from the applied field. This high energy electron may collide with a bound electron thereby resulting in a pair of free electrons. In the presence of a high field these electrons acquire sufficient energy to produce two more free electrons. Repetition of this process increases the number of free electrons, and since it is only free electrons which can acquire energy from the field, the avalanche can lead to very high local energy dissipated into the lattice, causing its disruption after a sufficient number of generations.

The critical number of generations was calculated as ~40 by Sietz [25], therefore the critical ionization rate per unit length α_c is given as:

$$\alpha_c = 40/d \tag{6.5}$$

where d is the material thickness. The field dependence of α at constant pressure may be expressed as:

$$\alpha = A' \exp(-B'/E) \tag{6.6}$$

where A' and B' are constant. B' depends on both the energy for impact ionization and mean free path for phonon collision.

Fowler–Nordheim field emission process [26] at the cathode is invoked to take care of the injection carriers which take part in the electron avalanching in the applied field, leading to failure. The expression for current J reaching the anode may be shown to be:

$$J = \frac{k_1 E^2}{\phi} \cdot \exp\left[-\frac{k_2 \phi^{3/2}}{E} + \alpha d\right] \tag{6.7}$$

where k_1 and k_2 are constants and Φ is the energy barrier for electrons escaping from cathode. At the instant of breakdown, the exponent term becomes zero. Therefore breakdown field E_b can be given as:

$$E_b = \frac{k_2 \phi^{3/2}}{\alpha d} \tag{6.8}$$

Similarly, conduction current effects can also take place due to field-dependent emission from traps in the bulk of the insulator (Pool–Frenkel effect [15,18]).

It is interesting to note that since the development of an avalanche is dependent on the availability of an initiating electron, a statistical time lag is also predicted by this theory. The field for avalanche breakdown is extremely well defined experimentally in crystalline systems. For example, in Zener diodes the current may rise by an order of magnitude for about 0.1% rise in the field. Kitani and Arii [27] have extensively studied time lags to breakdown in polymer films using nanosecond pulses, and consider that initiating electrons may be provided from the electrode or the bulk, depending on the material and the temperature.

The electron avalanche breakdown is purely electronic breakdown process usually observed in the low-temperature region for polymers. Typical characteristics of avalanche breakdown are a negative thickness dependence and a positive temperature dependence of the electric strength and the breakdown time lag of <1.0 ns [28].

6.4.2 Thermal Breakdown

When an insulation is stressed, heat is generated within it because of conduction currents and dielectric losses due to polarization. In general, the conductivity (σ) increases with temperature, conditions of instability are reached when the rate of heating exceeds the rate of cooling and the specimen may undergo thermal breakdown. To obtain the basic equation for thermal breakdown, let us consider a dielectric cube of face area $A(m^2)$.

Assume that the heat flow in the x-direction is as shown in Figure 6.11, then the

$$\text{Heat flow across face (1)} = KA \frac{dT}{dx} \quad (6.9)$$

where K is the thermal conductivity of the material.

$$\text{Heat flow across face (2)} = K \frac{dT}{dx} + KA \frac{d}{dx}\left(\frac{dT}{dx}\right) \Delta x \quad (6.10)$$

The second term in equation (6.10) represents the heat input into the block. Hence

$$\text{Heat flow/volume} = K \frac{d}{dx}\left(\frac{dT}{dx}\right) = \text{div}(K \text{ grad } T) \quad (6.11)$$

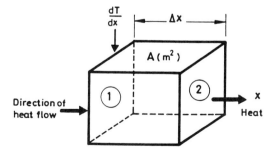

Figure 6.11 Heat input and output from a cubical specimen.

The conservation of energy requires that heat input into the element must be equal to the heat conducted away, plus the heat used to raise the temperature T of the solid, or heat generated = heat absorbed + heat lost to surroundings, hence:

$$\sigma E^2 = C_v \frac{dT}{dt} + \text{div}(K \text{ grad } T) \tag{6.12}$$

where C_v = thermal capacity of dielectric, σ = electric conductivity, T = temperature and t = time over which heat is dissipated.

Experimental data on breakdown with temperature shows that the voltage at which thermal breakdown can occur depends on the time of voltage application. Therefore two limiting cases can be considered: impulse thermal breakdown and steady-state thermal breakdown. Both of these represent breakdown that takes place on a broad front across the dielectric. Later we will also consider breakdown in filament through the dielectric.

Case I. Impulse Thermal Breakdown

In this case, the heat builds up very rapidly, therefore the heat lost to surrounding can be neglected. Therefore, equation (6.12) reduces to:

$$\sigma E^2 = C_v \frac{dT}{dt} = C_v \frac{dT}{dE} \frac{dE}{dt} \tag{6.13}$$

To obtain breakdown field E_b in time t_b, we apply a ramp function field; then

$$E = \left(\frac{E_b}{t_b}\right) t \tag{6.14}$$

Also the conductivity σ shall vary as:

$$\sigma = \sigma_o \exp\left[-\frac{W}{k_b T}\right] \tag{6.15}$$

where W = the activation energy to surmount the potential barrier, σ_o = conductivity at ambient temperature T_o and k_b is the Boltmann constant.

Substituting equations (6.14) and (6.15) into (6.13) yields

$$\int_0^{E_b} \frac{t_b}{E_b} \frac{\sigma_0}{C_v} E^2 \, dE = \int_{T_o}^{T_b} \exp\left(\frac{W}{k_b T}\right) dT$$

When breakdown occurs, then $T_o \to T_c$, i.e., the critical temperature and $W \gg k_b T$. Solution of this equation gives the value of E_b as:

$$E_b = \left[\frac{3C_v k_b T_o^2}{\sigma_o W t_b}\right] \exp\left[\frac{W}{2k_b T_o}\right] \quad (6.16)$$

This shows that under impulse thermal stress the applied critical field is related inversely to the time of the application of field; moreover it is approximately independent of the critical temperature.

Case II. Long-Time (Steady State) Thermal Breakdown

For this case we assume the dielectric slab is contained within big electrodes, as shown in Figure 6.12. In this case the metal electrodes will act as a strong heat sink. Therefore heat produced in the middle of the slab at a temperature T_i will be conducted toward the electrodes after a sufficiently long time lag. Therefore in this case we can neglect the "heat absorbed" term ($C_v dT/dt$) in equation (6.12) and the expression reduces to

$$\sigma E^2 = \text{div}(K \text{ grad } T) \quad (6.17)$$

At breakdown voltage, $E = E_b$. Since $\sigma E = J$ and $E = -dV/dx$, therefore $J = -\sigma dV/dx$. Substituting this value in equation (6.17) yields

$$J \frac{dV}{dx} = K \frac{d}{dx}\left(\frac{dT}{dx}\right)$$

Integrating both sides, we get

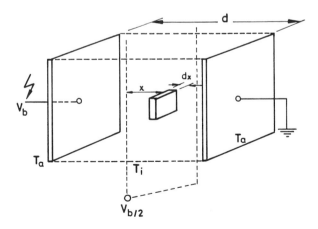

Figure 6.12 Arrangement for testing a solid dielectric under steady thermal stress.

$$\int_o^{V_x} JdV = K\int_o^x \frac{d}{dx}\left(\frac{dT}{dx}\right)\cdot dx$$

$$\times \left.\frac{V_x^2}{2}\right|^{V_b/2} = \int_{T_i}^T K\rho_v dT$$

where $\rho_v = 1/\sigma$

$$V_b^2 = 8\int_{T_i}^T K\rho_v dT \tag{6.18}$$

Critical conditions are reached when $T_i = T_b$, i.e., the critical temperature at which breakdown sets in. This shows that breakdown voltage under steady-state temperature conditions for a thick dielectric slab is independent of its thickness. Furthermore, materials of higher thermal conductivities are essential for a better performance. High thermal conductivity, K, helps to conduct unwanted heat away from hot spots within the dielectric more efficiently. Generally, a 10°C drop in hot spot temperature results in doubling the theoretical life of insulation [29].

The relation (6.18) does not hold for thin sheets of thickness d. A theoretical analysis yields an approximate relation in this case as [30]:

$$E_b = C\sqrt{d} \tag{6.19}$$

where C is a constant which depends on thermal conductivity.

The losses are much higher under AC voltages than under direct fields. Consequently, V_b is generally lower under alternating fields. Thermal breakdown is a well-established mechanism, therefore the loss factor ($\varepsilon \tan \delta$) is an essential parameter for application of dielectrics. Experimental data illustrates that the materials in which loss factor increases with temperature give rise to a quick thermal runway.

Case III. Filamentary Thermal Breakdown

In thin dielectric films the breakdown does not usually occur on a broad front across the insulation area, as assumed earlier, but generally occurs at weak spots. The temperature of a weak spot (with higher local σ, lower K, field-enhancing inclusions or the like) reaches the critical temperature before the rest of the insulation. Several experiments have been reported in support of such a filamentary thermal breakdown. Figure 6.13 illustrates localized heat generation in LDPE film using a two-dimensional visual thermograph [31]. Prebreakdown current measurements are also carried out to confirm this mechanism. The breakdown field can be related to current density and the time as [32]:

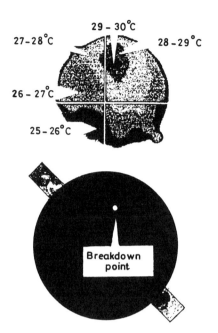

Figure 6.13 Thermograph of PE film before breakdown and breakdown point at 24.5°C. (From Ref. 31 © IEEE, 1990.)

$$J = \frac{k_b T_o^2 C_v D}{\phi E_b t_b} \qquad (6.20)$$

where D is the material density and t_b is the time measured just before the breakdown, after the application of voltage. To account for the current in a cylindrical filament of radius r_f, substitution of $I_f = \pi r_f^2 J$ turns equation (6.20) into

$$r_f = \left[\frac{\phi E_b I_f t_b}{\pi k_b T_o^2 C_v D} \right]^{0.5} \qquad (6.21)$$

values of E_b, I_f and t_b prior to breakdown are measured experimentally, and for a fixed material a plot of r_f versus $t_b^{0.5}$ has indeed been verified [32].

6.4.3 Electromechanical Breakdown

When solid dielectrics are subjected to high electric fields, failure occurs due to electrostatic compressive forces (electrostriction) which can exceed the mechanical compressive strength. If the initial thickness of the specimen is d_o, which is compressed to a thickness d under an applied voltage V, then the electrically developed compressive stress is in equilibrium if:

$$\varepsilon_o \varepsilon_r \frac{V^2}{2d^2} = Y \ln\left[\frac{d_o}{d}\right] \qquad (6.22)$$

where Y is the Young's modulus. Usually, mechanical instability occurs when $d/d_o = 0.6$ or $d_o/d = 1.67$. Consequently, from equation (6.22), the highest apparent electric stress E_m before breakdown is given as:

$$E_m = \frac{V}{d_o} = 0.6 \left[\frac{Y}{\varepsilon_o \varepsilon_r}\right]^{0.5} \qquad (6.23)$$

The above equation is only approximate as Y depends on mechanical stress. When a material is subjected to high stress the theory of elasticity does not hold well and plastic deformation has to be considered. Figure 6.14 shows plot of experimental data obtained for various samples of polyethylene [33]. It is clear that the curve (solid line) plotted based on equation (6.23) does not give a good fit to the measured values.

Recently, Dissado and Fothergill [34] refined this theory based on the concept of fracture mechanics and proposed a new mechanism in which filamentary-shaped cracks propagate through a dielectric, releasing both electrostatic energy and electromechanical strain energy stored in the material due to the applied electric field. This mechanism has analogy with

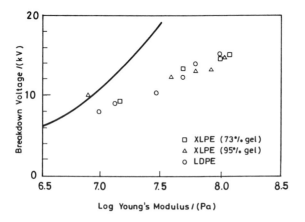

Figure 6.14 Breakdown voltage as a function of Young's modulus for various polymeric sheets. (From Ref. 33.)

Solid Dielectrics

conventional mechanical crack propagation in brittle materials, whereby crack propagates spontaneously if the strain energy released is greater than that required to overcome the toughness of the material (G).

Let σ' = stress and ε_s = mechanical strain, then Young's modules $Y = \sigma'/\varepsilon_s$. The strain energy density can be written as $\sigma'^2/2Y$. During a crack an extra area is created, which is produced by energy per unit area, called toughness (G). Therefore, for a crack to develop the condition is

$$\frac{\sigma'^2}{2Y} > G \quad (6.24)$$

In filamentary electromechanical breakdown mechanism, each spark filament is envisaged as a crack, as shown in Figure 6.15. At the hemispherical tip of the filament there will be an enhanced electrostatic field due the applied voltage. This will give rise to electrostatic energy density $W_e = 1/2\varepsilon_o\varepsilon_r E^2$.

Similarly, mechanical stress σ' induced by the electrostatic field = $1/2\varepsilon_o\varepsilon_r E^2$ resulting in a strain energy density per unit volume (W_m) of $\sigma'^2/2Y = \varepsilon_o^2\varepsilon_r^2 E^4/8Y$ is induced, therefore total strain energy (W) released (per unit volume) is made up of electrostatic (W_e) and electromechanical (W_m) components. Therefore, $W = W_e + W_m$.

Let the length of the tubular crack = dl, then its volume = πr_f^2 dl:

$$W = \left[\frac{1}{2}\varepsilon_o\varepsilon_r E^2 + \frac{\varepsilon_o^2\varepsilon_r^2 E^4}{8Y}\right] r_f^2 dl \quad (6.25)$$

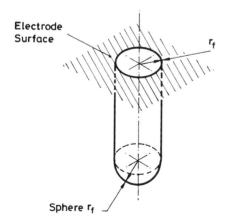

Figure 6.15 A filamentary crack emanating from an electrode.

If we consider the usual breakdown field of $\sim 10^9$ V/m and $Y = 3.10^7$ Pa (for polyethylene), we find that the contribution of W_m in the electrofractal process for polymeric insulation is much larger than W_e. Hence, the above relation reduces to

$$W = \frac{\varepsilon_o^2 \varepsilon_r^2 \pi r_f^2}{8Y} E^4 dl \qquad (6.26)$$

When a crack occurs it has to overcome the crack surface energy W_s and crack deformation energy W_f, therefore $W > (W_s + W_f)$ where

$$W_s = 2\pi r_f G dl \qquad (6.27)$$

$$W_f = \pi r_f^2 Y dl \qquad (6.28)$$

Combining equations (6.27) and (6.28), we get electromechanical breakdown criterion as

$$E^4 \frac{\varepsilon_o^2 \varepsilon_r^2 \pi r_f^2}{8Y} > (2\pi r_f G + \pi r_f^2 Y) \qquad (6.29)$$

or

$$E_b = \left[\frac{8Y(2G + Yr_f)}{\varepsilon_o^2 \varepsilon_r^2 r_f} \right]^{1/4}$$

The value of r_f depends on what initiates the crack. For example, it may be an impurity particle, microvoid, protrusion or extending element of electrical tree. The value of r_f never exceeds 10 μm. For polyethylene, $G \geq 6500$ J · m^{-2} and $Y = 3 \times 10^7$ Pa, then $2G \gg (Y \cdot r_f)$, which means $W_s \gg W_f$. Thus

$$E_b = \left[\frac{16GY}{\varepsilon_o^2 \varepsilon_r^2 r_f} \right]^{1/4} \qquad (6.30)$$

so

$$E_b \propto Y^{0.25}$$

The results of Figure 6.14 have been replotted in Figure 6.16 based on equation (6.30). It is clear that these give a good fit on this theory.

The speed of the crack propagation process is given by $(Y/D)^{0.5}$, where D = density. In polyethylene, $D = 930$ Kg m^{-3}, which will result in a maximum speed of 180 ms^{-1}. However, actual speeds measured in PE are much higher, i.e., 380 ms^{-1} (negative point) and 1700 ms^{-1} (positive point) [35]. This implies that the electromechanical process only plays role in

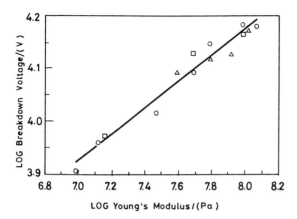

Figure 6.16 A log-log plot of data taken from Figure 6.14. Slope of the line = 0.25. (From Ref. 34.)

tree propagation once a faster electronic process has already generated low density region.

Experimental evidence shows that, in practice, the failure of solid dielectrics is a complicated process which often involves combination of several breakdown mechanisms. Figure 6.17 depicts schematically the tem-

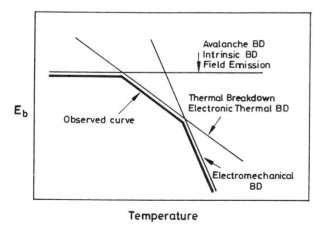

Figure 6.17 Schematically illustrated temperature dependence of various electrical breakdown processes operating in polymers. (From Ref. 21.)

perature dependence of the various breakdown processes operating in polymers. The minimum E_b–temperature curve shown is the usually observed behavior in polymers and includes the influence of various factors such as applied voltage (impulse, DC, AC), electrode conditions (material, metal-insulator interface, edge effects) and the dielectric itself involving impurities, defects, etc., together with secondary effects like space charge and the local heating [21]. In thicker insulation over a service life time, the breakdown is likely to be due to the progressive degradation caused by thermal aging, partial discharges and treeing. Also, mechanisms such as edge breakdown, chemical erosion, tracking and surface discharges continuously contribute toward the degradation of insulation. All of these mechanisms and their deleterious impact on insulation are dealt with in detail in chapters 8 and 9.

REFERENCES

1. T. J. Gallagher and A. F. Pearmain, *High Voltage Measurement, Testing and Design*, John Wiley and Sons, New York, 1983.
2. W. T. Shugg, *Handbook of Electrical and Electronic Insulating Materials*, IEEE Press, New York, 1990.
3. ASTM Standard D-351, Natural Muscovite Block Mica and Thins Based on Visual Quality, 1990.
4. R. C. M. Pompilli and C. Mazetti, IEEE Elect. Insul. Magazine, Vol. 8, No. 1, pp. 28–32, 1992.
5. G. Graham and S. Szaniszeo, IEEE Elect. Insul. Magazine, Vol. 11, No. 5, pp. 5–12, 1995.
6. E. Bauer, H. Karner, K. H. Muller, and P. Verma, CIGRE Report 22-11, Paris, France, 1980.
7. R. V. D. Huir and H. C. Karner, 6th ISH. New Orleans, 1989, Paper No. 21.03.
8. E. A. Cherney, IEEE Elect. Insul. Magazine, Vol. 11, No. 6, pp. 8–14, 1995.
9. J. W. Chang and R. S. Gorur, IEEE Trans. on Dielectrics and Elect. Insul., Vol. 1, No. 6, pp. 1039–1046, 1994.
10. R. J. Arhart, IEEE Elect. Insul. Magazine, Vol. 9, No. 6, pp. 11–14, 1993.
11. R. G. Hougate and D. A. Swift, IEEE Trans. on Power Delivery, Vol. 5, No. 4, pp. 1944–1955, 1990.
12. E. A. Cherney, IEEE Elect. Insul. Magazine, Vol. 12, No. 3, pp. 7–15, 1996.
13. R. W. Sillars, *Electrical Insulating Materials and their Applications*, Peter Peregrinus Ltd., London, England, 1973.
14. R. Arora and W. Mosch, *High Voltage Insulation Engineering*, Wiley Eastern Ltd., New Delhi, India, 1995.

15. R. Bartnikas and R. M. Eichhorn (Eds.), *Engineering Dielectrics Vol. IIA, Electrical Properties of Solid Insulating Materials: Molecular Structure and Electrical Behavior*, ASTM Publication, Philadelphia, 1983.
16. C. Wagner and W. Schottky, Zeitschrift fur Physikalische Chemie, Vol. 11, Part B, pp. 163–210, 1930.
17. A. Adler, *Physics of Electronic Ceramics*, Part A. L. L. Hench and D. B. Dove (Eds.), Marcel Dekker, Inc., New York, 1971.
18. J. J. O'Dwyer, *The Theory of Electrical Conduction and Breakdown in Solid Dielectrics*, Clarendon Press, Oxford, England, 1973.
19. H. St. Onge, H. Bartnikas, R. Braunovic, M. de Tourreil, M. Duval, EPRI Research Report No. EL-938 (RP-933-1), EPRI, Palo Alto, California, 1978.
20. C. R. Aschcraft, R. H. Boyd, J. Polymer Sci., Vol. 14, pp. 2153–2193, 1976.
21. M. Ieda, M. Nagao and M. Hikita, IEEE Trans. on Dielectrics and Elect. Insul., Vol. 1, No. 5, pp. 934–945, 1994.
22. A. von Hippel, J. Appl. Phys. Vol. 8, pp. 815–832, 1937.
23. H. Frohlich, Proc. of the Royal Society of London, Vol. A-160, pp. 230–241, 1937.
24. H. Frohlich, Proc. of the Royal Society of London, Vol. A-188, pp. 521–532, 1947.
25. F. Seitz, Physical Review, Vol. 76, pp. 1376–1381, 1949.
26. R. H. Fowler and L. Nordheim, Royal Society Proceedings, Vol. A119, pp. 173–181, 1928.
27. I. Kitani and K. Arii, IEEE Trans. on Elect. Insul. Vol. 17, pp. 571–576, 1982.
28. K. Arii, K. Kitani and Y. Inushi, Trans. of Japan Inst. of Elect. Eng. Japan, Vol. 94-A, pp. 251–258, 1974.
29. G. C. Zguris, IEEE Elect. Insul. Magazine, Vol. 4, No. 6, pp. 26–29, 1988.
30. M. M. Khalifa (Ed.), *High Voltage Engineering: Theory and Practice*, Marcel Dekker, Inc., New York, 1990.
31. M. Nagao, T. Kimura, Y. Mizuno and M. Kosaki, IEEE Trans. on Elect. Insul., Vol. 25, No. 4, pp. 715–722, 1990.
32. T. Mizutani, I. Kanno, M. Hikita and M. Ieda, IEEE Trans. on Elect. Insul. Vol. 22, No. 4, pp. 432–477, 1987.
33. M. Hakita, Japan J. Appl. Phys., Vol. 24, pp. 988–996, 1985.
34. L. A. Dissado and J. C. Fothergill, *Electrical Degradation and Breakdown in Polymers*, Peter Perigrinus Ltd., London, England, 1992.
35. Y. Yamada, S. Kimura and T. Sato, Proc. 3rd Int'l. Conf. on Conduction and Breakdown in Solid Dielectrics, Trondheim, Norway, pp. 87–91, 1989.
36. T. Tanaka and A. Greenwood, *Advanced Power Cable Technology*, CRC Press, Boca Raton, Florida, 1983.

7
Vacuum Dielectrics

7.1 INTRODUCTION

Compared to other dielectric materials, an ideal vacuum has, in principle, the highest possible dielectric strength, because there are no carriers to conduct the current. In a vacuum below 10^{-2} Pa, less than 3×10^{12} molecules per cubic centimeter are present and the length of the free path is in the order of meters. In such a vacuum, an electron may cross the electrode gap of few centimeters without any collision with particles. Therefore, in the absence of the multiplication process, a vacuum behaves as an ideal insulator. Nevertheless, in actual practice the existence of metallic and insulating surfaces within the vacuum and the presence of adsorbed gases and oil vapors contaminates the vacuum. Therefore, the vacuum has a breakdown threshold, since the electric field intensity ultimately generates charge carriers by literally pulling them from surrounding electrodes and contaminants. However, with a proper design, vacuum insulation can still exceed the breakdown strength of most dielectrics. This is why vacuum insulation has found widespread use where very high electric field intensities are needed.

Applications of vacuum insulation include electron microscopes, X-ray tubes, particle accelerators and electrical insulation in space. Its most important application in power systems is in the form of a vacuum interrupter, which is the central unit of a vacuum circuit breaker. Both the vacuum and SF_6 circuit breakers have increased their share in the power utilities worldwide and this trend is expected to grow even faster in the 21st century.

7.2 PREBREAKDOWN ELECTRON EMISSION IN VACUUM

When voltage across a small gap (<2 mm) encapsuled in vacuum is steadily increased to higher values, a relatively small steady current begins to flow which has been found to consist mainly of electrons. For longer gap spacings (>10 mm), small pulses of the order of microcoulomb charge values (microdischarges) can also occur independently or superimposed on the quasi-steady current. With a further increase in voltage, these microdischarges disappear and give rise to a steady current. For both small and larger gaps, a subsequent increment in voltage causes breakdown of the gap. With the interelectrode distance of 2 ~ 10 mm, breakdown may be caused by both the prebreakdown phenomena acting together or separately [1].

From decades of experimental work it has been established that the field-assisted electron emission process originates at the metallic protrusions on the electrodes and finally leads to breakdown of the gap when the field is sufficiently enhanced to a critical value. However, recent experimental evidence has explained a new mechanism, according to which electron emission originates from insulating/semiconducting surface oxides or impurity particles adhering to electrodes as well. Therefore, a form of nonmetallic emission mechanism has also been used to explain prebreakdown conduction [2]. Both of these mechanisms are the achievement of meticulous experimental work carried out almost from the onset of this century.

7.2.1 Electron Emission Mechanism from Metallic Surfaces

Several mechanisms can produce electron emission from metallic surfaces under different experimental conditions. These are thermionic emission, Schottky emission, Fowler–Nordheim cold field emission, photoelectron emission and secondary emission processes [3]. Today the most widely accepted model of electron emission is the cold field emission model that was originally presented by Fowler and Nordheim [4]. According to this model, the current density J from a field emitting sharp electrode as a function of electric field E at the tip is given as:

$$J = AE^2 \exp\left[-\frac{B\phi^{1.5}v(y)}{E}\right] \quad A/m^2 \quad (7.1)$$

where A = $(1.54 \times 10^{-2})/[\phi t^2(y)]$, B = -6.831×10^9 and ϕ = work function of the metal.

The expressions t(y) and v(y) are slow varying functions which are frequently regarded as constants. The above equation can also be expressed as:

$$\log\left[\frac{J}{E^2}\right] = -\log\left[\frac{1}{A}\right] - \frac{B\phi^{1.5}v(y)(1/E)}{2.3026} \qquad (7.2)$$

Since A, Φ and essentially v(y) are constants, a plot of log J/E^2 against the reciprocal of E produces a straight line having a negative slope. This straight line is often used to assess the applicability of Fowler–Nordheim relation to the experimental data. However, in the case of large area electrodes, the protrusions and uneven surfaces become emitters of unknown geometry. Therefore, to take account of these difficulties, Alpert [5] modified equation (7.2) to:

$$\log\left(\frac{I}{V^2}\right) = -\log\left[\frac{1}{A'A\beta^2}\right] - \frac{B\phi^{1.5}v(y)[d/V]}{2.3026\beta} \qquad (7.3)$$

Here A' is the electrode area and d is the gap length. Obviously J has been replaced by (I/A'), whereas the modified field is written as a product of local field enhancement factor β and the average field E (= V/d). The actual emitting area is replaced as $A'\beta^2$. Equations (7.1)–(7.3) have been experimentally confirmed by many investigators. If the enhancement factor β of the electric field due to protrusion is high enough and the critical threshold value of macroscopic field is exceeded in the gap, the emitting site explodes within 2 μs [6]. Field-assisted collisional ionization process then becomes active in the metal vapor thus formed. If sufficient charge carrier multiplication takes places, then the breakdown of the vacuum gap through this ionized vapor cloud becomes inevitable.

Other sources of field electron emission have been explained to occur from oxide films and trapped positive surface charges, absorbed particles on cathode surface, local hot spots and field-assisted thermionic emission from low work function regions and cathode microprojections [3]. However, under clean conditions, the prebreakdown current has been established to be due to only the field emission.

7.2.2 Electron Emission Mechanism from Nonmetallic Surfaces

Prebreakdown electronic currents and pulses have also frequently been found to originate from nonmetallic surfaces [7]. These are associated with some form of insulating/semiconducting oxide layers on the surfaces, or of impurity concentration. Microinclusions present on electrode surfaces can stimulate strong electron emission and can thus reduce breakdown voltage of the vacuum gap [8].

Large area electrodes in vacuum are generally contaminated with oxides, adsorbates and dust. Oxidation in vacuum takes place much more rapidly than under normal pressure. At a pressure of 10^{-3} Pa, the oxide layer on an electrode is formed within 60 ms. Halbritter [9] has shown that an oxide layer thickness of 5 nm and adsorbates of 10 nm are common. In the presence of the high electric field, these layers undergo chemical changes due to the impact of electrons, photons and ions. Presence of adsorbates and dust enhances the field emission of electrons, whereas oxides and adsorbates are found to enhance the secondary electron emission.

Electrode surfaces in vacuum, also have an adsorption layer which consists of atoms and molecules from gases and vapor of the pumping oil. Dominant species of adsorbates have been investigated and are found to consist of H_2, He, O_2, N_2, CO/N_2, H_2O, CO_2 and certain hydrocarbons. Under a high electric field, these adsorbed particles are released from surfaces due to charge exchange phenomena and cause a week discharge known as "microdischarge" [10,11]. To account for such electronic emission from insulating structures, a "field induced hot electron emission (FIHEE) mechanism" has been proposed by Latham [3]. It involves a composite of metal-insulating emission regime to explain the physical origin of prebreakdown burst current.

At present, it is a well-established fact that prebreakdown electron emission can also originate from submicron particulate structures that are randomly placed on the metallic electrode surfaces. These emission sites become a potential source of breakdown as soon as they become unstable in the presence of electric field.

7.2.3 Microdischarges

As explained earlier, the field electron emission is caused by the electric field at microprotrusions of cathode surface or at nonmetallic contaminants of the cathode. Beside the prebreakdown field current, the other phenomenon linked with electrode surfaces in long gaps is the onset of low power pulses called microdischarges. These are self limiting with durations of 0.1 to 100 ms, frequency of 0.1 to 100 s^{-1} and amplitudes of ≤ 10 mA [1]. They may be caused by (1) small particles of electrode material that are pulled out from one electrode and strike the other, (2) the beam of electrons from a cathode that can vaporize a small quantity of material from the surface of anode or of the cathode, or (3) positive and negative ions ejected from contaminant films on electrode surfaces through an ion exchange mechanism.

Microdischarges occur at critical voltage thresholds that depend on the state of electrode contamination. At a given pressure, the frequency of occurrence of microdischarges increases with increase in voltage, until the eventual occurrence of breakdown.

7.3 FACTORS AFFECTING BREAKDOWN VOLTAGE IN VACUUM

Electric strength of vacuum is defined in different ways. In the case of vacuum insulated switchgear, it is the value of the voltage to cause the first breakdown that is important. However, when a gap is caused to breakdown repeatedly, the breakdown voltage increases with the number of breakdown shots until it reaches a steady or "conditioned" value. This value is often taken as the breakdown strength of the vacuum gap. Figure 7.1 illustrates this characteristic. It has mainly been attributed to the removal and/or smoothing out of microprotrusions or due to the change in the work function of the electrodes because of adsorption processes [13]. In general the relative effects of various parameters affecting the breakdown voltage of the vacuum gap are always determined based on its conditioned value. These various factors include electrode separation, electrode material, residual pressure and the type of the applied voltage waveform.

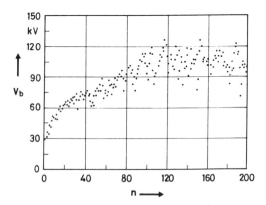

Figure 7.1 AC breakdown voltage V_b as a function of number of sparks (n). Gap spacing = 3.00 mm. (From Ref. 12 © IEEE, 1993.)

7.3.1 Effect of Electrode Separation

In a vacuum gap, as the voltage is raised above the onset of prebreakdown phenomena, various conduction processes intensify, as described earlier. In small gaps (<2 mm) and with extremely clean electrode surfaces, bright spots of light appear at the anode. At still elevated stresses, these hot regions that emerge due to impinging of electron beam from cathode can start radiating thermally, eventually leading to spark breakdown. The breakdown voltage V_b is linear function of gap spacing d such that:

$$V_b = kd \qquad (7.4)$$

where k is a constant. For such gaps the breakdown stress is relatively high and is of the order of 10^8 V/cm. Linear relation indicates that field emission of electrons plays a dominant role in the breakdown process in this region.

In longer gaps (>1 cm), field emission currents are low while the breakdown still occurs with a localized spark channel. However, in this case breakdown stress is considerably decreased, e.g., for a gap of 1 cm it is of the order of 10^4 V/cm only. It obeys a nonlinear relationship with gap length:

$$V_b = kd^n \qquad (7.5)$$

where, for a given electrode surface conditions, k and n are constants for a given range of d. In this case, generally the microsized particulate structures lying on the electrode surfaces are believed to dominate the breakdown process [15].

7.3.2 Effect of Electrode Material

The breakdown strength of a vacuum insulated gap is also a strong function of electrode material. Contact material exerts a strong influence on nearly every phase of the behavior of vacuum gap because the discharge depends entirely on the contact metal vapor, whereas the contact properties are directly influenced by the microstructure and the physical properties of the contact metal. Several investigators have shown that there exists a critical field for each material, at which breakdown occurs due to field resistive heating of the protrusions [3]. Table 7.1 gives a partial list of electrode materials and the values of the breakdown voltages associated with these electrodes when they are placed 1 mm apart. It clearly illustrates that the correct choice of electrode material is of vital importance if a high insulating capability is to be achieved for a given gap design.

Table 7.1 Experimentally Measured Critical Breakdown Voltage Across 1.0-mm Vacuum Gap

Electrode material	Voltage (kV)
Stainless steel	179
Chrome plated copper (baked at 500°C)	143
Chrome plated copper (unbaked)	89.4
Nickel	89.5
Aluminum	57
Silver	27

Source: Ref. 14.

Electrode geometry with regard to size and shape also plays a significant role. The larger the area of the electrode, the lower is the breakdown voltage. Similarly, electrodes of dissimilar material exhibit a strong polarity effect.

Surface finish is yet another factor associated with electrodes, which can be manipulated to get higher breakdown voltage. Besides this, there have been investigations into the effectiveness of the evaporated metal films as well as the insulating films which can give improvements in breakdown voltages by a factor of 2 to 3. The increase in breakdown voltage with insulating film has been attributed to the suppression of field emission by dielectric film.

7.3.3 Effect of Pressure

The influence of residual gas pressure on the breakdown voltage depends on the gap length. For small gaps (≤ 1 mm), the improvement of the vacuum beyond 10^{-2} Pa shows only insignificant changes in breakdown voltage [16]. However, longer gaps exhibit a definite pressure effect and an anomaly is observed such that with the increase in pressure, the breakdown also increases at first, but then decreases sharply with subsequent increase in its value. This phenomena has been confirmed by several investigators. Figure 7.2 illustrates this effect [17].

7.3.4 Effect of Voltage Waveform

For its application in power systems, broad area electrodes and longer gaps (>1 cm) are required in vacuum interrupters. However, most of the results reported for power frequency applications give breakdown data in gaps

Vacuum Dielectrics

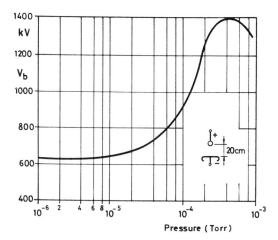

Figure 7.2 Effect of pressure on the breakdown voltage in a stainless steel sphere-plane electrode gap of 20 cm. (From Ref. 17.)

that are confined only to a few millimeters [16,18]. Comparatively little work has been done under both AC and lightning/switching impulses in longer gaps. Under AC, due to the reversal of polarity under each half cycle (making of anode region to cathode under each polarity switchover), the breakdown voltage level remains almost the same as reported for DC voltage [15]. Contrary to that, the breakdown values in 1.5 cm gaps under standard lightning impulse (1.2/50 μs) have recently been reported to be 30–40% higher than AC breakdown values (see Figure 7.3) [20]. It is also demonstrated that processes with time lags of >50 μs play an important role in the breakdown of vacuum gaps of >10 mm [20,21].

7.4 BREAKDOWN MECHANISMS

As a result of intense theoretical and experimental investigations over the last several years, different theories have been put forward in an attempt to explain quantitatively the mechanism that initiates breakdown in vacuum. Generally, these theories fall into four main categories [22].

7.4.1 Clump Theory

This theory postulates that breakdown is initiated by the removal of a charged particle or aggregate of material (clump) from one of the elec-

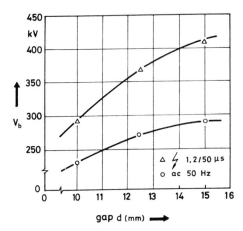

Figure 7.3 Breakdown voltage under AC and impulse voltage stress in a stainless steel electrode gap. (From Ref. 20.)

trodes, which is then accelerated across the gap to the opposite electrode. The impact energy causes a localized temperature rise that is high enough to free some other material from the electrode surface as a result of the impact. This mechanism was forwarded by Cranberg [23]. He postulated that when the energy deposited on impact by microparticles reaches a critical value, the breakdown occurs. A consequence of this hypothesis is that breakdown voltage is proportional to the square root of the gap length. This hypothesis has been reasonably substantiated as a contributing but not as a sole cause of vacuum breakdown [24].

7.4.2 Interaction Theory

According to this mechanism, an electron present in the gap would be accelerated by the voltage and impinge on the anode, where it will produce positive ions and photons due to ionization of absorbed gases. They, in turn, would be accelerated back to cathode, resulting in emission of secondary electrons. These would then be accelerated to anode, releasing further positive ions and photons. When such a chain reaction of electrons, ions and photons reaches a critical level, then the breakdown of the gap is triggered. A qualitative explanation of this mechanism is as follows:

Let A_o = average number of electrons produced by one electron
B_o = average number of secondary electrons produced by one of these electrons

C_o = average number of photons produced by one electron
D_o = average number of secondary electrons produced by a photon

The breakdown will occur if the coefficient of production of secondary electrons exceeds unity, i.e., $(A_o B_o + C_o D_o) > 1$.

7.4.3 Cathodic Theory

Cathodic theory assumes field emission of electrons from protrusions at the cathode, as explained earlier as due to Fowler–Nordheim mechanism. These protrusions can give rise to an enhanced electric field (βE) of the order of 200, which, when introduced in the Fowler–Nordheim equation, predicts sufficient current density to contribute to vacuum breakdown. Microprotrusions on the cathode surface are heated, evaporated or exploded, due to the emitted current, liberating the cathode material, which enters the interelectrode region and is subsequently ionized. This results in what is known as cathode-induced breakdown.

7.4.4 Anodic Theory

Anodic theory assumes involvement of both cathode and anode in the breakdown process. It assumes the existence of a beam emitted from the cathode that impinges on the anode causing a local rise in temperature and release of gases and vapors. Additional electrons ionize the atoms of gas and produce positive ions. At the cathode, the effect of these ions is twofold: (1) increased primary electron emission due to space charge formation and local field enhancement, and (2) electron emission by bombardment of the cathode surface. This process continues until sufficient gas is generated to give rise to a low-pressure discharge in the stressed gap.

Of all these theories, the interaction theory (see section 7.4.2) has been ruled out as applied to clean vacuum gaps and electrode surfaces. Similarly, no single theory seems to explain all breakdown events reported in the literature. Depending on gap geometry, electrode material, and voltage waveform, one can establish a situation in which a clump, cathode or anode process will be the primary cause of breakdown. However, a major volume of the present and past work reported in the literature tends to support the field-emission-initiated vacuum breakdown theory [24]. This field emission may be followed by, or may occur together with, other mechanisms discussed here. At low stress values, evidence supports a heating of the emission areas followed by melting of the emitter. At higher stresses, explosive destruction of the emitter occurs. In either case, the liberation of metal leads to plasma production in the near-cathode space. This copious production of electrons by the plasma may produce an anode plasma through deposition of energy at the anode surface. However, this anode plasma is not produced under all experimental conditions. In all cases the produced plasma can be further ionized to breakdown condition by electrons streaming from the cathode emission sites [24,25].

7.5 ARC INTERRUPTION IN VACUUMS

7.5.1 Vacuum Arc

An electric arc is basically a type of discharge between two electrodes. It differs from the general gaseous discharge, since in this case (1) the electrode current density is extremely high, of the order of kiloamperes per cm^2; (2) a greater part of discharge current near the cathode is carried only by the electrons and not by the ions formed during the ionization of medium; and (3) the potential difference between electrodes is small, i.e., only of the order of a few tens of volts. In the vacuum, arcing is established when the current carrying electrode contacts separate, melting or explosive vaporization of the last metallic point of contact occurs initially only with a single melting point on the cathode, known as the cathode spot. The cathode spot is a small, limited region of high temperature and pressure from which ions, electrons and neutral particles are emitted. At this spot, the current density is $>10^8$ A/cm^2, whereas surface temperature is close to the boiling point of the contact material. The peripheral areas of this spot represent an intense source of neutral metal vapor, which is ionized in the form of a discharge cone with its apex at the cathode. Contrary to arc discharge in a gaseous medium, more than 90% of the total current in the vacuum arc is transported by the electrons, whereas positive ions cause a neutralization of the negative space charge produced by the electrons.

If the arc current exceeds a limit that is found to be a function of contact material, the cathodic melted point splits into a number of parallel cathode spots. Studies show that without the influence of external magnetic field, these spots move in random directions around the entire cathode surface at high speeds, while the adjacent spots repel each other. This process leads to further production of copious metallic vapor. This type of discharge is called "diffused arc discharge" and is shown in Figure 7.4a. During diffused discharge, only the cathode is the active electrode, whereas the anode forms a condensation surface for vapor particles and charge carriers. Thus it functions merely as a collector. At current levels >10 kA (with cylindrical copper electrodes), the arc is constricted at the anode as shown in Figure 7.4b. Anode spots are formed while cathode spots con-

Vacuum Dielectrics

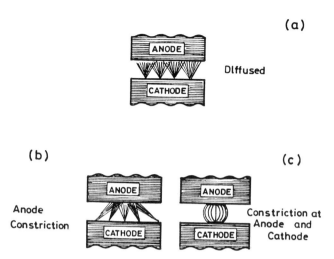

Figure 7.4 Schematic diagrams of various modes of the vacuum arc: (a) diffused arc; (b) arc constriction at anode; and (c) arc constriction at both electrodes (anode and cathode).

verge to a single spot. The constricted vacuum arc possesses in it a high-pressure metal vapor and also causes a voltage drop at the anode. The constriction of arc at both electrodes as shown in Figure 7.4c forms a thermal column, and in this case erosion of anode becomes inevitable. Therefore, in the case of high short-circuit current interrupters, local overheating of electrode contacts must be avoided. This is achieved by constant forced motion of the arc spots. For this purpose, spiral electrodes are used as shown in Figure 7.5 [39]. The current path induced by this spiral shaping of contacts generates a magnetic field whose tangential force vectors act on the spots forcing them to move around on the contact surfaces at a high velocity. This action not only avoids burning of the contacts, but also helps in the handling of large interrupting currents. The question of arc contraction and arc motion and the role of the magnetic field is a subject of continuing research [28].

7.5.2 Vacuum Gap After the Arc

In AC circuit breakers, the arc is interrupted at a current zero. At current zero, the interelectrode space is quickly deionized, and the space surrounding the electrodes is filled with a residue consisting of desorbed gas, neutral electrode vapor and plasma. The process, the time in which this residue

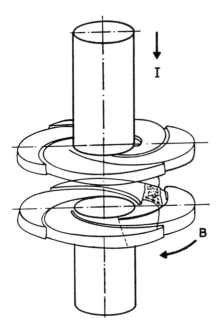

Figure 7.5 Spiral contacts of an interrupter. (From Ref. 39.)

decays and the result of the vacuum gap regaining its dielectric strength are referred to as an arc recovery phenomena.

If one considers the stresses on the interrupter contacts after the current zero, one can see that power frequency voltage exists at the generator side contacts, whereas the voltage of the faulted resonant circuit exists on the load side contacts. The vector sum of these voltages forms the stress on the contact gap. This total is known as transient recovery voltage (TRV). The rise of the dielectric strength in the contact gap (dielectric recovery voltage) must take place more rapidly after quenching of the arc than for recovery voltage, otherwise "restriking" of the arc will take place. Under real power network fault conditions, the TRV consists of many high-frequency oscillations, one superimposed on the other. This results in a very high rate of rise of recovery voltage (RRRV), and thus dictates the stress factor such that at any instant during arc interruption, it should not exceed the dielectric recovery of the vacuum gap.

It therefore becomes apparent that both the maximum RRRV and the peak TRV are important parameters for circuit breaker specifications and testing. In general, the dielectric recovery speed of separating contacts in an interrupter highly depends on (1) the prior thermal stresses on the contact gap, which are controlled by the arc voltage, the breaking current and the arcing time, and (2) the arc quenching medium, which also plays a significant role. In Figure 7.6, tests with nitrogen reflect the conditions in a compressed air breaker, and tests with hydrogen approximately reflect the conditions in minimum oil circuit breakers. It is clear that dielectric recovery takes place slowly in circuit breakers in which an arc quenching medium is used [29]. Contrary to that, the dielectric recovery in vacuum is accomplished within ~10 μs [30]. In practice, single units have shown the rate of rise of recovery voltage of the order of 20 kV/μs, which is at least an order of magnitude higher than in other types of interrupters. Similarly, the arcing time in a vacuum is extremely small. Interruption is usually accomplished at the first current zero after the contacts part. Arc duration is therefore always much less than one cycle.

7.5.3 Current Chopping and Contact Material

Below a certain minimum current, a metal vapor arc is interrupted prior to its sinusoidal zero. This phenomena is called current chopping [29]. It can lead to generation of high overvoltages on the power network or other devices connected across the vacuum circuit breaker (VCB), caused by the magnetic energy still trapped in the circuit's inductance. Studies have shown that chopping is not only dependent on the breaking current but also on the contact material [32,33]. This shows that at each instant during arc interruption, there must be exactly as much metal vapor available as is

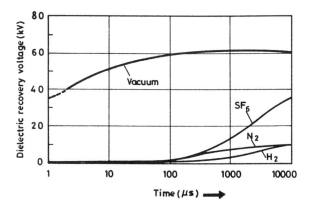

Figure 7.6 Dielectric recovery voltage as a function of time elapsed after arc extinction at current zero (rms current = 1600 ampere). Gas pressure 1.0 bar without switching media flow. Electrode spacing = 6.35 mm. (From Ref. 29.)

required to conduct the current then flowing, otherwise the current will be interrupted with a high rate of di/dt, resulting in a buildup of unnecessary voltage oscillations in the inductive circuits (high Ldi/dt) connected on the load side. While the interrupter may successfully withstand this voltage without breaking down, the voltage may cause failure in other parts of the connected circuit.

During the 1960s, tungsten alloys were used as contact material because they possess a high melting point. A great amount of energy is required to vaporize this material. With small currents approaching natural zero, this energy is no longer available, thus causing a forced quenching of the arc (its chopping). This problem impeded the introduction of VCBs in the market. However, in the 1970s this problem was solved with the introduction of a low-surge-type contact material consisting of a copper-chromium (CuCr) alloy. It combines good electrical properties with arc erosion and good welding resistance [34]. With these contacts the average value of chopping current has been contained to around a 3.0 A level [32,35]. More recently, a silver-tungsten carbide (AgWC) contact material has been introduced [33,36]. In this case, the base contact material is WC, and the high pressure metal component is silver. With these contacts the small current arc can be maintained to just before the current reaches its natural zero, because the vapor from highly molten silver metal is injected into the interelectrode space. In this case, the value of chopping current lies close to that of puffer type SF_6 breakers and minimum oil circuit breakers [30]. Other studies show that in some circuit conditions, even SF_6 breakers can have higher chopping currents than those exhibited by vacuum interrupters [37].

There are, however, some high surge impedance circuits where surge protection is required in parallel with VCBs [38]. For repeated switching of medium voltage motors using vacuum contractors, a high vapor pressure material such as bismuth is added to the CuCr base to lower the value of the chop current, which is generally not severely affected by repeated switchings [26,34]. Table 7.2 compares the chopping currents of various materials [28].

7.6 VACUUM CIRCUIT BREAKER

As discussed earlier, the vacuum is an excellent insulator and arc extinguisher. Extensive research and development effort over the past three decades in vacuum arcs, arc interruption processes and high voltage design have lead to the production of vacuum circuit breakers (VCB) that exhibit an outstanding interruption performance. Their short-circuit interruption

Table 7.2 Average Chopping Current of Pure Metals and Some Alloys at 45 A, 50 Hz and 400 V

Cu	18 A
Cr	6.5 A
W	9 A
Ca	8 A
Bi	0.3 A
Sb	0.4 A
Cu Cr25	6 A
Cr Cr50	5 A
Cu Cr25 Sb9	3.9 A
Cu Cr25 Zn10	3.2 A
W Cu30	6.2 A
W Cu30 Sb2	2.8 A
WC Ag 40	1.7 A

Surge impedance = 1 kΩ.
Source: Ref. 28 © IEEE, 1993.

capability has been pushed to currents of 50 kA and voltages of up to 36 kV. The applications of VCB over the past two decades have demonstrated its strength over the other technologies. It now enjoys more than 50% of the world market share, which is continuously growing [26]. Both, the vacuum and SF_6 technologies provide advantages over minimum oil and magnetic air circuit breakers. The user can employ either SF_6 or vacuum interrupter in metal clad switchgear and have circuit protection that will satisfy the required operational requirements.

7.6.1 Construction of Vacuum Interrupter

Central to the design of a VCB is the vacuum interrupter. Figure 7.7 illustrates the internal components of a typical vacuum interrupter. The ambient gas pressure within the evacuated envelope is kept in the range of 10^{-6}–10^{-4} Pa. Under normal conditions the contacts are closed and electrodes are in contact with each other.

When the bellowed contact moves away from the stationary contact, the arcing is established as a result of vapor evaporated from the local hot spots on the surfaces of the contacts. There is a continuous supply of this metal vapor, which is distributed on the contact surfaces, whereas the remaining vapor condenses on the surrounding metal vapor condensation

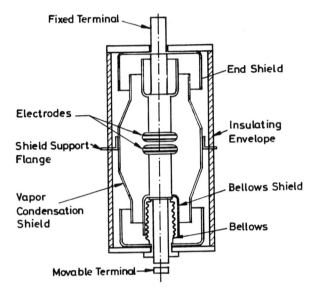

Figure 7.7 A typical vacuum interrupter. (From Ref. 26.)

shield. The latter is usually isolated from both contacts and serves to protect the glass or the ceramic external insulator from the deposition of the vapor. When the current wave reaches its natural zero, the vapor production stops and the original dielectric strength of the vacuum is rapidly restored, which is above the transient recovery voltage applied on the contacts due to the external circuit. While the contacts remain isolated, they are held externally by the insulating envelope. The metal bellow enables the moving contact stem to carry out its mechanical stroke, which varies according to the rated voltage of the interrupter.

External insulating envelopes are generally made of glass or metallized aluminum-oxide-based ceramics, which permits them to be permanently brazed to metal. Therefore, there are no replaceable seals, because the interrupter has a permanent sealed construction. Field data gathered by almost all manufacturers of vacuum interrupters have confirmed that the probability of vacuum loss is negligible [26]. Moreover, if these units are not mishandled, their service life with full current interrupting ratings is much longer than 20 years.

7.6.2 Limitations of VCB

Figure 7.8 shows the dielectric strength of different insulating mediums. For a gap of ~7.0 mm, the vacuum interrupter has higher breakdown

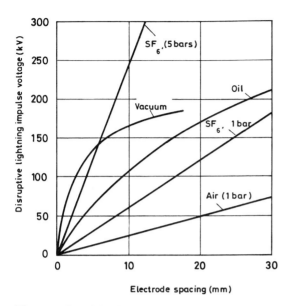

Figure 7.8 Lightning impulse withstand capability of various insulating media as a function of gap spacing in homogeneous field. (From Ref. 19.)

strength than the SF_6 breaker operating with 5.0 bar pressure. With the increase in contact gap, the breakdown strength of SF_6 gas increases linearly, whereas in a vacuum it tends toward saturation. For a gap of 16 mm, the corresponding strength of VCB is 180 kV, which is slightly higher than basic impulse level (BIL) of 170 kV recommended for 36 kV power frequency operations. It is clear that VCB has considerable advantage over an SF_6 breaker in the medium voltage range. To go for higher voltage levels, therefore, two or more of such interrupters will be required in series. This is always not very economical. It can thus be deduced that VCB is advantageous up to rated voltages of 36 kV, but beyond that level, the SF_6 breaker takes the lead.

7.6.3 Merits and Demerits of VCB

Merits

1. In a vacuum interrupter, the electrical contacts remain unaffected by the ambient internal environment, so there is very little change in its contact resistance throughout its life. In SF_6 breakers, on the other hand, arc by-products continuously react with internal com-

ponents, and if there is an ingress of moisture, it will produce hydrofluoric acid, which will react with contacts and glass filled insulators, which can affect the equipment integrity. Generally, VCB remains pollution free.
2. VCB is self contained and does not need gas or oil filling. There is no need for an auxiliary air system or oil handling system.
3. The necessary maintenance is modest, thus it is economical in the long term.
4. There is no risk of explosion or leakage of toxic byproducts.
5. VCBs have silent operation due to short stroke and light mechanical loadings and shorter contact parting time. Moreover, a smaller number of components are needed.
6. Since contact erosion is small, the normal current can be interrupted up to 30,000 times, and rated short-circuit current interruptions up to 100 times are generally quoted—more than all other types of circuit breakers.

Demerits

1. In the event of loss of vacuum, the unit cannot be repaired at the site.
2. In some applications where low magnetizing currents are involved—such as furnace transformers, reactors and some high voltage motors—the additional surge suppressors are required in parallel with the VCB in each phase [32]. These are installed as close as possible to the load and are also permissible in the switchgear panels.

REFERENCES

1. R. V. Latham, IEEE Trans. on Elect. Insul., Vol. 23, No. 1, pp. 9–16, 1988.
2. W. Ziomek and H. M. Grzesiak, IEEE Trans. on Elect. Insul., Vol. 28, No. 4, pp. 481–487, 1993.
3. R. V. Latham, *High Voltage Vacuum Insulation*, Academic Press, London, England, 1981.
4. R. H. Fowler and L. Nordheim, Royal Society Proceedings, Vol. A119, pp. 173–181, 1928.
5. D. Alpert, J. Vacuum Sci. and Technol., Vol. 1, No. 2, pp. 35–50, 1964.
6. E. Dullni, IEEE Trans. on Elect. Insul., Vol. 28, No. 4, pp. 454–460, 1993.
7. G. A. Farrall, IEEE Trans. on Elect. Insul., Vol. 20, No. 5, pp. 815–841, 1985.
8. B. M. Cox, J. Phys. D, Vol. 8, No. 17, pp. 2065–2073, 1975.
9. J. Halbritter, IEEE Trans. on Elect. Insul., Vol. 20, No. 4, pp. 671–681, 1985.

10. M. Wurtz, H. Adam and W. Walcher, *Theory and Practice of Vacuum Technology*, Freid Viewveg and Sohn, Braunsweig, Germany, 1989.
11. V. A. Nevrovskiy, J. Phys., Vol. 14, pp. 215–220, 1981.
12. J. Ballat, D. Konig and U. Reninghaus, IEEE Trans. on Elect. Insul., Vol. 28, No. 4, pp. 621–627, 1993.
13. M. G. Danikas, IEEE Trans. on Dielectrics and Elect. Insul., Vol. 3, No. 2, pp. 320–321, 1996.
14. J. B. Birks (Ed.), *Progress in Dielectrics*, Vol. 7, Heywood Books, London, England, 1967.
15. V. N. Maller and M. S. Naidu, *Advances in High Voltage Insulation and Arc Interrupter in SF_6 and Vacuum*, Pergamon Press, Oxford, England, 1981.
16. R. Hackam and L. Altech, J. Appl. Phys., Vol. 46, pp. 627–636, 1975.
17. K. W. Arnold, Proc. 6th International Conference on Ionization Phenomenon in Gases, 2, 101, 1963.
18. C. C. Erven, IEEE Trans. on PAS, Vol. 91, pp. 1589–1596, 1972.
19. R. Hawley, "Vacuum as an Insulator", Vacuum, Vol. 10, pp. 32–41, 1961.
20. F. Schneider, H. C. Karner and M. Gollar, Proc. 6th International Symposium on High Voltage Engrg., New Orleans, 1989, Paper 48.07.
21. H. G. Bender and H. C. Karner, IEEE Trans. on Elect. Insul., Vol. 23, No. 1, pp. 37–42, 1988.
22. H. C. Miller, IEEE Trans. on Elect. Insul., Vol. 26, No. 5, pp. 949–1043, 1991.
23. L. Cranberg, J. Appl. Phys., Vol. 23, No. 5, pp. 516–522, 1952.
24. J. E. Thompson, Proc. 6th International Symposium on High Voltage Engrg., New Orleans, 1989, Paper 48.01.
25. G. A. Meayats and D. I. Proskurovsky, *Pulsed Electrical Discharges in Vacuum*, Springer Verlag, Berlin, Germany, 1989.
26. P. G. Slade and R. W. William, Power Technical International, pp. 171–175, 1993.
27. F. G. Rowland, Electronics and Power, Vol. 21, pp. 496–499, 1975.
28. K. Frohlich, H. C. Karner, D. Konig, M. Lindmayer, K. Moller and W. Rieder, IEEE Trans. on Elect. Insul., Vol. 28, No. 4, pp. 592–596, 1993.
29. T. V. Armstrong and P. Headley, Electronics and Power, Vol. 20, pp. 198–201, 1974.
30. E. Dullni and E. Schade, IEEE Trans. on Elect. Insul., Vol. 28, No. 4, pp. 607–620, 1993.
31. J. M. Lafferty, *Vacuum Arcs, Theory and Application*, John Wiley and Sons, New York, 1980.
32. D. L. Swindle, IEEE Trans. on Industrial Applications, Vol. 20, No. 5, pp. 1355–1363, 1984.
33. E. Kaneko, K. Yokokura, M. Hommn, Y. Satoh, M. Okawa, J. Okutami and I. Ohshima, IEEE Trans. on Power Delivery, Vol. 10, No. 2, pp. 797–803, 1995.
34. P. G. Slade, Proc. 18th International Conference on Electrical Contacts, 1992.
35. S. S. Rao, *Switchgear and Protection*, Khanna Publishers, Delhi, India, 1993.

36. R. P. P. Smeets, E. Kaneko and I. Ohshima, IEEE Trans. on Plasma Sci., Vol. 2, No. 4, pp. 439–445, 1995.
37. J. D. Gibbs, D. Koch, P. Malkin and K. J. Cornick, IEEE Trans. on Power Delivery, Vol. 4, No. 1, pp. 308–316, 1989.
38. J. F. Peridne and D. Bheeavanich, IEEE Trans. on Industrial Applications, Vol. 19, No. 5, pp. 679–686, 1983.
39. ABB-Calor Emag, Germany, Cat. VD4-Vacuum Circuit Breaker, drawing no. V0003 Sp/E.

8
Composite Dielectrics

8.1 INTRODUCTION

The insulation requirements of a power network rarely consist of a single material. The use of two or more insulating materials becomes necessary due to design considerations or due to practical difficulties of fabrication. These different materials may be in parallel with one another such as an air gap in parallel with solid insulation, or oil in parallel with pressboard. Similarly they may be in series with one another, such as laminates. Apart from these, filled or reinforced insulating materials may have microscopic volumes of another phase of different materials present in the bulk, e.g., in polyethylene or in a mixture of two granulated substances in epoxy resins or impregnated solids, etc. Therefore, all of these forms of insulation are composite in nature.

In certain cases the behavior of a composite insulation system can be predicted by the behavior of individual components. But in most cases, the system as a whole has to be considered. The performance of such a system can be evaluated only by considering the factors such as: (1) the stress distribution at different parts of the insulation system, (2) breakdown characteristics at the surface, as they are effected by the boundaries of the composites, and (3) partial discharge products and/or chemical aging products of one component which may react with the other component. Similarly the economic life of the composite system will have to be considered, since under long-term operation, several types of breakdown mechanisms (other than electronic, electromechanical or thermal breakdown mechanisms, as explained in chapter six) may be in progress either independently or in combination. This chapter is concerned with composite dielectric systems that are generally encountered on a high voltage power network.

8.2 DIELECTRIC PROPERTIES OF COMPOSITES

Figure 8.1 illustrates slabs of two dielectrics placed perpendicular (in series) to the axis of an applied AC electric field E. If we assume that neither one of the dielectrics affects the properties of the other and no moisture, gas inclusion, etc., exist, then the ratio of electric fields in the two mediums varies inversely with their respective dielectric constants, i.e., $E_1 \varepsilon_1 = E_2 \varepsilon_2$.

Similarly, under a direct applied voltage the respective field ratio will vary directly with the resistivity ratios of the two mediums. Since the composite system shown in Figure 8.1 can be considered as two capacitors in series, the voltage distribution across the upper dielectric of thickness d_1 can be given as:

$$V_1 = \left[\frac{V}{1 + (\varepsilon_1 d_2 / \varepsilon_2 d_1)} \right] \qquad (8.1)$$

where V is the total voltage applied across the composite and d_2 is the thickness of the lower dielectric. If the upper dielectric is air ($\varepsilon_1 = 1$) in series with a solid insulating sheet or barrier, and suppose $d_1 = 4d_2$ and $\varepsilon_2 = 3$, the evaluation of stress using equation (8.1) shows that the stress in the air gap has increased by 15% due to the presence of the insulating barrier. This stress in the air gap will further increase if the barrier thickness is increased and will approach 300% of the stress value without the

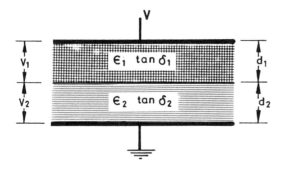

Figure 8.1 Two-layer composite dielectric.

Composite Dielectrics

barrier when d_2 is the only marginally smaller than d, the total gap length. If the air gap is marginally designed then there is a chance that it will breakdown and full voltage will appear across the barrier which is then likely to fail. Therefore, in such a situation, one must ensure that the air space is capable of withstanding the increased stress caused by the presence of the solid insulating material.

The other commonly used composite dielectric is generally a two-phase mixture of either reinforced fibers with resins, or inorganic filler in a polymeric resinous matrix as shown in Figure 8.2. Thin cavities may be present due to inefficient manufacturing process in such systems. The interface therefore exists around every minute fiber or particle of filler and cavities. The difference in coefficients of thermal expansion and volume shrinkage can create free volumes in the bulk. Similarly, many of the reinforcing materials have great affinity to absorb moisture. If a chemical bond does not exist at the interface of solids, then water molecules can easily diffuse into the composite and form water layers [1]. Presence of water will reduce the DC resistivity and the dielectric strength, whereas it will slightly increase the permittivity of the composite. However, the dissipation factor will suffer an appreciative increase. This implies that the dielectric properties of a composite measured at a certain time mainly depend on the history of the specimen.

The dielectric constant of a composite is generally based on the dielectric constants of the individual components and their volume fractions ($\bar{v}$). If a composite is formed of only two components, neither of them

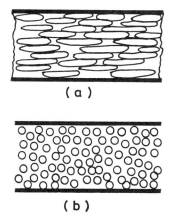

Figure 8.2 Internal bonding of composite mixtures: (a) mica platelets and (b) parallel rods of fiberglass.

affects the properties of the other and provided that moisture, gas inclusions or similar other factors are absent, then the value of ε and tan δ for the composite can be given as:

$$\varepsilon = \frac{\varepsilon_1 \varepsilon_2}{\varepsilon_1 \bar{\nu}_2 + \varepsilon_2 \bar{\nu}_1} \tag{8.2}$$

$$\tan \delta = \frac{\tan \delta_1 + (\varepsilon_1 \bar{\nu}_2 / \varepsilon_2 \bar{\nu}_1) \cdot \tan \delta_2}{1 + (\varepsilon_1 \bar{\nu}_2 / \varepsilon_2 \bar{\nu}_1)} \tag{8.3}$$

Here the prefix numbers represent the individual values of the two components used to formulate a composite. The representation of such a composite as a simple two-layer dielectric is no doubt a simplification of a complex problem. A full account of various theories and experimental results related to this topic have been given by van Beek [1] and Bartinkas and Eichhorn [2].

Interfacial Polarization in Composites

If an insulating material is composed of two or more different phases that contain dispersed macroscopic impurity regions, then on the application of an electric field a space charge builds up at the microscopic interfaces as a result of differences in conductivities and permittivities of its individual components. This is known as interfacial polarization. It will lead to increased dielectric losses and also cause field distortion in the material [2,3].

When one of the composite components has significant ionic conductivity, then mobile charges (usually impurity ions) diffuse under the influence of the applied field across the more conducting component up to the interface of the less conducting component. At this new interface they will become stationary and thus build up a surface charge, as shown in Figure 8.3. This effect will remain there until the applied field reverses, as in the case of alternating voltage. In the case of DC voltage and longer stressing time, a much larger space charge buildup continues until a very significant reverse electrical field develops from this surface charge, eventually arresting the current flow to a value that is determined by the less conducting component. This type of polarization is very common in composite insulation, since such interfaces are distributed throughout its structure. A recent study [4] has established that the use of DC voltage testing of polymeric cables is harmful for their life, which depends on several factors; the most important of these is due to the space charge buildup in semicrystalline structure of crosslinked polyethylene. The high concentrations of traps are found at the crystalline-amorphous interfaces [5]. The study

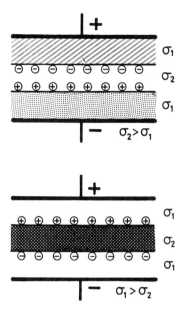

Figure 8.3 Interfacial polarization in a composite dielectric.

of space charge accumulation in polymeric cables and solid insulation is a topic of current research worldwide [6,7].

8.3 EDGE BREAKDOWN

In practical systems electrodes may be embedded on the surface of insulation as shown in Figure 8.4. It is clear that at the edges of electrode a

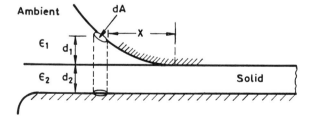

Figure 8.4 Breakdown of solid specimen due to edge effect.

composite insulation is formed, since the ambient medium becomes in series with the insulating slab under test. Assuming homogeneous field, consider elementary cylindrical volume of area dA spanning the electrode at distance x, then on application of voltage V across the electrodes, its fraction V_1 [as given in equation (8.1)] shall appear across the ambient air medium. For most of solid dielectrics used in power apparatus, $\varepsilon_r = 2 \sim 3$. Hence there will be substantial enhancement of the field in the air medium. The stress will further increase as x is decreased. Consequently, the ambient medium will breakdown at relatively low voltage as d_1 becomes very small. The charge at the tip of the resulting discharge will make the electrode arrangement highly nonuniform. Therefore, local breakdown at the tip of discharge impinging at the solid interface will result in a discharge channel which will erode its surface. Occurrence of many such breakdown channels will result in the formation of a tree in the slab which will extend step by step through the whole thickness, ultimately causing a complete breakdown of the slab insulation. However, it should be noted that the tree-like pattern of discharge is not limited specifically to the electrode edge effect but may be observed in other failure mechanisms in which nonuniform stresses predominate. For details, see Chapter 9.

8.4 CAVITY BREAKDOWN

Solid insulations, and to a lesser extent, liquid dielectrics, contain voids or cavities within the medium or at the boundaries between the dielectric and the electrodes. Similarly, in amorphous polymers, there are vacant spaces or holes between the molecules which are called free volume. In polyethylene, below the glass transition temperature T_g, the free volume is approximately 2.5% of the total volume, which increases with temperature above T_g[8]. Similarly, epoxy resin composites are prepared by mixing reinforcing fillers. Generally there exists between them a chemical or quasi-chemical adsorption-type bonding at the interface. But if care is not taken in fabrication, many areas of interface may have no bonding but a gap of varying thickness. These voids are generally filled with medium of lower dielectric strength and lower permittivity. Hence the electric strength in the voids, as shown earlier, will be higher than that across the dielectric. Therefore, even under normal working voltages the field in the voids may exceed their breakdown value, and breakdown may occur in these small entities.

Figure 8.5 shows a cross-section of a dielectric of thickness d, containing a cavity in the form of a disc of thickness t, together with its corresponding "abc" equivalent circuit. In this circuit the capacitance C_c corresponds to the cavity, C_b corresponds to the capacitance of the dielec-

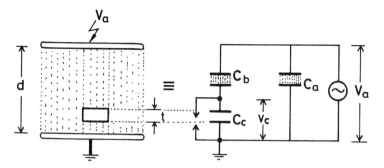

Figure 8.5 A cavity in a dielectric and its "abc" equivalent circuit.

tric that is in series with C_c and C_a is the capacitance of the rest of the dielectric. For $t \ll d$, which is usually the case, and assuming that the cavity is filled with a gas, the field strength across C_c is given by the expression $E_c = \varepsilon_r E_a$, where ε_r is the relative permittivity of the dielectric.

For the simple case of a disc-shape dielectric in solid shown in Figure 8.5, the discharge inception voltage can be expressed in terms of the cavity breakdown stress. Assuming that the gas-filled cavity breakdown stress is E_{cb}, then treating the cavity as series capacitance with the healthy part of the dielectric, we may write:

$$C_b = \frac{\varepsilon_o \varepsilon_r A}{d - t}$$

$$C_c = \frac{\varepsilon_o A}{t}$$

The voltage across the cavity is:

$$V_c = \frac{C_b}{C_c + C_b} V = \frac{V}{1 + \frac{1}{\varepsilon_r}\left(\frac{d}{t} - 1\right)}$$

Therefore the voltage V_i across the dielectric which will initiate discharge in the cavity will be given by

$$V_i = E_{cb} t \left\{ 1 + \frac{1}{\varepsilon_r}\left(\frac{d}{t} - 1\right) \right\} \tag{8.4}$$

Generally, a cavity in a material is often nearly spherical, and for such a case the internal field strength can be given as:

$$E_c = \frac{3\varepsilon_r E}{\varepsilon_{rc} + 2\varepsilon_r} \tag{8.5}$$

where ε_{rc} is the dielectric constant of the cavity, and E is the average stress in the dielectric. Since $\varepsilon_r \gg \varepsilon_{rc}$, therefore $E_c \approx 3/2\ E$. Hence, the stress inside the cavity becomes much greater than the average stress in the dielectric.

The sequence of breakdown under sinusoidal alternating voltage is illustrated in Figure 8.6. The dotted curve qualitatively shows the voltage that would appear across the cavity if it did not breakdown. During the positive half cycle, as V_c reaches the value V_i^+, a discharge takes place, the voltage V_c collapses and the gap extinguishes. The voltage across the cavity then starts again increasing until it reaches V_i^+, when a new discharge occurs. Thus several discharges may take place during the rising part of the applied voltage. Similarly, during the negative half cycle of the applied voltage, the cavity discharges as the voltage across it reaches V_i^-. In this way, groups of discharges originate from a single cavity and give rise to positive and negative current pulses on the positive and negative cycles of the voltage wave. The discharge in such an insulating medium can be measured with partial discharge (PD) detectors. For details refer to section 12.6.

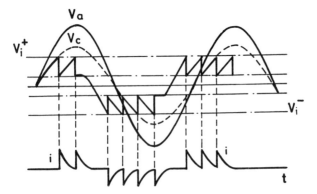

Figure 8.6 Sequence of cavity breakdown under alternating voltages.

When the gas in the cavity breaks down, the surface of the insulation acts as cathode and anode. Some of the electrons impinging upon the anode are sufficiently energetic to break the chemical bonds of the insulation surface. Similarly, bombardment of the cathode by positive ions may cause damage to the internal surface of the cavity by increasing the surface temperature and producing local thermal instability. Additional chemical degradation may result from active discharge products, e.g., O_3 or NO_2, formed in air, which may cause insulation deterioration. Whatever the deterioration mechanism operating, the net effect is a slow erosion of the material and a consequent reduction in the thickness of the solid insulation. Partial discharges are therefore harmful to the insulation, regardless of the size of the void in which the discharges take place [9]. However, a larger void is more dangerous than a smaller one, since the larger void not only discharges at a lower applied stress, but also dissipates more energy in each discharge [10], and thus causes more severe insulation damage. A survey has revealed that void size > 50 μm is the main cause of the high voltage cable-insulation failure [11].

8.5 BREAKDOWN DUE TO SURFACE EROSION AND TRACKING

Polymeric insulators are preferred in many application as they are easy to fabricate in complicated shapes and are tough, light in weight and possess excellent dielectric properties. However, their service on high voltage networks is coupled with the processes of degradation due to erosion and tracking. When the discharges occur on the insulation surface, the erosion takes place initially over a comparatively larger area. The erosion roughens the surface and also causes pitting, which aids accumulation of contamination and finally gives way to tracking. Erosion and pitting is also promoted due to the "edge mechanism" explained earlier.

Tracking is the formation of a permanent conducting path, usually carbon, across the surface of insulation, resulting from degradation due to continuous discharges and erosion. In service, the surface of insulation is progressively contaminated. The contamination takes up moisture from the atmosphere and the wet layer of pollution provides a continuous, conducting path between the HV electrode and the ground. The surface resistance decreases considerably in the presence of pollution and moisture. Low resistance, in turn, leads to high surface leakage currents and high power dissipation, causing significant loss of moisture from the surface. This loss is not uniform, and leads to the formation of dry bands. When a dry band is formed, the flow of surface leakage current is interrupted. The inductance of the system generates HV surges so that the effect is similar to a circuit breaker in which the contacts sustain the system voltage plus a high frequency transient reaching up to twice the supply voltage. In the case of dry banding, almost all of the applied voltage is concentrated across these dry bands and electrical discharges occur, causing the insulation surface to reach high local temperatures at the arc root, leading to gradual erosion and the formation of carbonaceous residues which become the focus of further action. Complete breakdown of the insulation surface generally follows when a conducting carbon path propagates to the extent that the remaining insulation is incapable of withstanding the system voltage.

A number of international standard testing methods have been established in order to evaluate the relative tracking resistance of different insulating materials [12]. At present, the generally accepted method is the liquid contaminant inclined plane tracking method, described in IEC-587 (1984) and ASTM-D 2303 (1984) [13,14]. The IEC method offers two test procedures for the time to track (1) under a constant voltage, and (2) under a stepwise voltage application. The IEC method is preferred because it has been observed that two materials with the same initial tracking voltage (ITV) but having a different resistance to tracking (due to filler content) may give very different times to track [15]. Another method widely used in many European countries and in the United States is described in the ASTM Test Method D 495 [16]. In this method, voltage is applied between two chisel-shaped tungsten rods as shown in Figure 8.7. No contaminant is used in this test method. The numerical value of the voltage that initiates tracking is called the "tracking index" and is used to quantify the surface properties of the dielectric under test.

8.6 CHEMICAL AND ELECTROCHEMICAL DETERIORATION AND BREAKDOWN

Since insulating materials are either composed of differing chemical substances or come in contact with materials of different composition, chemical reactions between various materials are inevitable. Application of continuous electric stress in the presence of high temperature may act as catalyst and enhance the rate of chemical reactions. Thus such composites may undergo chemical deterioration leading to reduction in electrical and mechanical strengths. The most deleterious effects are caused by O_2 in the presence of ultraviolet radiation and contact with moisture/water.

In the presence of air (O_2), materials such as rubber and PE undergo oxidation, giving rise to surface cracks. During electric discharge in air, nitrous oxide (NO) is produced, which in the presence of moisture forms

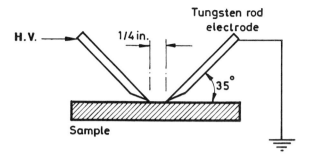

Figure 8.7 Dry arc electrode assembly.

nitric acid (HNO_3) which can degrade the dielectric. Similarly, in the presence of moisture, paper, cotton tape, cellose, etc., deteriorate due to hydrolysis.

Individual composite material such as epoxy resins, may have extensive interface area between the two mixed solids throughout the composite (see Figure 8.2). In such cases, thermal expansion coefficient of the two solids may have significant difference. For example, glass in reinforcing fibers has a linear thermal expansion coefficient (α) of about 0.5 to 0.7 × 10^{-5} cm/cm/°C, while the resins used in composites with the glass have $\alpha = 2 \sim 9 \times 10^{-5}$ cm/cm/°C, which is nearly tenfold higher [17]. With an increase of temperature, this may result into breakage of the interfacial chemical bonds, thus leaving behind cavities or microcracks. Since most of the reinforcing materials have a great affinity for water, water molecules can diffuse into these free sites and replace them with water layers. Presence of such a water film along the interface will further accelerate the process of electrochemical deterioration. Moreover presence of such a film will increase the electrical conductivity and the dielectric losses.

Similarly, water in contact with insulator surface (or absorbed moisture) has been found to decrease significantly the surface resistivity. Study on the cellulose-based laminates has shown a decline of about 65% in the dielectric strength, in the presence of 90% humidity in the air [17]. In case of composites containing hydrophilic fibers, if both sides of one of the internal fibers extends to the wet external surface, it will form an internal wet path. The conduction along the internal electrolyte path can develop sufficient heat to vaporize the water at the hottest point, thereby breaking the electrical circuit and forming a small arc. This will thus result into carbonized tracks. This mechanism is similar to external tracking, but in this case there is less cooling by evaporation and convection, so the internal arcs may develop at lower voltage gradients. Such an internal arc tracking has been found to cause carbon tracks in resins, which do not normally carbonize under external arcing conditions [17].

It should be particularly recognized that the power frequency short-time dielectric strength values usually quoted are not usable in service, since all organic resin composites decline very significantly in dielectric strength with the time of voltage application. The time to failure seems to vary as an inverse exponential of the electric stress, as shown in Figure 8.8. The decline in strength is due to partial discharges occurring internally or externally at the surface. There is also a decline in dielectric strength with thermal aging, which degrades the organic resins and the chemical bonds of the organic reinforcement [18].

8.7 MATERIALS OF OUTDOOR INSULATORS

Porcelain and glass have been the most widely used materials for outdoor HV insulation applications. Extensive service experience has shown that

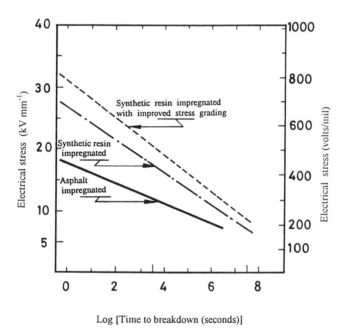

Figure 8.8 Voltage endurance of mica insulation. (From Ref. 2 © ASTM, reprinted with permission.)

these materials are very reliable and cost effective for a majority of outdoor HV applications. However, since the early 1960s, alternative materials, namely polymers, have emerged and presently they are being used extensively for a variety of outdoor HV applications. The chief advantages of polymers are their light weight, superior vandal resistance and better handling of mechanical shock loads.

Initially, polymeric insulators were considered as replacements for porcelain and glass for special applications such as areas with a high incidence of vandalism, urban locations with limitations on right of way, and areas of severe contamination problems. However, many difficulties were encountered regarding their performance in actual service during the fist two decades of operation. Typical of these were tracking and erosion of sheds, chalking and crazing of sheds, which lead to increased contamination collection; arcing and flashover; bonding failures and electrical breakdown along the rod-shed interface, corona splitting of sheds and water penetration due to hot line water washing which also resulted in electrical failures [21]. However, through continuous effort and research in polymeric material development and improved manufacturing technology, much improved versions of insulators have been introduced. Their comparable performance with porcelain and glass have lead to their widespread use even for routine outdoor HV insulation applications.

Today, polymeric line insulators are in use on lines operating up to 765 kV. However, they are more popular on transmission levels from 69 through 345 kV. A worldwide survey carried out by CIGRE [19] explains that there are several thousand polymeric insulators in service at all distribution voltage levels. An EPRI survey of polymeric insulators in the United States has reported that 78% of the utilities reported good performance, 18% reported acceptable, and only 4% were found unsatisfied with the performance of polymeric insulators.

8.7.1 Basic Polymeric Insulator Design

The basic construction of polymer insulators for overhead line applications consists of a core, weathersheds and metal end fittings. Two types are in common use. The suspension/dead-end type (Figure 8.9a) is used where line loads subject the insulator core to tension forces; the post type (Figure 8.9b) is used where line loads subject the core to appreciable bending forces in addition to the tension forces. The only significant differences are in the design of the attachment hardware and in the size of the core, which is much larger for posts [10].

The core of a nonceramic insulator has the dual burden of being the main insulating part and of being the main load-bearing member, be it in

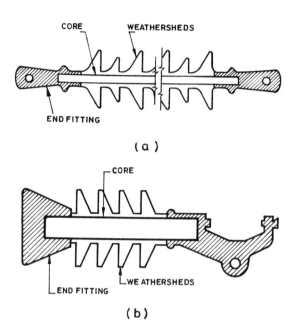

Figure 8.9 (a) Suspension/deadend-type and (b) post-type polymer insulators. (From Ref. 10 © IEEE, 1991.)

suspension, cantilever or compression. For suspension and line post insulators, the core consists of axially aligned, glass fiber-reinforced resin containing 70–75% by weight of E-type fiberglass. The fiber diameter ranges from 5 to 20 μm. The resin system can be of polyester or epoxy and the rod is formed by the pultrusion process. Although epoxy resin is considered to be the better of the two, because of lower cost the core used today is usually polyester resin.

The end seal is considered to be the most important element of the design of a nonceramic insulator. Field failures have occurred due to brittle fracture of the fiberglass rod due to breach of the end seal, thereby allowing the rod to come into contact with atmospheric pollutants and moisture. Tracking of the fiberglass rod leading to failure has also been observed in nonceramic insulators.

Nonceramic insulator end seals have three basic types: glued, friction and boned types. Glued-type seals that are made using a sealant material such as an RTV silicone rubber applied between the sleeved core and the hardware are not permanent, generally because of poor adhesion. Friction-type seals in which the sleeved core fits into the hardware are quite effective, as long as the dimensional tolerances are maintained; they do not cause any problems, provided that no movement of the fitting occurs. End seals that are made by molding the sleeved core material onto the end fitting are by far the best because of the physical bond obtained during molding [21,22].

8.7.2 Development of Composite Materials

Sheds made from various nonceramic materials compounded for electrical applications are shaped and spaced over the rod in various ways to protect the rod and to provide maximum electrical insulation between the attachment ends. It is quite clear that with such a diversity of constructions possible, the performance of nonceramic insulators depends on the choice of materials, the design and the construction of the insulator. A variety of polymer insulating materials has been developed for overhead electrical insulation. This includes PTFE (Teflon), epoxy resins, polyethylene, instant set polymers based on urethane chemistry, polymer concretes, various co-polymers, ethylene-propylene elastomers and room temperature and high temperature vulcanizing silicone elastomers. Each material offers particular characteristics. However, only the elastomeric materials have shown success in outdoor electrical insulation applications, with silicone meeting all of the requirements for long-term performance in practically all environments.

Today, only three classes of materials are in any significant use: epoxy resins, hydrocarbon elastomers and silicone elastomers. Properties of these have been described in detail in section 6.2.5. The polymers have the ability to interact with pollutants and reduce the conductance of this pollution layer [23,24,48]. This is illustrated in Figure 8.10. The important characteristic of this polymeric insulator which controls the conductance is due to hydrophobicity or water repellency of its surface. It has been observed that the hydrophobicity is maintained in silicone rubber materials even after many years in service, and it is this attribute that is responsible for the superior contamination performance of silicone rubber family of materials when compared to other polymers. The recovery of hydrophobicity is mainly due to (1) a diffusion process whereby low molecular weight (LMW) polymer chains with only a fraction ($\sim 20\%$) from within the bulk of material migrate to the surface, where they form a thin layer of silicone fluid, and (2) reorientation of surface hydrophillic groups away from the surface. These processes are temperature dependent. Higher temperature causes more rapid recovery. Moreover, regeneration of the LMW chain

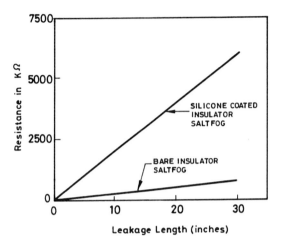

Figure 8.10 Surface resistance of bare and silicone-coated porcelain insulators under salt fog conditions (1.0 inch–25.4 mm). (From Ref. 23 © IEEE, 1991.)

occurs even after the initial supply of LMW polymer chains has been depleted, but this recovery from within the main material causes weakening of its mechanical properties [49].

8.8 OIL-IMPREGNATED INSULATION

The use of oil-impregnated cellulose paper in high voltage equipment has persisted to the present day despite the competition from a variety of synthetic materials. Although synthetic materials such as polypropylene (PP) possess lower tan δ, lower moisture absorption and increased dielectric strength, their capability of impregnation and mechanical strength are much inferior to that of cellulosic paper, generally known as "kraft paper." However, new materials that are in fact a sandwiched composite of paper and PP films are now available. Similarly, a variety of impregnants are also currently in use that range from low-cost petroleum oil to specially synthesized products.

8.8.1 Liquid–Solid Film Interaction

Impregnation can be regarded as a physical interactive process controlled predominantly by the surface tension force. The rate determining factors include the viscosity of the impregnant, its surface tension and the contact angle with the dielectric. A smaller contact angle (θ) between the liquid and the solid causes rapid penetration of the liquid from the insulation surfaces into the pores of the solid. The surface tension force is basically equal to $\eta \cos \theta$, where η is the surface tension of the liquid and is therefore dependent on the nature of the impregnant. In other words, impregnation, which is a very important process to fill the cavities of solid films, is dependent on the cavity size and the nature of the impregnant. On the other hand, excessive fluid absorption by the film results in its swelling, which produces several disadvantages, such as drop in electrical strength and reduction in tensile strength.

Chemical interaction between solid and liquid dielectrics is generally confined to various leaching processes whereby low molecular weight or ionic components migrate from the solid to the liquid or vice versa. The presence of ionic contamination in the liquid can be detected by an increase in conductivity or loss angle, as shown in Figure 8.11. This increase can, in turn, result in thermal limitations on AC equipment, and can also lead to further chemical deterioration at a rate which will depend on current

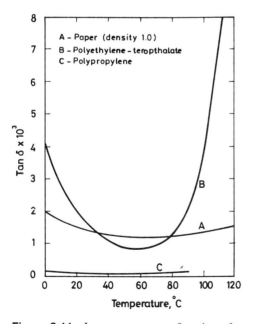

Figure 8.11 Loss tangent as a function of temperature for paper and various synthetic polymers. (From Ref. 44.)

density and chemical activity of ions liberated at the electrodes. It can also be enhanced by the presence of moisture and other contaminants that can dissociate in the dielectric. Electrochemical deterioration is a major cause of failure in liquid-impregnated power capacitors [25].

8.8.2 Effect of Multiple Layers

The simplest form of composite insulation system consists of two layers of the same material. This system is extensively used especially in cable and capacitor construction, where advantage is taken of the fact that two thin sheets have a higher electric strength than a single sheet of the same total thickness. Research work carried out on thin sheets has illustrated that this phenomenon is mainly due to the loss of energy from the discharges or partial breakdown channels at the interfaces. A discharge having penetrated one layer could not enter the next layer of material until the spot on the interface centered on the channel had been charged to a potential which could produce a field comparable with that of the channel at the level in question.

8.8.3 Application in HV Cables

In practice, cables have to experience mechanical stresses such as bending that could cause wrinkles or creases at the butt joints. It has been demonstrated experimentally that the formation of such deformation reduces breakdown strength. Therefore, cable insulation requires impregnated insulation with high mechanical and electrical strength. Impregnated paper is used in the form of tapes. This is lapped onto the cable conductor and then impregnated; it may or may not be pressurized. Such cables exist in two categories, self-contained oil-filled (OF) cables and pipe-type OF (POF) cables. The bulk of the underground cable systems installed in United States are high-pressure, POF-type cables. In contrast, self-contained OF cables are commonly used in Europe and Japan [8].

Experiments have shown that the thinner the paper sheet, the higher its dielectric strength. This arises from the resulting reduction in the oil-gap length, and increases the barrier effect and the impermeability. The barrier effect is caused by the arrangement of the stacked fine fiber. Figure 8.12 shows a linear relationship between impulse breakdown strength and the logarithm of paper thickness [8].

Impregnated-paper insulation in cable consists of simple layers of paper tapes. Each tape from 50 to 200 μm in thickness is wrapped helically around the conductor. The resulting insulation is comparatively homogeneous. One of the most important features of pressurized oil-impregnated

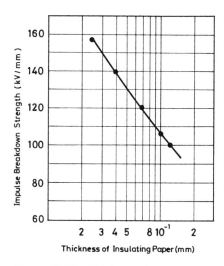

Figure 8.12 Thickness dependence of impregnated paper on impulse breakdown voltage. (From Ref. 8.)

paper is its high resistance to and/or freedom from corona or internal discharges. This provides the stability needed in cable insulation, especially for solid impregnated cables, and it is a main reason for its use in such applications.

Unfortunately, its electrical performance can be easily affected by a small amount of water. Figure 8.13 shows the variation of tan δ due to the addition of water. Thermal degradation is also enhanced by a small amount of water. For this reason, it is a standard practice to dry paper in a vacuum before oil impregnation and to protect a finished cable from water permeation with a lead sheath. In cables, since lower permittivity of insulation is required, mineral oils or alkylbenzene synthetic oil or their mixtures are used.

8.8.4 High Voltage Capacitors

Capacitors were introduced in HV power apparatus about 75 years ago when reliable and economical oil-impregnated paper was developed. Today HV capacitors are found in precision devices such as voltage dividers and in high kvar banks for power-factor correction on inductively loaded circuits. The dielectric is invariably an impregnated polymer of natural or synthetic origin. High voltage measuring divider systems generally contain a dielectric of mineral-oil-impregnated paper, but for power-factor correc-

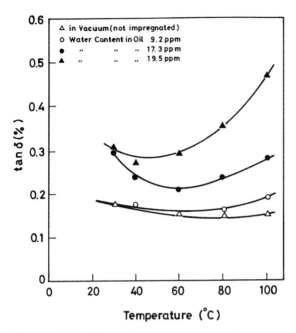

Figure 8.13 Effect of water content on loss tangent of oil-impregnated paper. (From Ref. 8.)

tion capacitors, the trend is towards an increasing use of polyproplyene film. The lower loss angle of dielectric materials in capacitors gives a reduction in dielectric heating and lower power consumption. Polypropylene film has been found to be a superior dielectric for capacitor application because of its reduced dielectric losses and higher dielectric constant. The volume of the dielectric in a capacitor is inversely proportional to the product of its ε_r and square of its working stress E, i.e., $\varepsilon_r E^2$ [47]. Therefore, use of PP film reduces both the size and weight of the capacitor. However, mixed paper/PP film dielectric also continues to be used in standard tank-type capacitors, since they enable a more complete impregnation of oils as compared with all-film-type capacitors. The latest designs of capacitors have loss tangents only one-tenth of the value for the early paper types.

The impregnants of power capacitors should have high dielectric strength, high dielectric constant, good insulation affinity with paper, chemical and thermal stability and adequate viscosity for efficient conduction of heat. Mineral oils with high aromatic content for high gas-absorbing

Composite Dielectrics 229

ability have traditionally been used. However, they have been replaced with a variety of synthetic oils such as di-isopropylnaphthalene (DIPN), mono isopropyl biphenyl (MIPB) and phenylxylyl ethane (PXE), because of their high gas-absorbing ability, high dielectric strength and dielectric constant. Dimethyl siloxane oil that is used as a replacement fluid in transformers instead of PCBs was found unsuitable for capacitor application because of its gas-evolving character under high electric stress. However, phenyl methyl silicone oil (PMS) possesses high gas-absorbing ability and is also used as nonflammable impregnant for HV capacitors. Blends of (BO) tricresyl phosphate (TCP) with DIPN and/or PXE are also being used as self-extinguishing capacitor impregnants.

Some modified aromatic hydrocarbons such as cumylphenylethene (CPE) and diphenylethane (DPE) have recently been introduced in Japan for capacitor impregnation [46]. Properties of some widely used high voltage capacitor impregnants are summarized in Table 8.1.

8.8.5 HV Bushings

Bushings are used at the interfaces between various power distribution equipments. The high voltage versions contain fluid- or resin-impregnated insulation which is operated at a fairly high stress level. The construction is typically a porcelain housing containing a paper element which is wound and lapped around a central conductor. Coarse textured papers are often used to facilitate drying and impregnation. If the bushing is integral with other equipment, a common impregnant is used, otherwise a good-quality mineral oil is suitable. A well-known design is the capacitor bushing, which contains concentric layers of paper dielectric interlayered with aluminum foils where each foil forms an equipotential screen. In service, bushings often experience the full extent of fast-rising transients and it is normal for the bushing insulation to be rated in terms of its ability to withstand impulse breakdown.

8.8.6 Impregnated Insulation in Transformers

During the past 30 years, many new types of synthetic varnishes and solid insulations have been used in transformers. Nevertheless, the mechanical stability and load-bearing properties of cellulose are still highly valued by design engineers. Much of the insulation system used in oil-filled transformers consists of oil-impregnated pressboard placed around the low and high voltage windings and supported by the iron core. The thin oil layers within the multilayer pressboard structure result in high dielectric strength, which increases with pressboard density. In addition, the oil passing be-

Table 8.1 Properties of Impregnants of High Voltage Power Capacitors

Property	Impregnating oil						
	MO	TCD[a]	DIPN	PXE	CPE	MIPB	PMS	BO
Specific gravity	0.88	1.40	0.96	0.99	0.96	0.99	0.97	1.07
Viscosity 30°C (cSt)	11.0	25.0	8.5	8.0	5.2	8.3	35	37
ε_r (60 Hz, 80°C)	2.18	5.20	2.48	2.51	2.44	2.55	2.65	4.61
Tan δ (60 Hz, 80°C)	0.01	0.04	0.005	0.005	0.005	0.005	0.08	0.61
Breakdown voltage (kV/2.5 mm)	70	65	80	80	80	80	58	60
Gas generating voltage (kV/mm)	46	—	70	70	72	71	—	60
Fire point (°C)	135	170	144	148	156	140	317	148
Pour point (°C)	−32	−17	−47.5	−47.5	−60	−50	−77	−40

[a]Use is restricted in many countries due to environmental concerns.
MO = mineral oil, TCD = trichlorodiphenyl, DIPN = di-isopropylnaphthalene, PXE = phenylxylyl ethane, CPE = cumylphenylethene, MIPB = mono isopropylbiphenyl, PMS = phenylmethyl dimethyl siloxane, BO = blend of DIPN or XPE and TCP.

tween the individual pressboard components of the insulation structure allows the required heat dissipation in order to achieve the necessary electrical and thermal performance.

In transformers, the average working electric stress is usually low and is typically less than one-fifth of the stress value used in impregnated paper cables and capacitors. This is essential, as there is a need to maintain a high impulse strength over complicated insulation geometry. Various techniques are used to determine the field distribution in transformers under both impulse and steady-state conditions. Similarly, thermal degradation of the fluid and paper has to be minimized and the operating temperature for the fluid is usually restricted to an upper limit of 60°C. The fluids which are chosen for use in larger power transformers generally have a high oxidation resistance and may contain inhibitors.

Aging of Paper/Pressboard Insulation

The lifetime of transformers is usually considered to be around 30 years. A substantial number of transformers were installed worldwide in the 1950s, a period of sustained economic growth. Such units are now anticipated to have reached the end of their life and therefore have been subjected to intensive investigation and diagnostics [26]. Various test procedures have emerged from these investigations.

Diagnosis of Lifetime by Degree of Polymerization

Insulating paper used in an oil-immersed transformer decomposes gradually due to heat dissipation. Figure 8.14 shows some of the decomposition products for cellulose paper. The number of cellulose molecules, m, indicates the degree of polymerization. It is usually 1000–1100 for the new transformer kraft paper. A critical value of end for lifetime estimation is considered to be around 500, as at that level the paper becomes extremely brittle and cracks upon bending. The degree of polymerization (DP) of a small amount of transformer paper sample is measured using IEC-450 [43]. Insulating paper located at the highest temperature point, such as the upper part of winding inside the transformer, is usually evaluated. The lifetime of the transformer is then predicted from equation (8.6) [27]:

$$D = D_o(1 - r)^{-t} \tag{8.6}$$

where D is the degree of polymerization, D_o is the initial degree of polymerization, r is a constant and t is the operation period.

Diagnosis by this method, however, has recently been subjected to criticism due to poor reproducibility of results. Investigations have shown that this method gives only a crude average of molecular chain size and is therefore a coarse measure of transformer degradation [28]. It is also re-

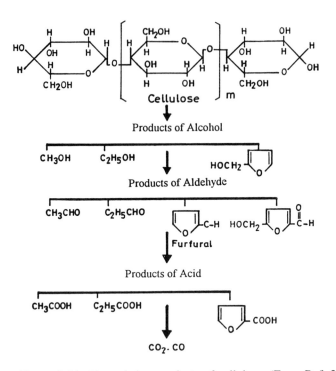

Figure 8.14 Degradation products of cellulose. (From Ref. 26 © IEEE, 1993.)

ported that due to long exposure to high temperature, although kraft paper becomes brittle and looses its tensile strength, yet the dielectric strength does not vary much with the reduction in degree of polymerization, as shown in Figure 8.15 [29]. This shows that transformer insulation is still good although it has in fact aged considerably, since it is ultimately the loss of tensile strength that determines the life expectancy of the insulation.

Diagnosis of Lifetime by Dissolved Gas Analysis

The diagnosis of oil-immersed transformers using the dissolved gas analysis (DGA) technique has been widely accepted internationally and is being used by laboratories, transformer users and manufacturers since the introduction of the IEC-599 [30]. This method is effective in the diagnosis of faults in transformers, such as arc discharge, partial discharge and overheating. The deterioration of paper insulation in this method uses the CO_2/CO ratio and other hydrocarbon gases dissolved in transformer oil, and not by directly sampling the paper or pressboard. According to this method, the formation of fault involving paper insulation is probable when

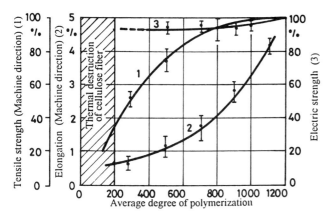

Figure 8.15 Tensile strength, elongation and dielectric strength as a function of an average degree of polymerization. (From Ref. 29 © IEEE, 1996.)

the CO_2/CO ratio is ≤3. It is based on the fact that the generation of CO increases faster than that of CO_2 as the decomposition temperature of cellulose increases. This method was first used by Tamara et al. [31] and is based on the fact that there exists a good correlation between the amount of dissolved CO_2 gas and the degree of polymerization of insulating paper in oil-immersed transformers.

The continuous use of this method has shown some difficulties in interpreting DGA results as well. Lifetime diagnosis is comparatively difficult for N_2-sealed-type oil-immersed transformers, because dissolved CO_2 escapes from oil into the nitrogen cushion. The concentration of CO_2 gas strongly depends upon the temperature at which the oil is sampled. It is often found that concentration of CO and CO_2 dissolved in transformer oil is higher in summer and lower in winter, which may result in misleading conclusions. Moreover, it is also found that CO_2, once decomposed from pressboard, can be reabsorbed by the paper if the operating temperature is below 80°C [32].

Diagnosis by the Measurement of Furfural Dissolved in Oil

As shown in Figure 8.14, insulating paper is decomposed into alcohol, aldehyde, acid and finally into carbon dioxide based on the degree of aging. At the aldehyde decomposition stage, 2-furaldehyde (commonly known as furfural) is generated. Abnormal operation conditions such as overheating and electrical discharge would generate CO_2 and furfural in abundance. Aging phenomena related to furfural have been reviewed in by Schroff and Stannett [33]. It was found that the amount of furfural in oil is proportional to the log (DP), i.e., the degree of paper polymerization. A method for furfural analysis has been proposed in IEC Publication 1198 [34], which utilizes high-performance liquid chromatography (HPLC). More recently, techniques such as "size exclusion chromatography" (SEC) of the paper to measure changes in its molecular weight distribution, which is related to the degree of polymerization, have also been introduced [35,36].

In fact, the whole are of transformer insulation aging, monitoring and life prediction has recently been reviewed [37,38], and a comparative study of different methods that are in common use shows that HPLC analysis of oil can potentially be better used to monitor in service degradation of paper. Normal levels of 2-furaldehyde are generally in the range of 100–1000 ppb. However, levels of about 1 ppm have also been measured in "normal" transformers, rising to 5 ppm in a transformer running hot and to over 10 ppm in transformer that was overheated due to failure in its cooling system. However, this is a subject of current research in a number of centers throughout the world.

8.9 FLEXIBLE LAMINATES

Flexible laminates are used extensively in electrical equipment. Often an application demands insulation that may have high dielectric strength and high mechanical strength at high temperatures. Usually a single dielectric does not possess all these characteristics. Therefore, two materials possessing different characteristics can be combined in the form of sheets so that one can take advantage of best of both types of materials.

8.9.1 Composite Laminate Classifications

Flexible laminates are manufactured by combining selected dielectrics with appropriate substances using various types of adhesives as bonding agents. Outer layers of these laminates may, in some cases, be further treated with various resinous compounds to provide surface or volume characteristics required for some applications. IEC Publication 626 [40,41] is a guide for the identification of these complex laminates.

Use is made of internationally recognized abbreviations for the various elements of these laminates as shown in Table 8.2. Further, a system of classification of the finished laminates has been developed which not only greatly simplifies their description (Table 8.3), but also provides a convenient and foolproof means for purchasing them internationally.

Composite Dielectrics

Table 8.2 Designation of Flexible Laminate Components

Form of component	Code designation	Nature of component	Code designation
Film	F	Cellulose acetate	CA
		Cellulose triacetate	CTA
		Polyethylene terephthalate	PETP
		Polyimide	PI
		Polycarbonate	PC
		Polypropylene	PP
Paper and non-woven fabric	P	Cellulose paper or presspaper	C
		Polyamide (aromatic) paper	PAa
		Polyethylene terephthalate	PETP
Woven fabrics	C	Cotton or viscose	C
		Glass fabric	G
		Polyethylene terephthalate	PETP
Adhesive	A	Pressure-sensitive, thermoplastic	T_P
		Pressure-sensitive, thermosetting	T_S

Source: Ref. 45 © IEEE, 1991.

Table 8.3 Laminate Descriptions

Performance	Properties	Parameters
Structural	Dimensional stability	Thickness, total and individual ply
	Deformation resistance	Water absorption
	Rigidity	Flexural modulus
Mechanical	Ability to withstand fabrication stress	Tensile strength elongation
	Resistance to service stresses, thermal expansion, etc.	Edge tear resistance
		Interply adhesion
Electrical	Insulation	Dielectric breakdown voltage
		Arc resistance
		Surface and volume resistivity

Source: Ref. 45 © IEEE, 1991.

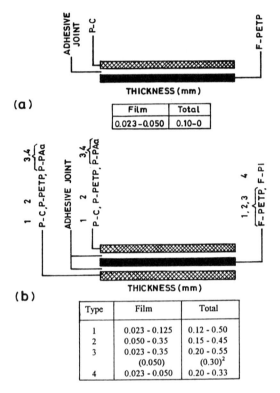

Figure 8.16 Combined flexible laminates, types and thickness range (abbreviations as per IEC 626-1): (a) double combination and (b) triple combination. (From Ref. 45 © IEEE, 1991.)

This classification system employs the abbreviations of Table 8.2, arranged sequentially in the order found in the actual laminate, with the various plies separated texturally by means of a slash (/). Thus, a laminate formed by bonding glass fabric to polyester film by means of a thermoplastic pressure-sensitive adhesive would be designated as:

C-G/A-Tp/F-PETP

Various types of combined flexible laminates comprising double, triple and quadruple plies are commercially used and are identified in the various specification sheets of IEC Publication 626 Part 3 [41]. Figure 8.16 shows a schematic representation of typical double- and triple-layer laminates. The outer layers may be coated or impregnated with selected resins that may be either fully cured to provide enhanced thermal capabilities or semi-

Composite Dielectrics 237

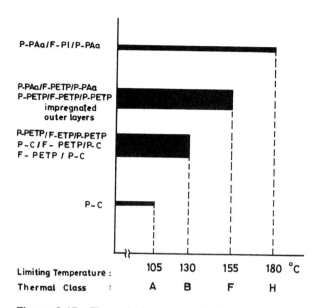

Figure 8.17 Thermal classes and limiting temperatures. (From Ref. 45 © IEEE, 1991.)

cured so that the laminates can be further cured by heating after insertion in the electrical apparatus, thus assuring improved mechanical strength and integrity of the unit.

8.9.2 Selecting Flexible Laminates for Thermal Requirements

At present there are no standard laboratory methods for evaluating the thermal endurance of flexible laminate insulation. Design engineers usually rely on materials known to perform satisfactorily under certain accelerated thermal conditions. IEC publication 85 [42] is a compilation of insulating materials which experience has shown to be suitable choices for specific temperature exposures.

Using IEC Publication 85 as a guide, and supported by available service data, the combinations shown in Figure 8.17 have been used to insulate electric equipment operating continuously at the temperatures indicated. It will be noted that application of selected resins to the outer layers of these laminates may in some cases upgrade the thermal capability. In the particular case of P-PETP/F-PETP/PETP, for example, such treatment has been shown to raise the insulation thermal classification from Class B

(130°C) to Class F (135°C). A commonly used saturant for this particular insulation is a suitable epoxy resin. It should be noted here that both the choice of resin and the degree of impregnation usually depend on the nature of the particular laminate, the needs of the apparatus to be insulated and the thermal capability sought.

REFERENCES

1. L.K.H. van Beek, "Progress in Dielectrics", J.B. Birks (Ed.), Chemical Rubber Co. Press, Ohio, Vol. 7, pp. 69–114, 1967.
2. R. Bartnikas and R.M. Eichhorn, *Engineering Dielectrics* Vol. 1, "Corona Measurement and Interpretation," ASTM Publications, Philadelphia, 1983.
3. C.J.F. Bottcher and P. Bordewijk, *Theory of Electric Polarization,* Elsevier Press, New York, 1979.
4. B.S. Bernstein, IEEE Elect. Insul. Magazine, Vol. 10, No. 4, pp. 33–38, 1994.
5. L.D. Dissado and J.C. Fothergill, *Electrical Degradation and Breakdown in Polymers,* Peter Perigrinus Ltd., London, England, 1992.
6. S.H. Lee, J.K. Park, J.H. Han and K.S. Suh, IEEE Trans. on Dielectrics and Elect. Insul., Vol. 2, No. 6, pp. 1132–1139, 1995.
7. D. Malec, R. Essolbi, H. Giam, B. Ai and B. Garros, IEEE Tran. on Dielectric. and Elect. Insul., Vol. 3, pp. 64–69, 1996.
8. T. Tanaka and A. Greenwood, *Advanced Power Cable Technology,* CRC Press, Boca Raton, Florida, 1983.
9. M.G. Danikas and T. Tanaka, IEEE Trans. on Dielectrics and Elect. Insul., Vol. 1, No. 3, pp. 548–549, 1994.
10. E.A. Cherney, IEEE Elect. Insul. Magazine, Vol. 7, No. 2, pp. 28–32, 1991.
11. T.D. Mintz, IEEE Trans. on PAS, Vol. 103, pp. 3448–3453, 1984.
12. A.S.G. Alghamdi and A.H. Mufti, Proc. 5th GCC-CIGRE Symposium on High Voltage Overhead Lines and Underground Cables, Qatar, 1994, Paper No. 12.
13. IEC Publication-587, "Standard Method for Evaluating Resistance to Tracking and Erosion of Electrical Insulating Materials used under Severe Ambient Conditions", IEC, Geneva, Switzerland, 1984.
14. ASTM Test Method D-2303, "Standard Test Method for Liquid Contaminant, Inclined Plane Tracking and Erosion of Insulating Materials", Philadelphia, 1984.
15. A.S.G. Alghamdi, D.W. Auckland, A.J. Risino and B.R. Varlow, IEEE Trans. on Dielectrics and Elect. Insul., Vol. 3, No. 3, pp. 439–443, 1996.
16. ASTM Test Method D495, "Standard Test Method for High Voltage, Low-Current, Dry Arc resistance of Solid Electrical Insulation", 1973.
17. R. Bartnikas and R.M. Eichhorn (Eds.), *Engineering Dielectrics Vol. IIA, Electrical Properties of Solid Insulating Materials: Molecular Structure and Electrical Behavior,* ASTM Publication, Philadelphia, 1983.

18. J.C. Botts, Proc. 5th Elect. Insulation Conference, Chicago, pp. 202–209, 1963.
19. CIGRE Committee 22, Sub-Working Group 03-01, Electra, Paris, France, No. 130, 1989.
20. H.M. Schneider and J.F. Hall, IEEE Trans. on Power Delivery, Vol. 4, No. 4, pp. 2214–2221, 1989.
21. E.A. Cherney, IEEE Elect. Insul. Magazine, Vol. 12, No. 3, pp. 7–15, 1996.
22. J.S.T. Looms, *Insulators for High Voltages*, Peter Peregrinus Ltd., London, England, 1988.
23. R.S. Gorur and T. Orbeck, IEEE Trans. on Elect. Insul., Vol. 26, No. 5, pp. 1064–1072, 1991.
24. S.H. Kim, E.A. Cherney and R. Hackam, IEEE Trans. on Power Delivery, Vol. 5, No. 3, pp. 1491–1499, 1990.
25. A.C.M. Wilson, *Insulating Liquids*, Peter Peregrinus Ltd., London, England, 1980.
26. T. Tanaka, T. Okamoto, K. Nakanishi and T. Miyamoto, IEEE Trans. on Elect. Insul., Vol. 28, No. 5, pp. 826–844, 1993.
27. C.R. Acker, Proceedings of IEEE-PES Winter Meeting and Tesla Symposium, IEEE, New Jersey, pp. 25–30, 1976.
28. M. Ali, C. Eley, A.M. Emsley, R. Haywood and X. Xiao, IEEE Elect. Insul. Magazine, Vol. 12, No. 3, pp. 28–34, 1996.
29. K. Giese, IEEE Elect. Insul. Magazine, Vol. 12, No. 1, pp. 29–33, 1996.
30. IEC-Publication 599, "Interpretation of the Analysis of Gases in Transformers and Other Oil Filled Electrical Equipment in Service", IEC, Geneva, Switzerland, 1978.
31. T. Tamara, H. Anetai, T. Ishii and T. Kawamura, Trans. of IEE Japan, Vol. 101A, No. 1, pp. 30–36, 1981.
32. H. Kan and T. Miyamoto, IEEE Elect. Insul. Magazine, Vol. 11, No. 6, pp. 15–21, 1995.
33. D.H. Shroff and A.W. Stannett, IEE Proceedings, Vol. 132, Part C, No. 6, pp. 312–319, 1985.
34. IEC-Draft Publication 1198, "Methods for the Determination of 2-Furfural and Related Compounds", IEC, Geneva, Switzerland, 1991.
35. J. Unsworth and F. Mitchell, IEEE Trans. on Elect. Insul., Vol. 25, No. 4, pp. 737–942, 1990.
36. M. Darveniza, T. Saba, D.J.T. Hill and T.T. Lee, Proc. 6th IEE Intl. Conf. on Dielectrics Materials, Measurements and Applications, Manchester, England, 1992.
37. A.M. Emsley and G.C. Stevens, IEE Proc. Part A, Vol. 5, No. 141, pp. 324–333, 1994.
38. A.M. Emsley, Polymer Degradation Stability, No. 44, pp. 343–349, 1994.
39. M. Schaible, IEEE Elect. Insul. Magazine, Vol. 3, No. 1, pp. 8–12, 1987.
40. IEC Publication 626 Part 1: Specification for Combined Flexible Materials for Electric Insulation, Definitions and General Requirements", IEC, Geneva, Switzerland, 1979.

41. IEC Publication 626 Part 3: "Specification for Combined Flexible Materials for Electric Insulation, Specifications for Individual Materials", IEC, Geneva, Switzerland, 1988.
42. IEC Publication 85: "Thermal Evaluation and Classification for Electric Insulation", IEC, Geneva, Switzerland, 1984.
43. IEC Publication 450, "Measurement of the Average Viscometeric Degree of Polymerization of New and Aged Electrical Papers", IEC, Geneva, Switzerland, 1982.
44. A Bradwell (Ed.), *Electrical Insulation*, Peter Peregrinus Ltd., London, England, 1983.
45. K. Giese, IEEE Elect. Insul. Magazine, Vol. 7, No. 4, pp. 27–30, 1991.
46. A. Sato, S, Kawakami, K. Endo and H. Dohi, CIGRE Conf., Paper 15-05, Paris, France, pp. 1–5, 1985.
47. Y. Yoshida and T. Muraoka, IEEE Elect. Insul. Magazine, Vol. 11, No. 1, pp. 32–45, 1995.
48. J.W. Chang and R.S. Gorur, IEEE Trans. on Dielectric and Elect. Insul., Vol. 1, No. 6, pp. 1039–1045, 1994.
49. R.S. Gorur, J. Mishra, R. Tay and R. McAfee, IEEE Trans. on Dielectrics and Elect. Insul., Vol. 3, No. 2, pp. 299–306, 1996.

9
High Voltage Cables

9.1 INTRODUCTION

Cables have been used in transmission and distribution networks since the early days of the electrical power industry. Generally, long-distance power transmission is carried out through overhead lines. However, transmission and distribution in densely populated urban areas mostly uses underground cables. Although significantly more expensive than the overhead lines, the cables are preferred in urban areas due to safety, reliability and aesthetical considerations. As a result of developments in insulating materials and manufacturing techniques, high voltage cable technology has improved significantly over the years. With a continuous increase in the overall length of cable networks, questions regarding reliability, failure modes and diagnostics of such cables have assumed greater significance. This chapter briefly discusses various aspects of high voltage power cables with emphasis on polymeric insulated cables, which are almost exclusively being used in distribution networks in many countries. The insulation testing of high voltage cables is discussed in Chapter 12.

9.2 CABLE MATERIALS

9.2.1 Conductors

Cables are constructed using a variety of materials for conductors, insulation, screening and armoring. The most common conductor materials are copper or aluminum of high purity (>99.5% pure), since the resistivity of such materials significantly increases with impurity content. Sometimes, sodium is also used as the conductor material [5]. In recent years, the possibility of using high-temperature superconducting materials for power applications is also being examined. The choice between aluminum and copper is normally based on resistivity (ρ), cost as well as mechanical and manufacturing considerations. Table 9.1 compares resistivity values for a few materials used in cables. For a given current rating, aluminum requires a larger conductor cross-sectional area than copper. Both solid and stranded conductors are used; the choice depends upon total cross-sectional area, flexibility and manufacturing considerations. For larger cross sections, stranded construction with alternate layers spiraled in opposite directions is generally preferred.

9.2.2 Insulation

The cable insulation usually consists of (1) impregnated paper, (2) synthetic polymers and (3) compressed gases. Early cables mostly used impregnated paper insulation. Next stage was the introduction of oil-filled paper insulation. At present, polymers are widely used. The polymers most often used are polyvinyle chloride (PVC) for low-voltage cables; polyethylene (PE), cross-linked polyethylene (XLPE) and ethylene propylene rubber (EPR) for medium-voltage cables; and XLPE and EPR for high-voltage cables. In addition, high-density polyethylene (HDPE), high-molecular-weight polyethylene (HMWPE), tree-resistant (or retardant) XLPE (TRXLPE) and terpolymer ethylene propylene diene monomer (EPDM) are also used for medium-voltage underground distribution (URD) system cables.

Table 9.1 Resistivity of Some Conductors at $\approx 20°C$

Material	ρ ($\mu\Omega$-cm)
Copper	1.73
Aluminum	2.83
Sodium	4.68
Lead	21.5
Steel	10.20

High Voltage Cables 243

Sometimes compressed gases such as SF_6, CCl_2F_2 (Freon-12) and N_2 are also used for cable insulation. Compressed-gas-insulated cables employing SF_6 have already been described in Chapter 4. Other types of compressed gas cables employ paper and compressed gas insulation [5].

The use of plastic insulated cables started in the early 1950s and now these cables are the most common choice in URD systems. However, at EHV and UHV levels, oil-filled paper-insulated (OF) cables are still extensively used. In recent years, XLPE insulated cables with mean AC working stresses approaching those used in OF cables (≈ 15 kV/mm) have been developed [1], and it appears that in the future such cables will become more common at transmission voltages as well.

9.2.3 Screens and Jackets

In paper-insulated cables, conductor screens in the form of lapped metallic (e.g., aluminum) foils or semiconductive carbon paper tapes are used to relieve the stress concentrations which arise due to conductor stranding. Moreover, a metallic sheath from a lead alloy or aluminum is also used for mechanical protection of the insulation which also helps to confine the electric stress within the insulation. Such a sheath also minimizes surface discharges, limits electrostatic and electromagnetic interference and prevents the ingress of moisture to the cable insulation. In polymeric insulated cables, conductor and insulation are provided with semiconducting screens. These screens have the following functions: (1) to equalize the electric field around the conductor periphery, (2) to maintain a contact between the conductors and the insulation throughout the expansion and contraction caused by load cycling and (3) to prevent extraneous damage to the insulation. Normally the two screens and the insulation are extruded simultaneously. The screens are formed from materials consisting of a polyethylene-based resin mixed with conductive carbon black and small amounts of antioxidants [1]. The level of impurities in the screens and their smoothness have a very strong influence on the cable performance. Supersmooth, extra-clean conductor screens can increase the cable's life [2].

The copper neutral wires or tapes that surround the insulation screen are usually covered with a polyethylene jacket for mechanical protection and to reduce the moisture intrusion into the cable insulation. It has been found that the life of URD cables can be increased by using a jacket of linear low-density polyethylene (LLDPE) [3]. Sometimes cables are jacketed with a semiconductive material. Such cables are finding increased applications when commonly shared tunnels are used for power and communication cables.

9.2.4 Armors

Armoring is required for cables subjected to mechanical stress. In such cases, a PVC sheath or a suitable bedding acts as a mechanical cushion and chemical insulation between the metal sheath and the armor. Armoring usually consists of steel tapes or steel wires. However, sometimes bronze wires are used instead to minimize hysterisis and eddy current losses in the armor. An outer layer of PVC sheath is normally provided to protect the armor against corrosion. Moreover, such a sheath is also used to inscribe information about the cable.

9.3 TYPES OF CABLES

Cables can be classified based on several parameters, such as:

1. Voltage rating of the cable, e.g., low voltage, medium voltage, high voltage and EHV cable.
2. Number of cores of the cable, e.g., single core, two core and three core.
3. Insulation of the cable, e.g., XLPE, PVC, EPR and oil-paper cable.
4. Oil-pressure level in case of oil filled cables, e.g., self-contained (OF) and pipe-type (POF) cable.
5. The presence or absence of the metallic shield over the cable insulation, e.g., shielded and nonshielded cable.

Furthermore, in the case of multicore cables, each core may be individually shielded thereby forming a coaxial cable or all three cores may have one common shield as in the belted cables, as in Figure 9.1. Usually, belted cables are restricted to voltage ratings of less than 33 kV. In general, shielding should be considered for nonmetallic covered cables operating at circuit voltages of above a few kV.

Generally, low and medium voltage URD cables are impregnated paper or polymeric insulated, single or multicore, and coaxial or belted cables. Figure 9.2 shows the cross section of a typical medium- and high-voltage polymeric-insulated three-core cable. For three-core construction, three cores (without armor and external jacket) are put together and enclosed within a common underarmor jacket, an armor and an outer jacket, in addition to necessary fillers, as shown in Figure 9.1b. Figure 9.3 shows typical cross sections of single- and three-core OF cables. Figure 9.4 shows typical cross-sectional details of a pipe type, oil-filled paper-insulated (POF) cable.

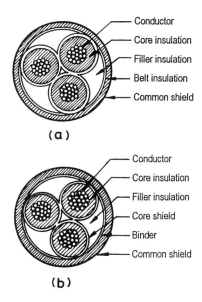

Figure 9.1 Three-core cables: (a) belted, (b) screened or H-type.

9.4 CABLE CONSTANTS

The most important cable constants include resistance, capacitance, inductance and insulation resistance which are briefly described next.

9.4.1 Conductor Resistance

The DC resistance of a conductor depends upon its resistivity, its length, its cross-sectional area as well as its temperature. The resistance, in terms of temperature, is given as:

$$\frac{R_2}{R_1} = \frac{T_o + T_2}{T_o + T_1} \qquad (9.1)$$

where R_1 = conductor resistance at temperature T_1 in °C, R_2 = conductor resistance at temperature T_2 in °C, and T_o = constant varying with conductor material and is equal to 234.5, 241 and 228 for annealed copper, hard-drawn copper, and hard-drawn aluminum, respectively.

The AC resistance is higher than the DC resistance due to skin effect. The difference between the two values depends upon the frequency and the conductor cross-sectional area [5]. Sometimes, if conductor area is

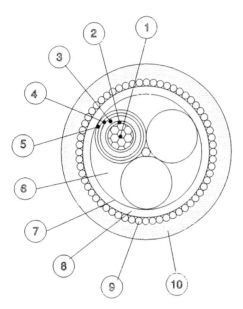

Figure 9.2 Typical medium and high voltage three-core cable cross-sectional details: 1 = copper or aluminum conductor, 2 = inner semiconducting screen (conductor screen), 3 = XLPE insulation, 4 = outer semiconducting screen (insulation screen), 5 = copper tape, 6 = filler, 7 = mylar tape or swellable material bedding layer, 8 = PVC bedding, 9 = steel wire armor, 10 = PVC or PE sheath.

large, segmented or hollow conductors are used to reduce the skin effect. In oil-filled cables, hollow conductors are preferred as these contain oil for better cooling and dielectric characteristics.

9.4.2 Cable Capacitance

Consider the single core coaxial cable of Figure 9.5. The capacitance of this cable is given by:

$$C = \frac{2\pi\varepsilon_0\varepsilon_r}{\ln(b/a)} \quad \text{F/m} \tag{9.2}$$

where $\varepsilon_0 = 8.854 \times 10^{-12}$ F/m and ε_r is the relative permittivity of the insulation. For three-core screened cables, each core will have a capacitance given by equation (9.2). For three-core belted cables, which can be modeled by the network of Figure 9.6, the capacitances between cores (C_c)

High Voltage Cables

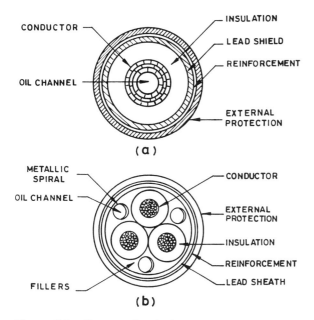

Figure 9.3 Cross-sectional views of typical oil-filled cables: (a) single core and (b) three core.

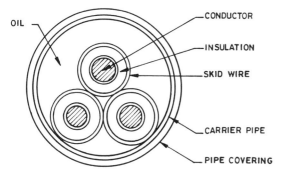

Figure 9.4 Oil-filled, paper-insulated, three-phase pipe-type cable.

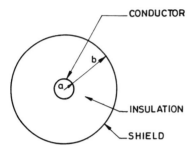

Figure 9.5 Cross section of a coaxial cable.

and from each core to the sheath (C_s) can either be measured or determined by numerical computation or approximate formulas [5–7].

9.4.3 Cable Inductance

For the coaxial cable of Figure 9.5 having a solid inner conductor and a perfect metal screen, the total inductance L is given as:

$$L = 2 \times 10^{-7} \left[\frac{1}{4} + \ln\left(\frac{b}{a}\right) \right] \quad \text{H/m} \quad (9.3)$$

For belted cables, the term (b/a) in equation (9.3) is replaced with the term S/a, where S is the spacing between two cores. For practical power cables, the inductance also depends on screening and armoring materials as well as the proximity of the cable to other conductors and ferrous objects, and

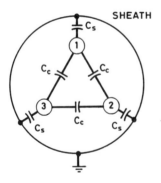

Figure 9.6 Capacitances of a three-core belted cable.

therefore its accurate calculation is difficult. However, it can be accurately determined from measurements [4].

9.4.4 Insulation Resistance

If the cables of Figure 9.5 have bulk insulation resistivity of ρ_i (in Ωm), then its per meter insulation resistance R_i is given as:

$$R_i = \frac{\rho_i}{2\pi} \ln(b/a) \quad \Omega m \quad (9.4)$$

Since ρ_i is usually very large for cable dielectrics (see Table 6.6), R_i is hundreds of MΩm for healthy high voltage cables. However, after a partial or complete insulation breakdown, R_i may decrease substantially.

9.5 ELECTRIC STRESS IN CABLES

The electric stress E in the coaxial cable of Figure 9.5 varies only radially and its value at a distance x from the cable center is given by:

$$E(x) = \frac{V}{x \ln(b/a)} \quad (9.5)$$

where V is the applied voltage across the cable insulation. The insulation next to the inner conductor surface is subjected to the maximum stress having a value of V/(aln(b/a)). If the conductor surface is not perfectly smooth, this value is increased due to the field intensification effect discussed in Chapter 1. Consequently, the maximum stress at a rough point will become fV/(aln(b/a)) where f is the field intensification factor given in Table 1.1 for some selected protrusion shapes. Such rough points can cause insulation deterioration and act as potential tree sites (for details see section 9.9) and significant efforts are being made by cable manufacturers to use smooth and defect-free conductor and insulation screens in order to reduce field intensifications.

It can be shown from equation (9.5) that for specified voltage and conductor radius values, the conductor stress is minimum when b/a = 2.781, i.e., the insulation thickness is ≈1.72 times the conductor radius. However, since this thickness depends upon the cable dielectric material used and the cable's voltage rating, the above condition for optimum stress design cannot always be realized in practice.

For belted cables, the electric stress is not merely confined to cable cores and its calculation requires use of simplifications or analysis using numerical methods. The charge simulation method can perform very accurate field calculations for such cables [8–10].

9.6 CABLE LOSSES

Cable losses include losses in conductor, insulation, sheath, screens and armors. Conductor losses (I^2R_{ac} losses) depend upon the rms current I and effective AC resistance of the cable conductor R_{ac}. Dielectric losses comprise of losses due to leakage through the cable insulation and losses caused by dielectric polarization under AC stresses. The net dielectric losses are given by equation (1.3) and depend upon cable voltage, its frequency as well as the permittivity and loss tangent of the cable dielectric material. Although for low and medium voltage power cables, dielectric losses are insignificant, these assume importance for HV and EHV cables. This is specially the case for oil paper cables. For instance, the dielectric losses in a 400 kV OF cable may reach about 20% of the total cable losses [4].

Generally, tan δ, which partially controls the dielectric losses, is significantly higher for oil-paper insulation as compared to XLPE insulation. For most of the dielectric materials used in cables, tan δ depends upon temperature, applied stress and supply frequency. For oil-paper insulation, tan δ is also strongly influenced by moisture content. Therefore, in high voltage cables, a moisture level of less than 0.05% is desirable in order to keep dielectric losses within acceptable limits [11]. Table 6.6 gives typical ε_r, tan δ and ρ_i values for some dielectrics used in power cables. The presence of voids and microcracks can also influence dielectric losses. Any partial discharge in such voids increases the effective tan δ value for the insulation. Consequently, when the applied voltage is raised above the discharge inception threshold, the dielectric losses exhibit a distinct increase. Similarly, impurities in the cable insulation and screening materials can also increase dielectric losses.

The AC current flowing along each cable conductor induces emf in the metallic sheaths of the cable. Without grounding, such sheaths would operate at a potential above the ground potential and can pose a safety hazard. Furthermore, it will accelerate degradation of the jacket and other materials, thereby affecting the cable's life and reliability. When the sheaths are bonded, circulating current flows in them causing power losses. However, for three-core cables such losses are negligible. In addition to circulating currents, eddy currents are also induced in sheaths of both single and multicore cables causing additional losses which usually are of small magnitudes.

High Voltage Cables

Single-conductor cable sheaths may be open or short-circuited. When the sheaths are short-circuited, they are usually bonded and grounded at every manhole. This makes sheath voltage equal to zero, but allows the flow of sheath currents. When the sheaths are open-circuited, no current flows but sheaths can assume certain potential. In order to keep such a potential within a safe value, i.e., 25 V in exposed positions, cable sheaths are usually cross-bonded. Figure 9.7 shows a typical cross-bonding and grounding arrangement to minimize sheath losses for single conductor cables. Alternate methods of operating sheaths in an open circuit include grounding the sheath at one point only, or bonding sheaths through impedances or transformers. The grounding conductor and its attachment to the sheath must be designed such that it can safely carry fault and circulating currents without overheating.

Losses also occur in armor and in any metallic pipe housing the cable. For the armors made of magnetic materials, additional losses due to magnetic hysterisis can be of significant magnitudes. Thus, all such losses must be considered in the design of a cable system.

9.7 CABLE AMPACITY

The ampacity or current carrying capacity of a cable is defined as the maximum current which the cable can carry continuously without the temperature at any point in the insulation exceeding the limits specified for the respectively material. The ampacity depends upon the rate of heat generation within the cable as well as the rate of heat dissipation from the cable to the surroundings. The rate of heat generation within a cable depends upon various losses (discussed in section 9.6), whereas the rate of heat dissipation depends upon the thermal resistances of different cable

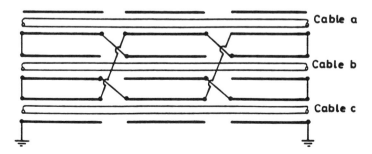

Figure 9.7 Cross bonding of single-conductor cable sheaths.

materials and the media surrounding the cable. Hence, the cable ampacity is influenced by numerous factors, such as the materials and dimensions of the cable conductor, insulation, shields and armors; the method of cable installation; the thermal characteristics of media surrounding the cable; the characteristics of other heat sources/sinks in the vicinity of the cable (e.g., other cables, hot/cold water and steam pipes); and general environmental conditions (e.g., solar radiation level, ambient temperatures) in the area where the cable is installed.

Neher and McGrath's method of calculations [12] forms the basis of ampacity related Association of Edison Illumination Companies (AEIC) specifications as well as IEC-287 [13]. For practical cables these calculations are fairly complex and are normally carried out using software packages [14] and numerical techniques such as the finite element method [15]. The calculations are based on the principle that the heat generated within a cable should be dissipated to the ambient without exceeding the maximum allowable conductor temperature T_c which depends upon cable's dielectric material. PE and XLPE insulated cables have $T_c \sim 75°C$ and $90°C$ under normal conditions and $90°C$ and $130°C$, respectively, under emergency operating conditions.

For a coaxial cable buried in a homogeneous earth, the thermal circuit of Figure 9.8 can be used to evaluate the cable ampacity. In Figure 9.8, W_c, W_d, W_s and W_a are, respectively, the conductor, the dielectric, the shield and the armor losses (W/m), whereas R_d, R_b, R_s and R_e are the thermal resistances of the cable dielectric, the bedding, the serving and the earth surrounding the cable, respectively, in m°C/W. T_c, T_s and T_e are the conductor, the sheath and the ambient earth temperatures, respectively. The conductor losses, $W_c = I^2 R_{ac}$, are calculated at the maximum operating temperature. W_s and W_a are expressed in terms of W_c by factors λ_1 and λ_2 such that $W_s = \lambda_1 W_c$ and $W_a = \lambda_2 W_c$. The cable ampacity is found by applying the thermal form of Ohm's law to the circuit of Figure 9.8. Consequently, we can write:

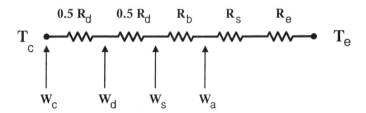

Figure 9.8 Thermal equivalent circuit of a single coaxial cable buried in earth.

High Voltage Cables

$$T_c - T_e = I^2 R_{ac}[R_d + (1 + \lambda_1)R_b + (1 + \lambda_1 + \lambda_2)(R_s + R_e)] + \Delta T_d \quad (9.6)$$

where ΔT_d is the cable conductor temperature rise caused by the dielectric losses and is given as:

$$\Delta T_d = W_d \left(\frac{R_d}{2} + R_b + R_s + R_e \right) \quad (9.7)$$

Thus, ignoring the skin and proximity effects, the ampacity of a coaxial cable is given by:

$$I = \left[\frac{T_c - T_e - \Delta T_d}{R_{ac}(R_d + (1 + \lambda_1)R_b + (1 + \lambda_1 + \lambda_2)(R_s + R_e))} \right]^{0.5} \quad (9.8)$$

For a coaxial cable buried in a uniform earth medium, the thermal resistances can be easily evaluated [12]. However, the evaluation of such resistances for belted cables and for nonhomogeneous multilayer earths are more difficult, and thus the subject of several reports [12–16].

9.8 PARTIAL DISCHARGES IN CABLES

Gas-filled cavities or voids are formed in the insulation or at the screens/insulation interfaces during manufacture, installation or operation. For example, voids may be formed in polymeric cables during the extrusion process. In paper-insulated cables, such voids may form during the impregnation cycle. Voids may also form in cables by the differential expansion and contraction of cable materials due to cyclic loading or short-circuit conditions.

Generally, voids have a higher electric stress as compared to the bulk of insulation. However, the gas inside a void usually has a lower breakdown strength as compared to the main insulation. When the electric stress in the void exceeds the breakdown strength of gas within the void, partial discharges (PDs) may occur. Such discharges, if sufficiently intense, will gradually degrade and erode the insulation, eventually leading to its breakdown. Therefore, the inception voltage for the onset of PDs, their magnitudes, their dependence on voltage, and methods of measurement and location of such discharges are of significant importance. Consequently, all solid dielectric high voltage power cables are tested for PDs.

For the coaxial cable of Figure 9.5, it can be shown that the theoretical discharge inception voltage (DIV) depends upon the radius of conductor; thickness and permittivity of the insulation; shape, size and location of the void; as well as the gas pressure inside the void [17,18]. Generally, the minimum theoretical DIV (V_{min}) can be expressed as [17]:

$$V_{min} = \frac{325}{KC} a \ln(b/a) \qquad (9.9)$$

where V_{min} is in volts, C (the cavity size is in meters) and the factor K = E_c/E_d depends upon cavity shape and ε_r of the insulation [see equations (1.14) and (1.15)]. In practice, cavity sizes are usually small and PD pulses may not be detectable in a reasonable amount of time due to a lack of initiatory electrons when the applied voltage is equal to V_{min}. Consequently, a somewhat higher applied voltage, a longer duration of applied stress or some external source of irradiation may be necessary to detect the PD initiation in small-size cavities.

When a PD occurs inside a cavity, the actual PD pulse thus generated has a duration of only a few nanoseconds. However, as the pulse propagates away from the source, it suffers frequency-dependent attenuations and, therefore, the signal detected outside the test object depends upon the nature of connection between the point of pulse generation and the external circuit. Moreover, for a given void size, the detected PD magnitude generally varies inversely with cable voltage rating (or cable size). Thus, the measurement sensitivity is the lowest for PD testing of the highest voltage cable where the room for measurement error is the minimum [19]. Standard methods exist for measurements and interpretation of PDs in power cables [21–23]. In addition to measurements of PD magnitudes, PD source location is also possible [24]. However, the accuracy of PD source location in power cables is limited by the high frequency attenuation of PD pulses as they propagate through the cable [20].

PDs cause cable insulation degradation through a variety of electronic, chemical and mechanical processes which have been the subjects of significant research in recent years. In many cases, PDs lead to initiation and growth of trees which ultimately cause failure of the cable. In addition to cavities, other defects which may ultimately lead to PD and tree initiation include protrusions and stress enhancements at sheaths, insulating or conducting contaminants in the insulation, microcracks and other defects. It is important to note that a PD test has limitations when used as a diagnostic tool for detection of electrical trees and some other defects [25]. Nevertheless, techniques to measure PD magnitudes of energized XLPE cable operating at EHV levels have been reported [26].

High Voltage Cables 255

9.9 TREEING IN CABLES

Treeing is an electrical prebreakdown phenomenon. This name is given to any type of damage which progresses through a stressed dielectric so that, if visible, its path resembles the form of a tree. Tree-like discharge patterns, sometimes leading to total breakdown of the insulation, have been observed for many years in oil-impregnated pressboard and in oil-impregnated paper insulated cables, Treeing can occur in most solid dielectrics including glass and porcelain but it is a serious problem in polymers, rubbers and epoxy resins, etc. However, since rubbers and resins are often pigmented or mineral filled, the existence of tree-like channels may go unnoticed in such materials.

Treeing may or may not be followed by complete electrical breakdown of the insulation; but in organic extruded dielectrics, it is the most likely mechanism of dielectric failure which is the result of a lengthy aging process [27,28]. Electric stress and stress concentration are always required for the initiation and growth of trees. Treeing can progress rapidly under high electric stresses in dry dielectrics by periodic partial discharges or more slowly in the presence of moisture at lower electric stresses without any detectable PD. Treeing can occur under DC, AC and impulse voltages.

The trees can be considered in two broad classifications: electrical trees and water trees. All trees are initiated at sites of high and divergent electric stresses and their growth rate may be aggravated by the presence of moisture, chemicals, contaminants and other defects in the dielectric as discussed next.

9.9.1 Electrical Trees

Electrical trees initiate and propagate due to high and divergent electric stress at metallic or semiconducting contaminants and/or voids, etc., by partial discharges occurring in a dry dielectric [28]. Such trees consist of hollow channels resulting from decomposition of dielectric material by the PDs. The tree shows up clearly in PE and other translucent solid dielectrics when examined with an optical microscope and transmitted light. Electrical tree channels are permanently visible and there is a great variety of the visual appearance of stems and branches of such trees as well as the circumstances in which initiation and growth of such trees occur. Figure 9.9 shows the typical appearance of some electrical trees [29]. Many names such as dendrites, branch type, bush type, spikes, strings, bow-ties and vented trees have been used in the literature to describe such trees. Trees which start to grow from within the insulation and progress symmetrically outwards from the electrodes are called bow-tie trees because of their ap-

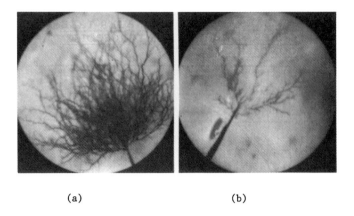

Figure 9.9 Typical patterns of electrical trees in PE: (a) bush-like and (b) tree-line patterns. (From Ref. 29 © IEEE, 1978.)

pearance. On the other hand, trees which initiate at an electrode (or semiconductive screen) insulation interface and progress towards the opposite electrode are called vented trees.

Access to free air is an important factor in the growth of a vented tree. Such trees are capable of growing continuously and long enough to bridge the electrodes or cause a dielectric failure. Bow-tie or nonvented trees do not have a free supply of air to support continuous PDs. Therefore, the growth of such trees is intermittent and discharge occurs with longer periods of extinction, which is believed to be due to an increased void pressure resulting from ionization. During the extinction period, gas pressure in the tree channel is reduced by diffusion and conditions become favorable for occurrence of another PD causing further growth of the tree. Usually vented trees do not grow long enough to bridge the entire insulation thickness or cause a failure [27,28].

There are two distinct periods in electrical treeing. The first is an incubation period during which no measurable PD can be detected, but at the end of which a tree-like figure is first observed. The second is a propagation period during which a tree-like figure grows in the insulation and significant PD magnitude can be measured. Usually, the PD level fluctuates over a wide range with the growth of an electrical tree, as shown in Figure 9.10 [30]. The incubation period depends upon the stress level and its distribution at the initiation site, the composition and properties of the dielectric and the environmental conditions. Generally, at low stress levels, cumulative processes are proceeding and eventually foster conditions

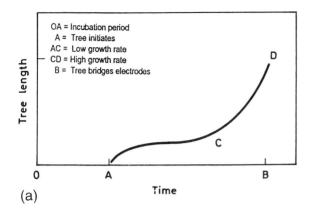

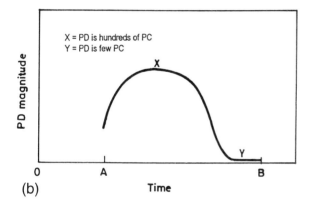

Figure 9.10 Typical evolution of PD level and electrical tree length. (From Ref. 25 © IEEE, 1992.)

which initiate treeing. The mechanisms which have been proposed to explain the initiation of electrical treeing include electron injection and extraction at the tree site, mechanical fatigue cracking due to periodic Maxwell stresses, heating, thermal decomposition and PDs in voids [27,28]. The mechanical effects including the buildup of strain, fatigue failure and fracture due to shock waves involved are believed to play an important role in the initiation and growth of electrical trees [30].

Aging of low density PE under uniform and divergent fields has shown that electrical trees do not develop at fields of up to 20 kV/mm. Local fields of ≈100 kV/mm are required for tree initiation. Such fields can occur at cable working voltages due to various stress enhancement mechanisms, as discussed earlier in Chapter 1. For example, a conductive protrusion with a 10 to 1 ratio of axis will have a local field at its tip which is about 50 times the macroscopic average field. The stress at which trees initiate in a given polymer depends upon the waveform, frequency, magnitude and time of test voltage, and on whether the voltage is applied continuously or is interrupted periodically.

After initiation the tree growth proceeds by a series of sporadic bursts of activity. Consequently, tree branching becomes more frequent and the rate of tree growth slows down. Two of the most important factors that influence tree propagation rate are the development of internal gas pressure due to PDs and the shielding effect of adjacent tree branches on the electric field. Tree channels are generally hollow where conducting carbon particles may also be found. The channel diameter can vary from a few up to several tens of microns.

9.9.2 Water Trees

The tree-like figures which appear in water-exposed polymer-insulated stressed cables are named water trees. Water treeing occurs in the presence of moisture. As compared to electrical treeing, water treeing usually starts at lower electric stress values and progresses more slowly without any detectable PDs.

Water (or wet) trees are different from electrical (or dry) trees. Contrary to electrical trees, water trees do not exhibit measurable PD levels of ≥0.1 pC. The propagation time of water trees is measured in years, whereas once initiated an electrical tree can very quickly propagate through the insulation, e.g., under the influence of a surge overvoltage. The appearance of the two types is usually different from each other as water trees do not exhibit much branching. However, sometimes the two types are difficult to distinguish. A practical method for differentiating between the two types is to examine them after drying the insulation. Unstained water trees become invisible after the insulation is dried while electrical trees are clearly visible even under dry conditions. However, it is possible to make water trees permanently visible by the use of methylene blue with a base or acidic rhodamine dye [27,28]. Unlike electrical trees, water trees do not consist of permanent hollow channels. Instead, water tree channels usually consist of fine filamentary paths between small cavities through which moisture penetrates under the action of a voltage gradient.

Similar to electrical trees, there are two basic types of water trees, namely, bow-tie trees and vented trees (Figure 9.11) [31]. Vented trees are

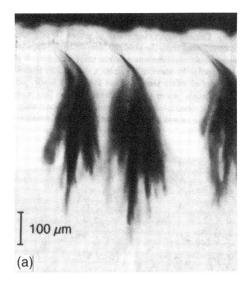

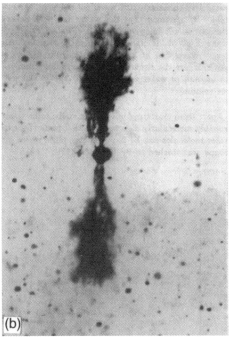

Figure 9.11 (a) Bow-tie water tree (total length = 200 μm) initiated from one of the impurities in the insulation, and (b) vented water tree bent near the insulation outer surface. (From Ref. 31 © IEEE, 1990.)

initiated at the insulation surfaces whereas bow-tie trees are initiated in the insulation volume. Both types have different growth behavior and pose different levels of danger. Both types grow from points having high electric stress values which are also moisture or moisture vapor sources. The moisture source may consist of condensed water or water vapor of ≈65–70% relative humidity. The water vapor may become available from external sources or may be contained in the dielectric during the cable manufacture.

The concentration of vented trees is often low compared to that of bow-tie trees, and at the beginning of their growth, the propagation rate of vented trees is normally lower than that of bow-tie trees. However, at a later stage, the opposite may be true since the growth of bow-tie trees is strongly reduced after a certain time and, therefore, their total length is restricted. Consequently, a bow-tie water tree is seldom the origin of cable breakdown. However, vented water trees usually have access to water and are capable of growing long enough to bridge the dielectric. Alternatively, such trees may grow long enough to reduce the effective insulation thickness below that required to support the electric stress, after which failure may occur by electrical treeing. The growth of electrical trees near or at the tip of water trees have often been observed [27–30].

Water trees contain water. If this water is evaporated, e.g., by heating, the tree channel becomes invisible. Usually the tree absorbs water again if the insulation is exposed to water or water vapor afterwards. Near the initiation spot, a vented tree column can contain up to 10% water. At a certain distance from the tree site, the water content may be up to 1–2% [31]. Though water trees weaken the dielectric, these do not totally damage the insulation and the tree channels exhibit properties of a poor dielectric material. There is a clear relation between the size of the water trees and the electric breakdown strength, as shown in Figure 9.12 [31]. It has been observed that water trees crossing the entire dielectric section do not cause immediate breakdown under service conditions. Such dielectrics often still have a breakdown strength above the service stress level of ~2 kV/mm, as evident from Figure 9.12. Thus, breakdown in cables containing even large water trees may be initiated by some kind of transient surge or temporary overvoltage. In order to determine the level of degradation of aged cables by water treeing, a characterization test has been proposed [32].

There is a hypothesis that undetectable PDs of ≤0.02 pC magnitude accompanied by light emission do occur during water treeing. Recently, it has been shown that long vented water trees could generate measurable PDs when excited with an AC voltage of moderate magnitude [33]. The tree propagation rate increases with applied voltage and is also influenced by supply frequency, ambient temperature, mechanical stress, water conductivity and the nature of salts and chemicals present in the water.

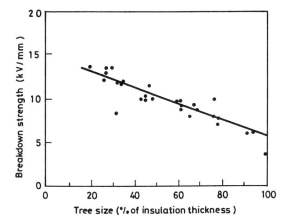

Figure 9.12 Relation between the mean breakdown stress level and the water tree size. (From Ref. 31 © IEEE, 1990.)

Though a definite mechanism of inception and growth of water treeing is not agreed upon, it appears that the capillary action, osmosis, Coulomb's forces, dielectrophoresis, thermal degradation, partial discharges and chemical degradation all play some part in water treeing. Various proposed mechanisms of water treeing are summarized in the literature [27,28, 32,34].

9.9.3 Electrochemical Trees

Electrochemical trees are the water trees which are stained permanently during their growth period due to the presence of certain minerals or ions in the water. Such ions penetrate the dielectric material under an applied stress. Examples of electrochemical trees include sulfide trees caused by H_2S and green, blue and brown trees caused by sulfur, iron, copper or aluminum ions [27,28]. Such trees are permanently visible, like electrical trees, and can have a color depending upon the chemistry of water soluble ions and the cable dielectric material.

9.9.4 Techniques to Suppress Trees

Although there is no definite theory to explain all aspects of treeing in dielectrics, it is generally accepted that the trees start from regions of high and divergent electric stresses. Such regions are usually caused by defects such as loose and/or rough semiconductive screens, cavities, defects and inclusions of foreign particles in the insulation. Therefore, the first step to suppress trees is to minimize such defects by employing suitable materials and manufacturing techniques. In recent years, smoother semiconducting screens with improved bonding to the insulation have been developed. In addition, the following four principles may be used to reduce treeing [27]:

    Principle I: to fill voids with some suitable material
    Principle II: to coat the internal surfaces of voids with semiconducting compounds
    Principle III: to relax locally intensified electric stresses
    Principle IV: to trap or decelerate high energy electrons

The treeing problems in oil-paper cables were overcome by careful drying and degassing of both the paper and the oil and by the development of OF and POF cables (principle I). In this case, the original gas in the void is replaced by a gas, a liquid or even a solid of either higher dielectric strength and/or higher permittivity in order to increase the discharge inception voltage and to reduce the PD magnitude. This is the basic philosophy of principle I.

Lowering the surface resistivity of a void by suitable additives can suppress internal discharges because, in this case, the void is virtually shorted out. Various additives having the required characteristics are available and used. When mixed with the dielectric, they bleed out to the surface of the insulation, including the void surfaces. There are two groups of additives; one has low resistivity by nature, while the other exhibits low resistivity only after being subjected to partial discharges [27].

Field grading with additives or voltage stabilizers is applicable to principle III. Voltage stabilizers act to soften sharp electrode profiles and reduce electric stress around them. The addition of such semiconducting materials increases tan δ, which in some cases may cause thermal runaway of the insulation and therefore can be harmful. One example of this principle is the replacement of the conductive tapes by extruded layers of semiconductive materials.

Principle IV is to add various organic materials to the dielectric which can quickly absorb injected electrons before they could react with the insulation. Cable manufactures have patented different additives in this category [27].

Cable performance against water treeing could be improved by reducing the tree initiation sites. In addition, water-tight construction and suitable curing methods are also useful. Moreover, metallic or nonmetallic water impervious sheath and radial and/or longitudinal barriers can also be used to minimize the availability of water for rapid tree growth. The use of laminate sheaths as moisture barriers is also being explored [35]. Furthermore, tree retardant additives may be introduced in the cable dielectric. References [27,28,31] provide more details about the role of such additives in suppressing water tree growth in cables. Very low density polyethylene (VLDPE) has shown some resistance to water treeing due to its crystallinity and the additives used [36].

9.10 CABLE AGING AND LIFE ESTIMATION

All cables are subjected to simultaneous electrical, thermal and mechanical stresses due to high voltages and high currents. In addition to aging of the cable dielectric under the above mentioned stresses, there are normal chemical changes occurring in the dielectric material as well. Since it is important to estimate the lifetime of a given dielectric from accelerated life tests, the usual voltage, frequency and temperature aging tests as applied to power cables are briefly outlined here.

9.10.1 Voltage Aging

Partial discharges and treeing can reduce the cable life appreciably. The material degradation by PDs is strongly influenced by two important factors: applied voltage and frequency. In the presence of cavities and PDs, the lifetime (t) of a dielectric is strongly affected by the applied voltage V and follows a relationship of the form:

$$V^n t = D_v \qquad (9.10)$$

where D_v is a constant and the voltage life factor, n, depends upon the dielectric material, method of cable manufacture, cable size and type of applied voltage. For cables, n is between 5 and 25 [1,27]. For medium voltage XLPE cables, n = 9 is considered suitable. Figure 9.13 shows a voltage life diagram of unscreened, three-phase, belted-type XLPE and PVC cables [37]. Equation (9.10) is used to predict cable life under voltage stress alone and it does not take into account other stresses which may result under actual service conditions.

9.10.2 Frequency Aging

When the applied voltage is held constant but supply frequency is increased, the insulation degradation due to partial discharges increases. Consequently, the time to failure is related to the frequency F by:

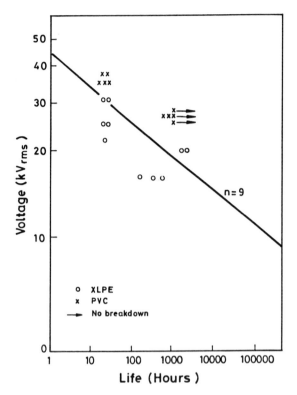

Figure 9.13 Voltage-life curve of unscreened three-phase belted, XLPE and PVC cables with PDs. (From Ref. 37 © IEEE, 1989.)

$$F^m t = D_f \qquad (9.11)$$

where D_f is a constant and the frequency life factor, m, is ~1 for surface discharges and for electrical tree inception. For void discharges and for tree propagation, m < 1 [27]. Combining both voltage and frequency accelerated aging, the dielectric life time can be expressed as:

$$t = D_{fv} F^{-m} V^{-n} \qquad (9.12)$$

where D_{fv} is the constant for simultaneous voltage and frequency aging. This equation can be used for life estimation based on accelerated life testing of cable samples and materials.

9.10.3 Thermal Aging

At elevated temperatures, insulation deteriorates more rapidly leading to a reduction in its useful life period. Since the cables operate at elevated temperatures, it is important to consider the effects of thermal stresses on cable aging as well. Dakin [38] proposed a chemical rate theory model for thermal aging and suggested that the internal chemical changes in a dielectric material depend upon the temperature. Such changes influence the dielectric properties and hence the expected life of the material. Consequently, the life is related to temperature by:

$$t = G \exp (H/T) \qquad (9.13)$$

where G and H are constants determined by the activation energy of the reaction, which influences the dielectric behavior, and T is temperature in °K. Thus, if the logarithm of life is plotted against $1/T$, a straight line is usually obtained. This is known as the Arrhenius relation and is used to estimate the thermal aging behavior of cables. Generally, if temperature is increased by 8–10°K, the life of a given cable insulation is reduced by about one-half. In order to use equation (9.13), some suitable parameter such as tensile strength, breakdown strength or tan δ, must be identified along with suitable end point values (e.g., life is reached when the dielectric strength is reduced to one-half of the original value) which are monitored at different temperatures to estimate the life.

9.10.4 Multifactor Stress Aging

Cables are usually subjected to electrical, thermal, mechanical, radiation, environmental and chemical stress aging simultaneously. With one aging stress at a time, the lifetime results are quite different from those obtained when several aging parameters are applied simultaneously. Therefore, considerable research effort is being exerted in understanding multifactor stress aging [39]. However, so far, no general quantitative model capable of predicting the aging behavior and life expectancy under multifactor stress aging of insulation exists. At present most of the accelerated aging studies concentrate on simultaneous voltage, temperature and environmental (water and chemicals) effects. Figure 9.14 shows a voltage-life curve for XLPE and PE cables with accelerated aging tests along with cables removed from service [40]. This figure shows that due to water treeing, the life of a cable is shortened considerably. However, without treeing, the cable's life points lie on the characteristic aging curve of the respective material as given by equation (9.10).

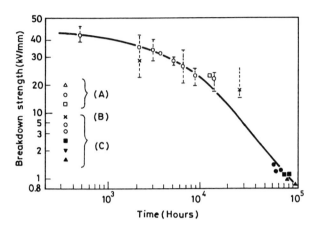

Figure 9.14 E-t curve of XLPE cable and cables with water trees. (From Ref. 40 © IEEE, 1976.)

9.11 CABLE ACCESSORIES

Limitations of manufacture, transportation and installation require cables to be jointed in the field when the total length exceeds a certain value. Moreover, terminations are required where the cables are connected to overhead lines or electrical apparatus. Thus, joints and terminations are an integral part of any cable system and should possess the same integrity as their associated cables. Various types of joints and terminations can broadly be classified as shown in Table 9.2 along with their potential applications [27].

9.11.1 Joints

Straight joints splice two cables of the same kind in a straight line. Besides the normal joints commonly used, there are other types such as insulated joints, stop joints, semistop joints and gas-stop joints. Insulated joints are constructed such that the conductors are joined while the sheaths are insulated from each other and are used for cross-bonding purposes (Figure 9.7). Beyond a certain optimum length, OF cables are sectionalized for the purpose of oil feeding and maintenance and stop joints are used in such cases. These connect cables electrically but block oil flow. In gas-filled cables, gas-stop joints perform a similar function. Semistop joints are used to permit or stop the flow of oil in POF cables. Heterojoints connect two

Table 9.2 Classification of Splices and Terminations

Joints	Comments
A. Straight Joints	
1. Normal joints	Taped and extruded cables.
2. Insulated joints	Shields insulated for cross bonding.
3. Stop joints	To stop oil flow in OF cables.
4. Semistop joints	To stop oil flow in POF cables.
5. Gas-stop joints	To stop gas flow in GF cables.
B. Terminal joints	
1. Termination in air	Used for taped and extruded cables. For taped cables, porcelain bushing is used. For extruded cables, porcelain, epoxy resin or elastomer mold bushings are used.
2. Termination in oil	Porcelain bushing are used.
3. Termination in SF_6	Epoxy-resin bushing are used.
4. Direct connection	With circuit breakers, transformers, etc.
C. Heterojoints	Used to connect OF-XLPE, OF-POF, cables, etc.
D. Branch joints	Used for making Y (or T branch) and X (or + branch).

Source: Ref. 27. Reprinted with permission.

different types of cable systems whereas branch joints are used when a cable is to be connected to multiple cable systems.

The basic concept of jointing of a single conductor cable is outlined in Figure 9.15. Here the joint is made up of (a) a conductor compression sleeve which joins together the conductors of the two cable ends, (b) a tapering down of the insulation on each cable, (c) the joint insulation which is applied over the conductor sleeve and (d) the stress of relief cone.

The tapering of the insulation is called stepping for taped cables and penciling for extruded cables. Stepping consists of a set of steps having risers and treads from the level of the conductor surface to that of the cable insulation surface. Penciling, however, gives a smooth surface. The insulation applied over the conductor sleeve should be well blended into the cable insulation so as to make the overall cable insulation as homogeneous as possible. Generally a tangential component of electric stress is introduced at the end of conductor sleeve in the tapered insulation. Usually the joint insulation is covered with a suitable protective layer. Since the joint insulation is normally built up to some diameter greater than that of the cable insulation, thermal discontinuities may arise at joints because the

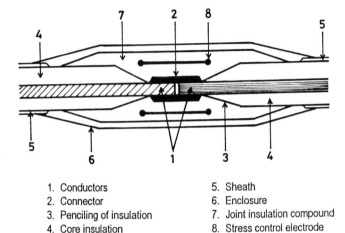

1. Conductors
2. Connector
3. Penciling of insulation
4. Core insulation
5. Sheath
6. Enclosure
7. Joint insulation compound
8. Stress control electrode

Figure 9.15 Basic components of a cable joint.

thermal resistance tends to be higher at the joints than in any other part of the cable. In addition to these simple elements, a joint should also have conductor and insulation shields and a sheath or a protective covering. The arrangement employed depends upon the type of the joint as well as the type of the cable. Due to the availability of premolded, heat-shrinkable components, the jointing of extruded cables is much simpler than that of the taped cables. Table 9.3 summarizes the types of splices available for extruded dielectric cables [27]. The detailed design principles and performance requirements of joints are discussed by Tanaka and Greenwood [27].

Table 9.3 Types of Extruded Cable Splices

Types	Insulating materials
Tape wrap splices	Self-adhesive tapes, pressure sensitive tapes or heat shrinkable tapes are used with silicone oil, etc.
Field molded splices	Involves taping, heating and cross-linking using EPR, PE vulcanizable tape or injection mold (PE).
Prefabricated splices	EPR, silicone rubber, epoxy resin and insulating oil are normally used in prefabricated splices.
Semiprefab splices	Uses EPR mold or tape and partial replacement of tape with mold.

Source: Ref. 27. Reprinted with permission.

9.11.2 Terminations

Terminations are required where cables are connected to overhead lines or other electrical equipment that may be air, oil or SF_6 insulated requiring the use of different terminations, as mentioned in Table 9.2. In some cases, direct cable connection to the apparatus is used instead. Figure 9.16 shows the basic structure of a single-core cable termination. It consists of a conductor lead-out rod, an insulation reinforcing layer with a stress relief cone, and a casing or a bushing. The end of taped cables is usually encapsulated with a porcelain bushing which acts as an external insulation. The space between cable core and the inner surface of the bushing may be filled with oil, SF_6 or other suitable insulation compound. The external insulation may be made of plastic tape, a porcelain bushing, an epoxy resin bushing or a rubber molded bushing. The end of the termination is often open to air for connection to an overhead line, but sometimes it is connected to an apparatus in oil or SF_6. Terminations in oil or SF_6 usually need less space. For SF_6 applications, epoxy resin bushings are preferred.

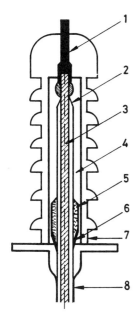

Figure 9.16 Basic components of a cable termination: 1 = conductor lead-out rod, 2 = penciling of insulation, 3 = cable conductor, 4 = insulating compound, 5 = insulation reinforcing layer, 6 = stress relief cone, 7 = bushing, 8 = cable sheath.

The electric field in a coaxial cable is purely radial and the axial stress is zero. At a joint or termination, the stress distribution is no longer completely radial as joints cannot, in general, be made without introducing an increase in diameters of both the conductor and the insulation shield. At these transitions, an axial or longitudinal component of stress is introduced. Similarly, at a termination, a longitudinal stress component is introduced between the high voltage terminal and the cable shield. This longitudinal stress is controlled by the accessory geometry, i.e., the profile of the stress relief cone and the pencling or stepping of the joint insulation. Normally stress relief cones are used to reduce stress concentrations that occur at the sheath's edge. For higher voltage ratings, even such a cone is not sufficient and a capacitively graded termination is usually preferred [27]. For such applications, capacitor bushings made of cylindrical laminates of metal foil and insulating paper are designed to get a uniform stress distribution.

Problems can arise at joints and terminations due to voids and due to the use of inhomogeneous dielectrics. Other problems such as corona degradation, surface flashover and tracking are caused by higher longitudinal and sheath edge stresses. Any such problem can lead to a premature failure of the respective cable splice. In addition, when a termination is made up of different materials, the different rates of aging, expansion/contraction or other inhomogeneous properties can also cause premature failure.

9.12 CABLE FAULT LOCATION

All types of cables are subjected to electrical faults. It is important to locate the point of such a fault as soon as possible. OF, POF, and gas-filled (GF) cables can also experience other types of faults which may be followed by electrical faults. Oil leaks in case of OF and POF cables and gas leaks in case of GF cables are examples of such faults. A brief summary of different fault location techniques is given in this section.

9.12.1 Detection of Oil Leaks

An oil-leak point can be located by visual inspection, by oil-flow behavior or by the frozen cable method. It usually involves judging the oil flow direction from axial variation of temperature on the pipe when part of the pipe is heated. When the oil leak is caused by external forces, the protective covering of the cable is damaged and it may also be earthed. In this event, it may be possible to utilize a ground point detection technique. Details of various methods for detection and location of oil leaks are reported by Tanaka and Greenwood [27].

9.12.2 Detection of Gas Leaks

Gas leak detection methods include (1) the bubble observation method, which is usually suitable for joint boxes and other cable accessories, and (2) gas flow, tracer gas and acoustic detection methods, which are suitable for main cables. The gas flow method estimates the position of a gas leak by measuring the pressure gradient caused by the leak. In the tracer gas method, a certain amount of tracer gas (e.g., a halogen gas) is injected in the cable and conditions are monitored along the cable route to detect the leak. Acoustic methods detect the leak through acoustic waves which are generated at the point of a gas leak [27].

9.12.3 Fault Location

Electrical faults in multiconductor cables may be classified as:

1. High- or low-resistance earth faults involving one or more conductors
2. Open-circuit faults
3. External flashover or self-healing faults

The cable fault location is accomplished in three basic steps. These are: (1) recognition of a fault condition, (2) estimation of the point of fault and (3) confirmation of the point of fault. Fault presence is indicated by the operation of relays. The faulty conductor and fault type are identified by simple tests at both cable ends. Such tests include measuring the conductor resistance and measuring the insulation resistances between the cable cores and between the cores and the ground. From such measurements, the type of fault, the faulted phase and the approximate fault zone are identified. Terminal measurement methods are used to measure the distance D from either end of the cable and then locate the fault more accurately. The distance D can be measured by either a pulse method or a bridge method. A DC bridge is suitable for low-resistance faults, while AC bridges are more appropriate for open-circuit faults such as broken conductors. Murray loop method is useful to find D for low-resistance single line to ground fault. However, it is not suitable for open conductor faults or for three-phase faults. If the fault has a high resistance to ground, usually a high voltage is first applied to "burn out" the fault and reduce the fault resistance. A Murray–Fischer bridge is useful to locate a fault when no healthy cable is available [27].

Reflection or pulse methods, which can be used for precise fault location for a variety of cable faults [27], are based on the behavior of traveling waves. For an open or a short circuit, cable impedance changes are very sudden and traveling waves are reflected back with or without a complete change in the incident pulse polarity. In pulse methods, a pulse is sent from one cable end and its reflection is monitored. The polarity of the reflected pulse, the time interval between the incident and the reflected pulse, and the velocity of wave propagation are used to estimate the fault location. Moreover, if a train of high-frequency pulses is sent along the cable and the resulting magnetic field is detected, there would be a detectable signal all along the cable route up to the point of the fault, beyond which it would cease. A high-resolution radar method to locate cable faults as well as underground splices and feed-through devices with good accuracy has been described by Banker et al. [41].

9.13 RECENT ADVANCES IN CABLE TECHNOLOGY

Present OF and POF cables are the outcome of fairly mature and proven technology. Recent years have seen increasing interest in the development of paper-polypropylene-paper insulation (PPP) for cable applications. Consequently, a 345-kV underground/underwater PPP insulated pipe-type cable was successfully deployed [42]. However, the major developments are taking place in the materials, manufacturing techniques, applications and diagnostics of extruded dielectric power cables. In the past two decades, XLPE and EPR have been extensively investigated in order to develop cable insulations with improved resistance to initiation and growth of electrical and water trees. Consequently, TRXLPE and EPR insulations have shown improved characteristics in this regard. Considerable improvements have also been made in the extrusion processes, curing methods and use of semiconducting screens [1,43]. In addition to improved smoothness of these screens and better bonding between the screens and the insulation, the number and size of impurities and other defects in the insulation are also being controlled.

The use of solid dielectric cables have been extended to EHV range and 500-kV, XLPE insulated cables have been in service since 1987 with good performance records [44,45]. EPR cables are also being manufactured for medium and high voltage applications. Moreover, XLPE cables are increasingly being used in DC and underwater applications. Along with such developments in extruded cables, accessories for such cables have also been developed for voltages of up to 500 kV [46].

In the medium voltage URD systems, utilities have gained from their earlier experience and have modified the cable specifications for improved performance against treeing and premature failures. Consequently, besides other factors, a water impervious jacket is recommended for URD cables as there is an overwhelming evidence that the use of moisture barriers on medium and high voltage cables can solve the problem of water treeing [3,47]. Metal-plastic laminates are also being developed for such applications [35]. For existing cables that have suffered degradation due to treeing, use of silicon fluids to extend cable life has been proposed [48].

A significant research effort has also gone into understanding the aging and degradation mechanisms as well as into developing diagnostic techniques. There is a general consensus that DC testing of service aged extruded dielectric cables can cause premature cable failures after the cables are returned to service, and the DC test should be replaced with a very low frequency AC test [49]. Impulse overvoltages can also affect the remaining life of extruded cables by aiding tree propagation. PD and leakage current methods have been proposed to monitor the deterioration of XLPE cable insulation [26,50].

An important area in which progress is desirable is cryogenic superconducting cables. Some of the cryogenic cables tested up to the present time have demonstrated an acceptable short-term performance and the operating parameters achieved should permit economical superconducting systems to be developed [51]. However, a considerable research and development effort will be required before such systems, using either conventional or high temperature superconductors, are commercially used. Further details of various aspects of high voltage cables can be found in cable handbooks, e.g., Bungay and McAllister [52].

REFERENCES

1. T. Fukuda, IEEE Elect. Insul. Magazine, Vol. 4, No. 5, pp. 9–16, 1988.
2. N.M. Burns, R.M. Eichhorn and C.G. Reid, IEEE Elect. Insul. Magazine, Vol. 8, No. 5, pp. 8–24, 1992.
3. G. Graham and S. Szaniszlo, IEEE Elect. Insul. Magazine, Vol. 11, No. 5, pp. 5–12, 1995.
4. M. Khalifa (ed.), *High Voltage Engineering: Theory and Practice*, Marcel Dekker, Inc., New York, 1990.
5. P. Graneau, *Underground Power Transmission: The Science, Technology and Economics of High Voltage Cables*, John Wiley & Sons, New York, 1979.
6. N.H. Malik and A.A. Al-Arainy, Int. J. Elect. Engineering Education, Vol. 25, No. 1, pp. 27–32, 1988.
7. T. Gonen, *Electric Power Transmission System Engineering: Analysis and Design*, John Wiley & Sons, New York, 1988.
8. N.H. Malik, IEEE Trans. on Elect. Insul., Vol. 24, No. 1, pp. 3–20, 1989.

9. N.H. Malik and A.A. Al-Arainy, IEEE Trans. on Power Delivery, Vol. 2, No. 3, pp. 589–595, 1987.
10. N.H. Malik and A.A. Al-Arainy, IEEE Trans. on Elect. Insul., Vol. 20, No. 3, pp. 499–503, 1988.
11. H.M. Ryan (ed.), *High Voltage Engineering and Testing*, Peter Peregrinus Ltd., London, England, 1994.
12. J.H. Neher and M.H. McGrath, AIEE Trans. on PAS, Vol. 76, pp. 752–772, 1957.
13. IEC Standard Publication 287, 2nd Edition, 1982.
14. G.J. Anders, M.A. El-Kady, R.W. Ganton, D.H. Horrocks and J. Motlis, IEEE Trans. on PWRD, Vol. 2, No. 2, pp. 289–295, 1987.
15. M.A. El-Kady, IEEE Trans. on PAS, Vol. 103, pp. 2043–2050, 1984.
16. B.M. Weedy, *Thermal Design of Underground Systems*, John Wiley & Sons, New York, 1988.
17. N.H. Malik, A.A. Al-Arainy, A.M. Kailani and M.J. Khan, IEEE Trans. on Elect. Insul., Vol. 22, No. 6, pp. 787–793, 1987.
18. J.C. Chan, P. Duffy, L.J. Hiiala and J. Wasik, IEEE Elect Insul. Magazine, Vol. 7, No. 5, pp. 9–20, 1991.
19. S.A. Boggs, IEEE Elect. Insul. Magazine, Vol. 6, No. 4, pp. 33–39, 1990.
20. S.A. Boggs, A. Pathak and P. Walker, IEEE Elect. Insul. Magazine, Vol. 12, No. 1, pp. 9–16, 1996.
21. ICEA Publication T-24-380, "Guide for Partial Discharge Test Procedures", 1980, USA.
22. IEC Publication 270, "Partial Discharge Measurements", IEC, Geneva, Switzerland, 1981.
23. IEC Publication 885, "Partial Discharge Tests" and "Test Method for Partial Discharge Measurements on Lengths of Extruded Power Cables", Parts 2 and 3, IEC, Geneva, Switzerland, 1987 and 1983.
24. F.H. Kreuger, M.G. Wezelenburg, A.G. Wiemer and W.A. Sonneveld, IEEE Elect. Insul. Magazine, Vol. 9, No. 6, pp. 15–23, 1993.
25. C. Laurent and C. Mayoux, IEEE Elect. Insul. Magazine, Vol. 8, No. 2, pp. 14–17, 1992.
26. G. Katsuta, A. Toyo, K. Muraoka, T. Endoh, Y. Sekii and C. Ikeda, IEEE Trans. on PWRD, Vol. 7, No. 3, pp. 1068–1074, 1992.
27. T. Tanaka and A. Greenwood, *Advanced Power Cable Technology*, Vol. I and II, CRC Press, Boca Raton, Florida, 1983.
28. R.M. Eichhorn, "Treeing in Solid Organic Dielectric Materials" in *Engineering Dielectrics*, Vol. IIA, R. Bartnikas and R.M. Eichhorn (eds.), ASTM Press, Philadelphia, 1983.
29. S. Grzybowski and R. Dobroszewski, Proc. IEEE Int. Symp. on Elect. Insul., pp. 122–125, 1978.
30. B.R. Varlow and D.W. Auckland, IEEE Elect. Insul. Magazine, Vol. 12, No. 2, pp. 21–26, 1996.
31. E.F. Steennis and F.H. Kreuger, IEEE Trans. on Elect. Insul., Vol. 25, No. 5, pp. 989–1028, 1990.

32. E.F. Steennis and A. van der Laar, Electra, No. 125, pp. 89–101, 1989.
33. M.S. Mashikian, R. Luther, J.C. McIver, J. Jurcisin and P.W. Spencer, IEEE Trans. on PWRD, Vol. 9, No. 2, pp. 620–628, 1994.
34. J.J. Xu and S.A. Boggs, IEEE Elect. Insul. Magazine, Vol. 10, No. 5, pp. 29–37, 1994.
35. K.E. Bow, IEEE Elect. Insul. Magazine, Vol. 9, No. 5, pp. 17–28, 1993.
36. A. Asano, T. Takahashi, K. Maeda and T. Niwa, IEEE Trans. on PWRD, Vol. 9, No. 1, pp. 553–558, 1994.
37. F.H. Kreuger, P.H. Morshuis and A.M. van der Laar, IEEE Trans. on Elect. Insul., Vol. 24, No. 6, pp. 1063–1070, 1989.
38. T.W. Dakin, Trans. AIEE, Vol. 67, pp. 113–122, 1948.
39. V.K. Agarwal, H.M. Banford, B.S. Bernstein, E. Brancato, R.A. Fouracre, G.C. Montanari, J.L. Paral, J.N. Seguin, D.M. Ryder and J. Tanaka, IEEE Elect. Insul. Magazine, Vol. 11, No. 3, pp. 37–43, 1995.
40. T. Tanaka, T. Fukuda and S. Suzuki, IEEE Trans. on PAS, Vol. 95, No. 6, pp. 1892–1900, 1976.
41. W.A. Banker, P.R. Nannery, J.W. Tarpey, D.W. Meyer and G.H. Piesinger, IEEE Trans. on PWRD, Vol. 9, No. 2, pp. 1187–1194, 1994.
42. J. Grzan, R.V. Casalaina, E.I. Hahn and J.O. Kansog, IEEE Trans. on PWRD, Vol. 8, No. 3, pp. 750–760, 1993.
43. K.B. Mueller, U. Trefow, B. Dellby and G. Hjalmarsson, IEEE Trans. on PWRD, Vol. 5, No. 4, pp. 1660–1668, 1990.
44. K. Ogawa, T. Kosugi, N. Kato and Y. Kawawata, IEEE Trans. on PWRD, Vol. 5, No. 1, pp. 26–32, 1990.
45. K. Kaminaga, M. Ichihara, M. Jinno, O. Fujii, S. Fukunaga, M. Kobayashi and K. Watanabe, IEEE Trans. on PWRD, Vol. 11, No. 3, pp. 1185–1194, 1996.
46. T. Kubota, H. Takaaki, H. Noda, M. Yamaguchi and M. Tan, IEEE Trans. on PWRD, Vol. 9, No. 4, pp. 1741–1759, 1994.
47. J.H. Dudas, IEEE Elect. Insul. Magazine, Vol. 10, No. 2, pp. 7–16, 1994.
48. P.R. Nannery, J.W. Tarpey, J.S. Lacenere, D.F. Meyer and G. Bertini, IEEE Trans. on PWRD, Vol. 4, No. 4, pp. 1991–1996, 1989.
49. H.R. Gnerlich, IEEE Elect. Insul. Magazine, Vol. 11, No. 5, pp. 13–16, 1995.
50. T. Nakayama, IEEE Trans. on PWRD, Vol. 6, No. 4, pp. 1359–1365, 1991.
51. E.B. Forsyth, IEEE Elect. Insul. Magazine, Vol. 6, No. 4, pp. 7–16, 1990.
52. E.W.G. Bungay and D. McAllister (eds.), *Electric Cable Handbook*, BSP Professional Books, Oxford, England, 1990.

10
Generation and Measurement of Testing Voltages

10.1 INTRODUCTION

High voltage testing is the final step in ensuring the dielectric quality of the developed insulation material. Similarly, the complete HV system or device is tested to ensure its integrity and performance. Chapter 12 will deal with various aspects of high voltage testing. This chapter discusses the basic circuits used for generating the test voltages and the traditional measuring techniques. The next chapter will outline more recent advances in high voltage measurement techniques. High test voltages normally require only moderate currents (from few mA up to a few A), and thus the generation schemes discussed in this chapter will generally fall in this category.

10.2 HIGH VOLTAGE DC GENERATION

High DC voltages (HVDC) are mainly used in scientific research, in testing equipment for HVDC transmission systems, and sometimes for high voltage cables and insulation testing. HVDC can be generated by either direct method, i.e., changing mechanical energy directly to HVDC, or by indirect method, i.e., changing mechanical energy to high AC voltages (HVAC) and then converting it to HVDC. The direct generation is obtained by van de Graff generators, but this method suffers from low kVA output, limiting its use to some special applications in physics. In the indirect method,

Generation and Measurement of Testing Voltages 277

HVAC is rectified using HV diodes and capacitors. Ripple factor (δV) and voltage drop (ΔV) are the two important parameters to be considered when dealing with rectification of HVAC to get HVDC. In reference to Figure 10.1, δV is defined as half the difference between the maximum and the minimum voltage values, i.e.:

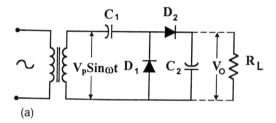
(a)

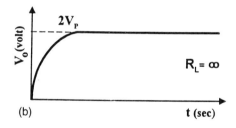
(b)

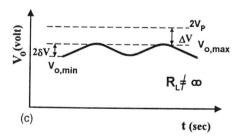
(c)

Figure 10.1 Voltage doubler: (a) circuit, (b) output voltage at no load and (c) output voltage at load.

$$\delta V = 0.5(V_{max} - V_{min}) \quad (10.1)$$

In most applications, the percentage ripple allowed is between 3% and 5%. The voltage drop ΔV is defined as:

$$\Delta V = V_{n \cdot L} - V_{F \cdot L} \quad (10.2)$$

where $V_{n \cdot L}$ and $V_{F \cdot L}$ are average values of no-load and full-load DC voltages, respectively. The voltage doubler circuit shown in Figure 10.1a is widely used to generate a DC voltage of $\pm 2\,V_p$ from an AC voltage of $V_p \sin \omega t$. The voltage waveshape at no load is shown in Figure 10.1b. The output voltage with load is depicted in Figure 10.1c, showing ΔV as well as δV. The maximum voltage across the diodes or the capacitor is $2V_p$. Since the costs of these components increase at much higher rates than the increase in their voltage rating, it is not economical to increase their voltage rating to generate higher voltages. Instead, two or more doubler circuits are connected in cascade to form a voltage multiplier circuit. Figure 10.2 shows a multiplier circuit consisting of three stages (n = 3) with maximum output voltage of $6V_p$. For such a circuit, δV and ΔV are given as [1]:

$$\delta V = \frac{I}{4fC} n(n+1) \quad (10.3)$$

$$\Delta V = \frac{I}{3fC}\left(\frac{2n^3}{5} - \frac{n}{2}\right) \quad (10.4)$$

where I = DC load current, f = AC supply frequency, C = stage capacitance and n = number of stages. It can be seen from the above equations that δV is proportional to n^2 while ΔV is proportional to n^3. Thus, the

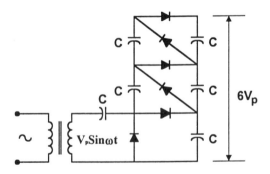

Figure 10.2 Multiplier circuit (n = 3) for HVDC generation.

number of stages in such a circuit is the limiting factor as it influences adversely the values of δV as well as ΔV.

The above limitations can be overcome by energizing each stage by a transformer comprising a tertiary low voltage winding which excites the primary winding of the next upper stage. In this way, the problems of δV and ΔV are solved; however, there are extra costs involved for the additional transformer windings. The limiting factor in such circuits employing cascaded transformers is the loading of the first stage transformer's primary which has to supply the current for all n stages. In addition to the above circuits, there are other HVDC generating arrangements like Deltatron circuit and voltage multiplier circuit with individual transformers for each stage [1].

10.3 HIGH VOLTAGE AC GENERATION

There are three main methods for HVAC generation, which are briefly described next.

10.3.1 Single Testing Transformer

This is similar to a single-phase step-up transformer with emphasis on the insulation of the HV winding. The windings are generally designed for low current ratings. The per unit impedance is kept less than 0.05 and the magnetizing current is also kept low to minimize the harmonics in the output voltage. Furthermore, care must also be taken to ensure that the field distribution over the HV winding is uniform. Single test transformers are economical for voltages of ≤ 300 kV. Above this level, a single test transformer becomes quite expensive due to the added insulation requirements and due to the burden in transportation and erection of large transformers. Thus, for higher voltages, cascaded transformers are the main choice.

10.3.2 Cascaded Transformers

The idea of cascading transformers is based on connecting the secondary windings of various transformers in series while their primary windings are energized in the normal manner. Figure 10.3 shows three stages of a cascaded transformer. The second- and third-stage transformers are isolated from the ground by insulators capable of withstanding voltages of V and 2V, respectively. Here the total output voltage is 3V and is taken across

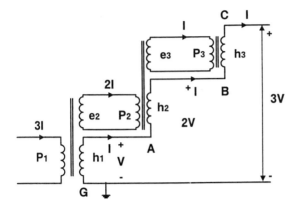

Figure 10.3 Cascaded transformer for HVAC generation.

the high voltage terminal of the third transformer and the ground. The two main limitations of increasing the number of stages in this scheme are the high total internal impedance and the high loading of the primary of the first stage. As can be seen from Figure 10.3, the current in the first transformer's primary winding is 3I. The net internal impedance of this circuit can be derived as follows.

The total reactive power is given as:

$$Q = I^2(X_{h_1} + X_{h_2} + X_{h_3}) + I^2(X_{e_3} + X_{p_3})$$
$$+ (2I)^2(X_{e_2} + X_{p_2}) + (3I)^2 X_{p_1} \quad (10.5)$$

If $X_{h1} = X_{h2} = X_{h3} = X_h$, $X_{e2} = X_{e3} = X_e$, and $X_{p1} = X_{p2} = X_{p3} = X_p$, then:

$$Q = 3X_h I^2 + 5X_e I^2 + 14X_p I^2 \quad (10.6)$$

If the whole arrangement is modeled by one impedance X_{eq} and one current I then:

$$Q = I^2 X_{eq} = I^2[3X_h + 5X_e + 14X_p] \quad (10.7)$$

Thus,

$$X_{eq} = 3X_h + 5X_e + 14X_p \quad (10.8)$$

In general, for an n stage cascaded transformer, the total impedance is given as:

$$X_{eq} = \sum_{j=1}^{n} [X_{hj} + (n-j)^2 X_{ej} + (n+1-j)^2 X_{pj}] \quad (10.9)$$

10.3.3 Resonance Circuit

One of the problems associated with the HV test transformer is the resonance of its inductance with the load capacitance. Figure 10.4a shows a simplified transformer equivalent circuit with single inductance L_T (it is assumed that $R_T = 0$). If the load is purely capacitive (C_L), such as a long cable, then the load voltage is given as:

$$V_L = \frac{V_0}{1 - \omega^2 L_T C_L} \qquad (10.10)$$

where ω is the supply frequency. Since the denominator is less than 1, $V_L > V_0$ and could be several times V_0 depending upon ω, L_T and C_L and could lead to the destruction or breakdown of the tested load. If this phenomenon is used in a controlled manner, output voltage V_L of up to 50 times V_0 could be obtained from a supply voltage V_0. This is achieved by using adjustable inductance L (where $L \gg L_T$) and a capacitance C such that $C \gg C_L$ (see Figure 10.4b). This type of circuit is called a resonance circuit and it is used to generate HVAC especially when the required output current is low. The main advantages of such a circuit are:

1. Less power requirement from the source (<10% of kVA required for testing)

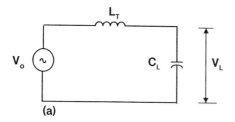

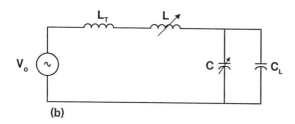

Figure 10.4 Series LC circuit: (a) simplified HV transformer equivalent circuit (R = 0) with capacitive load and (b) resonance circuit for HVAC generation.

2. Pure sinusoidal output waveforms due to resonance
3. No arcing in case of a short circuit since the condition of resonance ceases to exist with low shunt resistance of the arc
4. Simple and compact system
5. More than one resonance circuit can be cascaded to obtain even higher voltages

Such systems are very suitable for testing capacitive loads such as cables.

10.4 HIGH VOLTAGE IMPULSE GENERATION

It is well known that power system components are subjected to severe overvoltage due to internal switching surges or external lightning surges. Consequently, the integrity of the individual components, devices and subsystems must be checked through high voltage surge testing. High voltage surges that are simulated in laboratories are commonly known as high voltage impulses. An impulse is defined as a unidirectional voltage (or current) rising quickly to its peak value and then decaying slowly to zero. Thus, an impulse has two parts: a rising part, which is usually realized by charging a capacitor, and a decaying part, which is realized by discharging a capacitor. Figure 10.5a shows a capacitor C_2 which is charged through a resistor R_1 from a much larger capacitor C_1. In Figure 10.5b the two capacitors are discharged through a resistor R_2. Combining the two circuits and replacing the mechanical switch by a HV switch, e.g., a sphere spark gap, the simple impulse generating circuit of Figure 10.5c is obtained. The impulse voltage waveform can be modeled mathematically by a double exponential wave defined by the equation:

$$V_0(t) = A[\exp(-\alpha t) - \exp(-\beta t)] \quad (10.11)$$

Generally $\beta \gg \alpha$, where β corresponds to the charging or the rising portion of the voltage while α corresponds to the discharging or decaying portion of the waveform. Figure 10.6 shows the typical impulse waveshape, which is characterized by its front time t_f and the tail time t_t. Tail time or time to half of the voltage peak value (t_t) is the time between $t_0 = 0$ and the time where the impulse decays to half of its peak value. Front time t_f is defined as peak time (t_p) or [($t_{.9} - t_{.3}$)/0.6] for switching and lightning impulses, respectively. Equation (10.11) can be derived from Figure 10.5c using either time or frequency domain circuit analysis and is written as:

$$V_0(t) = \frac{V}{k(\beta - \alpha)} [\exp(-\alpha t) - \exp(-\beta t)] \quad (10.12)$$

where

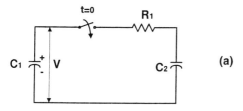

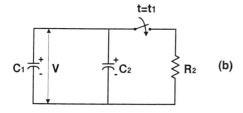

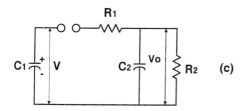

Figure 10.5 Principle of HV impulse generation: (a) charging a capacitor, (b) discharging capacitors and (c) basic impulse generating circuit.

$$\alpha, \beta = \frac{a}{2} \pm \sqrt{\left(\frac{a}{2}\right)^2 - b} \qquad (10.13)$$

$$a = \left(\frac{1}{R_1 C_1} + \frac{1}{R_1 C_2} + \frac{1}{R_2 C_1}\right), \ b = \frac{1}{R_1 R_2 C_1 C_2}, \ k = R_1 C_2 \quad (10.14)$$

and V = initial voltage across C_1. The impulse generator circuit can also be analyzed using computer software such as PSPICE, where the adjustment of parameters becomes simpler. If the circuit components are known, t_f and t_t can be found by sketching V(t) versus t or by solving equation (10.12) using the definitions of t_f and t_t. If $R_2 \gg R_1$, and $C_1 \gg C_2$, then t_f and t_t can be found approximately by the following formulas:

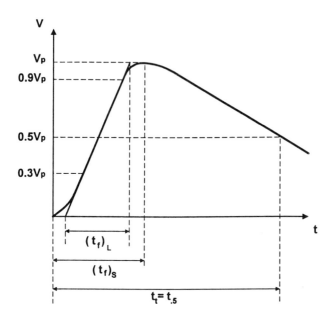

Figure 10.6 Impulse waveshape parameters. $(t_f)_L$ = front time for lightning impulses, $(t_f)_S$ = front time for switching impulses and t_t = time to half value.

$$t_f = 3R_1 \frac{C_1 C_2}{C_1 + C_2} \qquad (10.15)$$

$$t_t = 0.7(R_1 + R_2)(C_1 + C_2) \qquad (10.16)$$

These simplified expressions are based on the assumption that a capacitor can be fully charged in about $3\tau_c$, where τ_c is the changing time constant, i.e., $\tau_c = R_1 C_1 C_2/(C_1 + C_2)$. Furthermore, it is assumed that the capacitor is discharged to its half voltage when the elapsed time is $\approx 0.7\tau_d$, where τ_d is the discharge time constant and is given as $\tau_d = (R_1 + R_2)(C_1 + C_2)$.

When t_f and t_t are known and it is required to design the generating circuit elements, we need to find α, β, and then R_1, R_2, C_1 and C_2. Both exact as well as approximate methods are available. For either case, it is a normal practice to specify the values of two circuit elements and then calculate the values of the remaining two elements. Usually C_1 and C_2 are specified such that $C_2 \ll C_1$. C_2 can also be used as the voltage divider for measuring and recording the output voltage. The values of R_1 and R_2 can be found accurately by substituting the values of t_f and t_t in equation (10.12) and finding α and β using numerical techniques. Then the values of R_1 and R_2 can be evaluated by using equations (10.13) and (10.14) and assumed values of C_1 and C_2. The values of α and β and consequently R_1 and R_2 can also be found approximately by using the following formulas [2]:

$$\alpha = \frac{1}{R_1 C_2} \text{ and } \beta = \frac{1}{C_2 R_1} \qquad (10.17)$$

Furthermore, Table 10.1 can be used to find approximate waveshape or circuit parameters. When using Table 10.1 or equation (10.17) to find the circuit parameters, the calculated values can be verified by evaluating t_f and t_t and comparing them with specified values.

The single stage impulse generator circuit shown in Figure 10.5 is one of the several circuits in which either the position of R_2 is changed, or it includes some additional resistors or inductors [3]. When very high impulse voltages are needed, multistage impulse generators are used since the costs of a single stage circuit's elements become prohibitive as the voltage rating becomes high. The basic idea of a multistage circuit is to charge several stage capacitors in parallel and then discharge them in series. Figure 10.7a shows such a circuit; Figure 10.7b shows the same circuit during the stage capacitor charging period; and Figure 10.7c shows the same circuit during the discharge period. Figure 10.7d shows the equivalent single stage circuit which can be used to analyze the multistage circuit. In this circuit, only C_1 is distributed throughout all the stages. In other multistage impulse circuits, R_1 and R_2 can also be distributed throughout the different stages. However C_2 is normally left as a single unit which normally works as a voltage divider or represents the load capacitance. There are several details concerning the operation and control of this type of circuit. These can be found in manufacturer's manuals or in various references [1,2].

Such single and multistage impulse generation circuits are used to generate lightning and switching impulses. For some applications, special

Table 10.1 Impulse Parameters for Various Standard Waveshapes

t_f/t_t	$\frac{1}{\alpha} (\mu s)^{-1}$	$\frac{1}{\beta} (\mu s)^{-1}$	$\eta = V_p/nV$
1.2/50	68	0.4	0.96
1.2/200	284	0.38	0.99
170/1700	2200	43.5	0.91
250/2500	3160	62.5	0.9
650/2600	2500	250	0.75

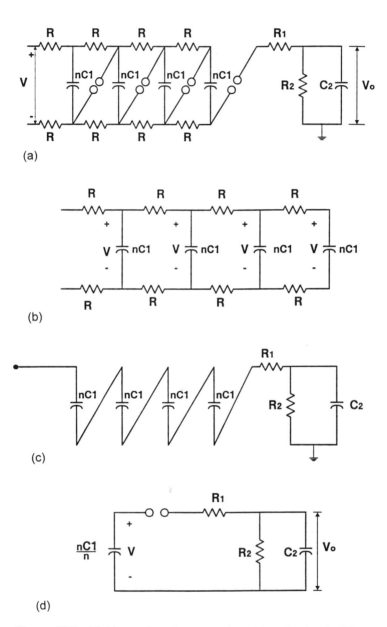

Figure 10.7 Multistage impulse generation: (a) main circuit, (b) stage capacitor charging part, (c) capacitor discharging part and (d) equivalent single stage circuit.

Generation and Measurement of Testing Voltages

kinds of switching impulses, such as impulses with very long tail times (in ms) or impulses with some controlled oscillations on their tail, are required. Table 10.2 shows a few circuits along with their output voltage waveforms and mathematical representations of such waveforms.

When a full wave surge occurs on a power network and a flashover takes place, e.g., across a bushing or an insulator, the voltage instantaneously falls to zero, resulting in a chopped wave. The voltage chopping can take place either on the front, at the peak or on the tail of a surge. To simulate a chopped surge wave, a rod-rod chopping gap is normally placed in parallel with the tested object. The distance of the chopping gap can be adjusted to control the width of the applied chopped wave during the

Table 10.2 Special Impulse Generating Circuits and Their Output Voltage Waveforms

Circuit diagram	Output Voltage waveform	Output Voltage formula in s domain
		$V_o(s) = \dfrac{V/K}{s^3 + as^2 + bs + c}$ $K = R_1 C_2$ $a = \dfrac{R_1}{L} + \dfrac{1}{C_2 R_2}$ $b = \dfrac{R_1 C_1 + R_2 C_2 + R_2 C_1}{L C_1 C_2 R_2}$ $c = \dfrac{1}{C_1 C_2 R_2 L}$
		$V_o(s) = \dfrac{V}{K} \cdot \dfrac{s}{s^4 + as^3 + bs^2 + cs + d}$ $K = L_2 C_2, \quad a = \dfrac{R}{L_2}$ $b = \dfrac{1}{C_1 L_2} + \dfrac{1}{C_2 L_2}$ $c = \dfrac{R}{C_1 L_1 L_1}, \quad d = \dfrac{1}{C_1 C_2 L_1 L_2}$ $R_1 = 0, L_2 \gg L_1, R_2 = R$
		$V_o(s) = \dfrac{Va}{s^2 + as + b}$ $a = \dfrac{R}{L}$ $b = \dfrac{1}{LC}$

chopped impulse testing. Triggered chopping gaps are often used to control the chopping time. Chopped impulse testing is required in some applications as discussed in Chapter 12.

10.5 NANOSECOND PULSE GENERATION

In Chapter 2 the gas breakdown under high voltage pulses of nanosecond duration was discussed. Due to an increasing number of applications of discharges under nanosecond voltage pulses, several electric circuits have been developed to produce such voltage pulses. The most common circuit used for this purpose is based on the discharge of a capacitor into a low inductance through a spark gap switch. However, modifications have been incorporated in this basic circuit by several investigators for achieving a better performance. Masuada [4] reported a simple circuit to generate nanosecond voltage pulses. A capacitor is rapidly charged and discharged utilizing a rotary spark gap switch. Rea [5] analyzed the characteristics of a nanosecond pulse generator circuit of Figure 10.8, which is based on Masuada's idea. In this case, the pulse rise rate can be controlled by R and L_i, while the pulse repetition rate is controlled by the rotation speed of the spark gap. To improve the circuit performance, a DC supply, V_{dc}, is added and coupled to the circuit through C. It is found that the intensity of the

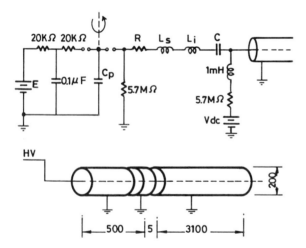

Figure 10.8 Nanosecond pulse generator circuit; dimensions are in mm. (From Ref. 5 © IEEE, 1995.)

corona discharge produced by voltage pulses of this circuit depends on total voltage magnitude, pulse voltage rise rate, value of C_p and the shape of corona electrode. Bajeu et al. [6] used a four-stage Marx generator with rotating spark gaps to generate nanosecond voltage pulses. The main problems encountered in this circuit are relatively longer rise times (tens to hundreds of nanoseconds), the energy loss in the sparking of the rotary gaps and the erosion of the spark gap. Kamase et al. [7] tested several types and shapes of materials for the rotary gaps and found that flat electrodes made from sintered tungsten with thorium dioxide coating had the longest life. Recently a specially made HV semiconductor switching arrangement was introduced to overcome the problems of mechanical switching where the pulse repetition rate can be increased to very high values [8]. A typical wave shape produced by this circuit is shown in Figure 10.9 where the rise time is around 35 ns.

The main applications of nanosecond pulsed corona are the removal of harmful impurities from air [5,6], purification of water from volatile materials [8] and the destruction of weeds and some other harmful insects and bacteria in the soil [9,10].

10.6 SPARK GAPS AS A VOLTAGE MEASURING DEVICE

10.6.1 Sphere Gaps

The sphere-sphere spark gap is one of the oldest and most reliable devices for high voltage measurements. A roughly uniform field distribution is achieved in the sphere-sphere gap when the gap length (d) is equal to or less than the sphere radius (D/2). It is well known that under known en-

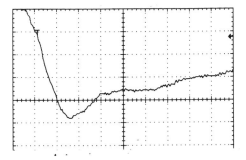

Figure 10.9 Nanosecond pulse voltage generated by ns pulse circuit using semiconductor switching: y-axis = 5 kV/div and x-axis = 20 ns/div.

vironmental conditions a uniform field air gap has a reasonably well-defined breakdown voltage. Thus, such a gap can be used for the measurement of the peak voltage value. Since the breakdown voltage characteristics of air gaps at standard atmospheric conditions are reasonably well defined, it is possible to establish tables (or figures) relating the gap spacing with the breakdown voltage. Table 10.3 lists the sparkover voltages of sphere-sphere gaps for various values of D and d at standard environmental conditions of temperature = 20°C, pressure = 1013 mbar, absolute humidity = 11 g/m^3 and clean air. For other atmospheric conditions, the sparkover voltage will vary and can be found using the procedure discussed in Chapter 3. Consequently, for a gap distance d between two spheres of diameter D, the actual sparkover voltage V_a when $P = P_1$, $T = T_1$ and $H = H_1$ is given as:

$$V_a = K_t V_t \qquad (10.18)$$

where K_t is the atmospheric correction factor described in Chapter 3, and V_t is the breakdown voltage value corresponding to d and D given in Table 10.3.

On the other hand, if one needs to adjust a gap spacing between two spheres of diameter D such that the spheres sparkover at a certain pre-specified voltage V_a, then first V_t is calculated using equation (10.8). Then from Table 10.3 the required value of d can be found which exhibits a sparkover voltage of V_t corresponding to the sphere diameter D. When $d \leq D/2$, the error in the voltage values measured by sphere gaps is within ±3% [15]. However for $D/2 < d < D$, the error is within ±5%. The breakdown voltage values are given for such large gaps in parenthesis in Table 10.3.

Measuring spheres can be arranged horizontally or vertically as shown in Figure 10.10. Moreover, for an accurate measurement scheme, there are certain requirements regarding the minimum clearances to the nearby earthed objects as well as to the ground. These requirements are specified in IEC 52 [11]. Table 10.4 summarizes the required minimum clearances. For uniform field air gaps, the breakdown voltage (in kV peak) can also be estimated by the Bruce formula [12]:

$$V_a = 24.22d + 6.08\sqrt{d.RAD} \qquad (10.20)$$

where RAD = relative air density (for details see Chapters 2 and 3) and d = gap length in cm.

For the measurements of lightning impulses, breakdown time lags can significantly influence the measurement accuracy when spark gaps are used as the voltage measuring device. Irradiation increases the number of free electrons needed for avalanche initiation. Studies show that external irra-

Generation and Measurement of Testing Voltages

Table 10.3 Peak Values of Breakdown Voltages of Sphere Gaps with One Sphere Grounded at Standard Atmospheric Conditions (P = 1013 mbar, T = 20°C)

Sphere Gap Spacing d(mm)	Peak Voltage kV — Sphere dia D in cms														
	12.5		25		50		75		100		150		200	
	a	b	a	b	a	b	a	b	a	b	a	b	a	b
10	31.7	31.7												
20	59	59	59	59										
30	85	85.5	86	86										
40	108	110	112	112										
50	129	134	137	138	138	138	138	138	138	138	138	138		
75	167	(181)	195	199	202	203	203	202	203	203	203	203	203	203
100	(195)	(215)	244	254	263	263	265	265	266	266	266	266	266	266
125	(214)	(239)	282	299	320	323	327	327	330	330	330	330	330	330
150			(314)	(337)	373	380	387	387	390	390	390	390	390	390
175			(342)	(368)	420	432	443	447	443		450	450	450	450
200			(366)	(395)	460	480	492	505	510		510	510	510	510
250			(400)	(433)	530	555	585	605	615		630	630	630	630
300					(585)	(620)	665	695	710		745	745	750	750
350					(630)	(670)	735	770	800		850	858	855	860
400					(670)	(715)	(800)	(835)	875	815	955	965	975	980
450					(700)	(745)	(850)	(890)	945	900	1050	1060	1080	1090
500					(730)	(775)	(895)	(940)	1010	980	1130	1150	1180	1190
600							(970)	(1020)	(1110)	1040 (1150)	1280	1310	1340	1380

Table 10.3 Continued

Sphere Gap							Peak Voltage kV Sphere dia D in cms							
Spacing d(mm)	12.5		25		50		75		100		150		200	
	a	b	a	b	a	b	a	b	a	b	a	b	a	b
700							(1025)	(1070)	(1200)	(1240)	1390	(1430)	1480	1550
750							(1040)	(1090)	(1230)	(1280)	(1440)	(1480)	1540	1620
800									(1260)	(1310)	(1490)	(1530)	1600	1690
900									(1320)	(1370)	(1580)	(1630)	1720	1820
1000									(1360)	(1410)	(1660)	(1720)	1840	1930
1100											(1730)	(1790)	(1940)	(2030)
1200											(1800)	(1860)	(2020)	(2120)
1300											(1870)	(1930)	(2100)	
1400											(1920)	(1980)	(2180)	
1500											(1960)	(2020)	(2250)	
1600													(2320)	
1700													(2370)	
1800													(2410)	
1900													(2460)	
2000													(2490)	

[a] For ac, ±dc, −ve lightning and −ve switching impulses.
[b] For +ve lightning and +ve switching impulses.

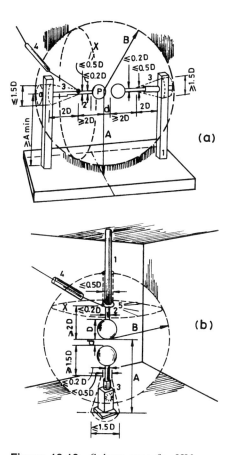

Figure 10.10 Sphere gaps for HV measurement: (a) horizontal arrangement and (b) vertical arrangement. 1 = insulating support, 2 = sphere shank, 3 = operating gear, 4 = high voltage connection with series resister, 5 = stress distributor, P = sparking point of HV sphere, A = height of P above ground, B = radius of space free from external structures, X = item 4 not to pass through this plane with a distance B from P.

diation is necessary for measuring low impulse voltages of <50 kV or for measurements using spark gaps having spheres of diameter ≤125 mm if an acceptable accuracy of ±3% is to be maintained [1,11].

A dust film on the surface of the unirradiated spheres may also influence the voltage measurement accuracy. Experiments show that the presence of dust film reduces the scatter of impulse breakdown voltages and significantly improves the measurement accuracy for lightning and switching impulses which can further be improved by providing enough external

Table 10.4 Clearances for Sphere Gaps

D (cm)	Value of A		Value of B (min)
	Maximum	Minimum	
up to 6.25	7D	9D	14d
10 to 15	6D	8D	12d
25	5D	7D	10d
50	4D	6D	8d
100	3.5D	5D	7d
150	3D	4D	6d
200	3D	4D	6d

A and B are shown in Figure 10.10; D = sphere diameter, d = gap length, d ≤ D/2.

irradiation [13,14]. However, for measurements of AC and DC voltages, the error caused by dust particles on the sphere surfaces may reach up to 6% [13]. Thus, the presence of dust particles on sphere electrodes should be considered when such gaps are used for high voltage measurements.

10.6.2 Rod Gaps

Rod-rod gaps are used for measuring HVDC with an error of less than 3% [15,16]. The used rods should have a square cross section with each side between 15 and 25 mm. For gap length in the range of 25 to 250 cm and for absolute humidity H such that $1 \text{ g/m}^3 < H/5 \leq 13 \text{ g/m}^3$, the breakdown voltage V_b (in kV) as function of gap spacing d (in cm) can be expressed as [16]:

$$V_b = 2 + 5.34d \qquad (10.21)$$

In this case, the applicable air density and humidity correction procedures are similar to those discussed in Chapter 3. Moreover, these apply only to the above specified gap lengths where the breakdown voltage–gap spacing relationship is linear. Beyond these limits, the rod gaps cannot be used for HVDC measurements, since the influence of environmental parameters especially humidity becomes nonlinear [17].

10.7 POTENTIAL DIVIDERS FOR HIGH VOLTAGE MEASUREMENT

For the human safety and for the protection of measuring instruments, potential dividers are used to decrease the high voltages. Two basic types of dividers, i.e., resistive and capacitive dividers are commonly used for

HV measurements. However each type has certain limitations when used for measurements of very high transient voltages. Such limitations arise from the stray capacitances and series inductances of divider elements. Table 10.5 shows the two basic dividers, their inherent problems when used for EHV and UHV measurements and the possible techniques to overcome such problems, in addition to the conditions which must be satisfied to ensure the divider transfer function is frequency independent. The divider is usually connected to a low voltage measuring instrument (e.g., peak voltmeter or oscilloscope) through a coaxial cable. In order to have an accurate measuring system, the divider, the cable, and the low voltage measuring and recording instruments must have adequate bandwidths suitable for the measurement of the required transient voltage. In addition there should be a complete matching between different components in order to avoid reflections and hence measuring waveform distortions. There are certain test procedures to determine the divider ratio, and to assess the divider suitability for high voltage and fast impulse measurements (see section 12.5).

10.8 OTHER HIGH VOLTAGE MEASURING DEVICES

Table 10.6 summarizes the basic devices used for high voltage measurements. The operating principles of some of these devices are described here.

10.8.1 Electrostatic Voltmeter

When a voltage (V) is applied across a pair of parallel plate electrodes of area A, separated by a spacing S and having a uniform electric field intensity, E, the energy stored W is given as:

$$W = \frac{1}{2} \varepsilon E^2 A S \qquad (10.22)$$

where ε = absolute permittivity of the medium between the plates. If the spacing changes by dS, then the change in energy dW is given by:

$$dW = \frac{1}{2} \varepsilon E^2 A dS \qquad (10.23)$$

Thus, mechanical force experienced by the plates is given by:

$$F = \frac{dW}{dS} = \frac{1}{2} \varepsilon E^2 A \qquad (10.24)$$

Table 10.5 Potential Dividers Used for HV Measurement

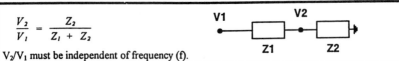

$\frac{V_2}{V_1} = \frac{Z_2}{Z_1 + Z_2}$ V_2/V_1 must be independent of frequency (f).	
Resistive divider	**Capacitive divider**
Ideally $\frac{V_2}{V_1} = \frac{R_2}{R_1 + R_2}$	**Ideally** $\frac{V_2}{V_1} = \frac{C_1}{C_1 + C_2}$
Problem With high V, R_1 becomes long and then stray cap C_S is not negligible, C_S in pF is given by: $C_S = \frac{2\pi \varepsilon_0 l}{\ln\frac{1.15l}{2r}}$ l = resistor length (m) and r = resistor radius (m). V_2/V_1 is dependent on f. It will cause distortion and errors.	**Problem** When increasing V, C_1 becomes long and series inductance (L) becomes important. V_2/V_1 is dependent on frequency. $\frac{V_2}{V_1} = \frac{C_1}{C_1 + C_2 - \omega^2 L C_1 C_2}$
Remedies A) Add capacitors in parallel with LV & HV sides. $\frac{V_2}{V_1} = \frac{Z_2}{Z_1 + Z_2}$ $Z_1 = \frac{R_1}{1 + j\omega C_1 R_1}$ $Z_2 = \frac{R_2}{1 + j\omega C_2 R_2}$ $\frac{V_2}{V_1} = \frac{R_2}{R_2 + R_1 \frac{1 + j\omega C_2 R_2}{1 + j\omega C_1 R_1}}$ $\frac{V_2}{V_1} = \frac{R_2}{R_1 + R_2}$ if $R_1 C_1 = R_2 C_2$ B) screen the resistor, i.e., make the stray capacitance large but across the whole divider.	**Remedy** Add series resistor to both LV & HV sides of the divider (damp the oscillations). $\frac{V_2}{V_1} = \frac{R_2 + \frac{1}{j\omega C_2}}{R_1 + \frac{1}{j\omega C_1} + R_2 + \frac{1}{j\omega C_2}}$ $\frac{V_2}{V_1} = \frac{1}{1 + \frac{C_2}{C_1}\left[\frac{1 + j\omega C_1 R_1}{1 + j\omega C_2 R_2}\right]}$ if $C_1 R_1 = C_2 R_2$ $\frac{V_2}{V_1} = \frac{1}{1 + \frac{C_2}{C_1}} = \frac{C_1}{C_1 + C_2}$ select $C_1 R_1 = C_2 R_2$

Table 10.6 High Voltage Measurement Systems and Devices

System/device	Type of voltage measured	Main advantages	Main drawbacks
1. Sphere gaps	Peak value of AC, DC and impulse	Simple, reliable, cheap	Sensitive to gap adjustment, need correction for environmental conditions. Continuous arc for AC and DC: Accuracy $\approx \pm 3\%$
2. Rod gaps 25 cm $< d <$ 250 cm	Peak value of HVDC	Similar to sphere gaps	Similar to sphere gaps. In addition, the influence of humidity outside the assigned range is not defined
3. Electrostatic voltmeter	AC (rms), DC (mean)	Extremely high input impedance, very accurate (error $< .25\%$)	Expensive and bulky especially for very high voltages
4. Peak voltmeter	Peak value of AC	Simple, can be constructed and calibrated in the laboratory	There are many sources of error (frequency, capacitors, diodes, LV meters)
5. Resistive divider, a cable and LV meter or an oscilloscope	AC, DC and impulse	Easily constructed in the lab, waveshape can be displayed. The frequency of AC, and the front and tail times of impulses can be measured	The presence of stray capacitance especially for EHV and UHV dividers and fast transient (see section 10.7)

Table 10.6 Continued

System/device	Type of voltage measured	Main advantages	Main drawbacks
6. Capacitive divider, cable and LV meter or an oscilloscope	AC and impulse	Similar to resistive dividers	The presence of series inductance especially for EHV and UHV dividers and fast transients can cause errors (see section 10.7)
7. Voltage transformer and LV meter	AC	Accurate	The transformer is costly especially for EHV and UHV ranges
8. Series impedance and LV meter	AC and DC	Simple, can be constructed in the lab	Not highly accurate especially for AC, loads the source, and the reading is sensitive to changes in temperature
9. Generating voltmeter	AC and DC	Does not load the supply, can also measure electric field directly	Requires separate drive, and needs constant calibration since any disturbance in position or mounting needs new calibration

Hence, the mechanical force experienced by the parallel plates can be used for voltage measurement purposes since it is proportional to E^2 or V^2.

Electrostatic voltmeters usually have one fixed plate; the other is moveable within a fraction of a millimeter in order not to disturb the original electric field too much. Since the mechanical movement is very small, its effect is amplified through electrical or optical means so that a reasonable scale can be obtained. Figure 10.11 shows the arrangement of plates and the use of light reflection for voltage measurement. The logarithmic scale can be used so it can give the values of V_{rms} directly. The detailed construction of electrostatic voltmeters vary depending on the manufacturer, the measured voltage range and the insulation medium used.

10.8.2 AC Peak Voltmeter

The 90° phase shift between the current and the voltage in a capacitor is utilized in peak AC voltmeter based on the Chubb and Fortesque circuit. Figure 10.12 shows the basic circuit which includes a standard capacitor C, two diodes and a low voltage ammeter. This ammeter will read the average value of the current I through the diode D_1 which is given as:

$$I = \frac{1}{T} \int_0^T C \frac{dV(t)}{dt} \cdot dt = \frac{1}{T} \int_0^{T/2} C dV(t) \qquad (10.25)$$

For an AC voltage, $v(t) = V_p \sin \omega t$. Therefore, equation (10.25) leads to:

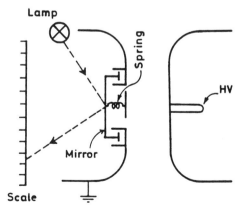

Figure 10.11 Basic principle of electrostatic voltmeter.

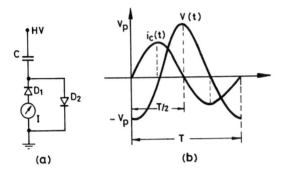

Figure 10.12 AC peak voltmeter: (a) circuit and (b) $i_c t$ and $v(t)$ waveshapes.

$$I = \frac{2C}{T} V_p \qquad (10.26)$$

Since $T = 1/f$, equation (10.26) can be rewritten as:

$$V_p = \frac{I}{2fC} \qquad (10.27)$$

Thus, by measuring I and knowing f and C, V_p can be determined. This method is accurate provided the positive and the negative half cycles of AC voltages are symmetrical and equal.

The above principle can be modified to have digital peak voltmeters for measuring AC as well as impulse voltages. This is achieved by passing the rectified current through a resistor and then transferring this voltage to a digital meter using (A/D convertor) and electronically controlled gates.

10.8.3 Generating Voltmeter

It is a variable capacitor electrostatic voltage generator which generates current I that is proportional to the electric field E or to the applied voltage V. Figure 10.13 shows a schematic diagram for a rotating-vane-type generating voltmeter. The measured high voltage (V) is connected to disc (D_h), while D_r is being rotated by a motor at a suitable constant speed, and is located at distance S from D_h; D_o is connected directly to the ground, while D_m is connected to the ground through an ammeter in order to measure the current I(t), which is given as:

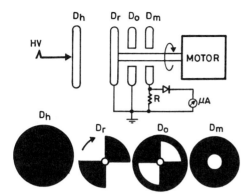

Figure 10.13 Rotating-vane-type generating voltmeter.

$$I(t) = C\frac{dV}{dt} + V\frac{dC}{dt} \quad (10.28)$$

For a DC voltage, dV/dt = 0. Thus, equation (10.28) becomes:

$$I(t) = V\frac{dC}{dt} \quad (10.29)$$

When the motor is at a standstill, dC/dt = 0 and no current flows. When the motor is rotating, C changes since the exposed area of D_m to electric field is changing with time from zero when it is completely covered, to A_m when the plates of disc D_m are completely exposed to E. Thus:

$$C(t) = \varepsilon\frac{A(t)}{S} \quad (10.30)$$

Hence,

$$I(t) = \frac{V\varepsilon}{S}\frac{dA(t)}{dt} \quad (10.31)$$

Figure 10.14 shows the variation of C(t) and I(t) with t. The variation of A(t) can be made linear or sinusoidal. This voltmeter can be used for AC voltage measurements as well provided the speed of the drive motor is half of the frequency of the measured voltage. For a four-pole motor, a speed of 1800 or 1500 rpm is suitable for measuring 60 or 50 Hz voltages, respectively. For peak value measurements, the phase angle of the motor must also be controlled so that V_p coincides with C_{max}.

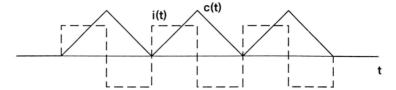

Figure 10.14 The interdisc capacitance and generated current as function of time of generating voltmeter.

Field measurement is also possible by using this meter provided the field between the HV disc and ground disc is uniform (other methods of field measurements are given in Chapter 11). Precautions must be taken to avoid breakdown between D_h and D_G. Thus, the voltage limit of such an instrument should be clearly identified.

10.9 MEASUREMENT OF CORONA AND GAP DISCHARGE CURRENTS

The fast current pulses resulting from corona and gap discharges are normally measured by placing a resistive shunt of low ohmic value between the grounded electrode and the earth. It must be ensured that the inductance of this shunt is very small. One method of generating smooth and true corona current pulses is the use of coaxial cylindrical electrode geometry where the high voltage is applied on the outer cylinder while the inner cylindrical (pipe) is grounded [18,19]. A small protrusion (corona point) is introduced in the center of the cylinder where it is isolated from the grounded pipe by a small (i.e., 50 Ω) resistor. A 50-Ω cable is used to carry the voltage drop across the resistor to the oscilloscope or any other measuring instrument. Figure 10.15 shows the schematic diagram for the above arrangement and the negative corona resulting from this corona point.

There are other ways of reducing the inductance of the detection resistance, such as placing several resistors in parallel to obtain the exact value of the required resistance but reducing the resultant inductance by paralleling the individual resistor's inductances. Figure 10.16 shows such an arrangement, where six 300-Ω resistors are arranged in parallel to give a 50-Ω shunt while reducing its inductance to 1/6th the value for individual resistor element [20,21]. This shunt can be used for detecting corona or

Generation and Measurement of Testing Voltages

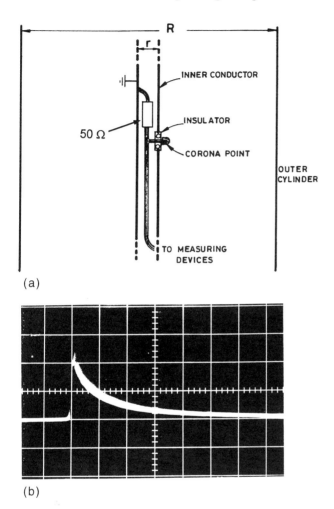

Figure 10.15 (a) Experimental setup for the generation and measurement of corona using cylindrical geometry. (b) Resulting corona current pulse (time scale = 20 ns/div).

gap discharge current. Figure 2.14 shows typical gap discharge current measured using such a detection resistor [21].

The suitability of the resistive shunts for measurements can be checked either by applying and measuring a known signal or by finding the spectral bandwidth of the measuring system, i.e., resistor, cable and the oscilloscope.

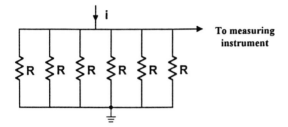

Figure 10.16 Current shunt resistors with low equivalent inductance used for corona and gap-type discharge measurements.

REFERENCES

1. E. Kuffel and W. Zaengle, *High Voltage Engineering Fundamentals*, Pergamon Press Ltd., Oxford, England, 1984.
2. M. Naidu and U. Kamaraju, *High Voltage Engineering* Tata McGraw-Hill, New Delhi, India, 1982.
3. M. Khalifa (ed.), *High Voltage Engineering Theory and Practice*, Marcel Dekker Inc., New York, 1990.
4. S. Masuada and S. Husakawa, "Submicrosecond Pulse Energization for Retrofitting Applications", Proc. 2nd Int. Conf. on Electric Precipitation, Kyoto, Japan, p. 613, 1984.
5. M. Rea and K. Yan, IEEE Trans. on Industry Applications, Vol. 31, No. 3, pp. 507–512, 1995.
6. G. Bajeu, I. Mustata, C. Lungu and G. Musa, Proc. IEEE Industry Application Society's Annual Meeting, Vol. 3, pp. 1583–1586, 1994.
7. Y. Kamase, M. Shimiza, T. Nagahama and A. Mizuno, IEEE Trans. on Industry Applications, Vol. 29, No. 4, pp. 793–797, 1993.
8. A. Al-Arainy, S. Jayaram and J.D. Cross, Proc. of 12th Int. Conf. on Cond. and Breakdown in Diel. Liquids (12-ICDL 96), Rome, Italy, pp. 427–431, 1996.
9. A. Mizuno and Y. Hori, IEEE Trans. on Industry Applications, Vol. 24, No. 3, pp. 387–394, 1988.
10. A. Mizuno, T. Inoue, S. Yamaguchi and K. Sadamoto, Proc. IEEE Industry Application Society's Annual Meeting, pp. 713–727, 1990.
11. IEC 52, "Recommendation for Voltage Measurement By Means of Sphere Gaps (One Sphere Earthed)", Geneva, Switzerland, 1960.
12. C. Wadhwa, *High Voltage Engineering*, Wiley Eastern Limited, New Delhi, India, 1994.
13. N.H. Malik, A. Al-Arainy and M.I. Qureshi, Journal of King Saud University (Engineering Sciences) Vol. 9, No. 1, 1996.
14. A.A. Al-Arainy, N.H. Malik and M.I. Qureshi, The Arabian Journal for Science and Engineering, Vol. 20, No. 3, pp. 495–509, 1995.

15. N. Allen, IEEE Trans. on Elect. Insul., Vol. 28, No. 2, pp. 183–191, 1993.
16. IEC 60-1, "High Voltage Test Technique", IEC, Geneva, Switzerland, 1989.
17. CIGRE Report #72, "Guide for the Evaluation of the Dielectric Strength of External Insulation," Working group 07 of Study Committee No. 33, CIGRE, Paris, France, 1992.
18. W. Janischewskyj and G.L. Ford, IEEE Int. Symp. on Electromagnetic Compatability, IEEE Publication No. 70C28-EMC, pp. 436–441, 1970.
19. W. Janischewskyj and A. Al-Arainy, IEEE Trans., Vol. PAS-100, No. 2, pp. 539–551, 1981.
20. R. Dobroszewski and W. Janischewskyj, IEEE Trans., Vol. PAS-100, No. 5, pp. 2695–2702, 1981.
21. A. Al-Arainy, "Laboratory Analysis of Gap Discharges on Power Lines," Ph.D thesis, University of Toronto, Toronto, Canada, 1982.

11
New Measurement and Diagnostic Technologies

11.1 INTRODUCTION

High voltage components and apparati are tested for a variety of reasons (see chapter 12). During such tests, measurements of voltages, currents, partial discharge levels, etc. are made to ensure that required test voltage and/or current waveforms and amplitudes have been applied and the tested equipment complies with the test requirements and specifications. Chapter 10 briefly reviewed the methods of generation of high test voltages and their measurements. Due to great advances made during the last two decades in the fields of computers, electronics and optics, interest in the applications of such advanced technologies for the purpose of high voltage measurements, testing and diagnostics have been growing. Consequently, electro-optical and digital techniques are finding increasing use in the area of high voltage, high current measurements and diagnostics. This chapter provides a brief introduction to such new measurement and diagnostic technologies.

11.2 DIGITAL IMPULSE RECORDERS

The conventional method of measuring high test voltages involves a suitable voltage divider, with a well-defined and frequency-independent divider ratio, and a suitable low voltage measuring instrument. For high DC and AC test voltages, a suitable analog or a digital reading peak voltmeter is normally sufficient since breakdown of most insulating materials is related to the peak instantaneous applied voltage. For transient or impulse voltages, an impulse oscilloscope or some other suitable recording device is used, since in addition to the peak amplitude the voltage waveform is also required. A storage oscilloscope or an oscilloscope equipped with a camera is usually used for this purpose. Alternately, a digital recorder may be used. Commercial storage oscilloscopes can record waveforms with rise times as short as a few nanoseconds and camera-equipped oscilloscopes can even record pulses of subnanosecond rise times. Moreover, the amplitude can be measured with $\approx 1\%$ accuracy.

Digital recorders have several advantages over conventional oscilloscopes. For example, waveform data can be easily recorded, stored, retrieved, processed, compared and transmitted to other locations at great speed and are, therefore, becoming increasingly popular for measurement and recording of transient voltages or currents. For HV testing, these recording instruments may be called upon to record standard lightning and switching impulses, chopped and steep fronted impulses, as well as very fast transients generated during the operation of SF_6 and vacuum circuit breakers, testing of surge arresters, insulators and breakdown of some insulating materials. Therefore, such recorders should not only be fast enough for required applications, but should also be accurate and reliable to capture random transients. For high voltage measurement and recording applications, the requirements for digital recorders are specified in the relevant IEC and IEEE standards [1,2].

11.2.1 Recorder Parameters

In a digital recorder, an analog-to-digital converter (ADC) samples the input signal at regular but discrete time intervals, which are defined by a clock. The samples are then converted to digital codes and stored in successive memory locations. Once recorded, the contents of the memory can be transferred to a computer for analysis, display and printout or can be displayed repetitively on a slow conventional oscilloscope with the aid of a digital-to-analog converter (DAC). Thus, there are two important parameters which control the performance limits of a digital recorder. These are the sampling rate (R) and the number of ADC bits (N) [3]. The amplitude resolution commonly known as quantization error or code bin width (W) is given as:

$$W = \frac{FSV}{2^N} \qquad (11.1)$$

where FSV is the allowable full scale voltage which is usually ≈100 V. Existing high speed digital recorders typically have 8, 10 or 12 bits, resulting in quantization errors of 0.39%, 0.1% and 0.025%, respectively. The sampling rate (R) defines how many times per second the input analog signal will be converted to a digital code. According to information theory, a sinusoidal signal of frequency F can be reconstructed if R = 2F. However, for single-shot recording of voltage surges which are not bandwidth limited and which provide only one opportunity to capture the waveform, it is desirable to have as high a sampling rate as possible. Figure 11.1 shows the errors caused by slow sampling rates when measuring the amplitude of a linearly rising surge voltage. Recorders with R ranging from 20 MSa/s (or megasamples per second) to 5 GSa/s are available. R is generally related to T_m, the time to be measured (e.g., front time of an impulse) by:

$$R \geq \frac{30}{T_m} \quad (11.2)$$

A third important specification of a digital recorder is the memory length and its speed, since the memory must be able to store all the magnitude codes as fast as generated by the ADC. The readout speed of the memory may be slower.

11.2.2 Measurement Errors

As discussed above, the number of ADC bits N and the sampling rate R set the theoretical limits on the precision of amplitude and time measurements by digital recorders. Such recorders normally employ ADC technologies such as scan converters, charge coupled devices, flash converters and ribbon-beam digitizers [3,4], each of which exhibits nonideal characteristics. In such recorders, dynamic measurement errors can occur which are quantified by differential nonlinearity (DNL), integral nonlinearity (INL) and aperture uncertainty [4]. DNL is the difference between the measured code bin width and the average code bin width (W) divided by the average code bin width. For an ideal recorder, all of the code bin widths are equal. However, DNL introduces a local variation in the code bin widths. INL gives an estimate of the deviation of the recorders quantization characteristics from its ideal counterpart. Thus, INL gives difference between the corresponding points on the measured quantization characteristic and the quantization characteristic calculated using the static scale factors. Aperture uncertainty is the standard deviation of the time intervals between adjacent instants of sampling. Internally generated noise is another source

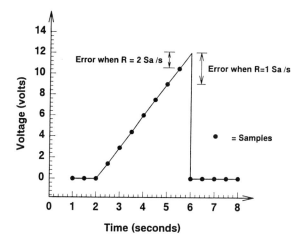

Figure 11.1 Error due to slow sampling rate of a digitizer.

of error with real digital recorders. High resolution recorders are generally more susceptible to such an error due to noise. For very accurate measurements, the recorder errors have to be quantified, and therefore, technical procedures and computer models have been proposed to assess the characteristics of digital recorders [5,6].

11.2.3 Technical Assessment

The traditional approach of testing the recorders uses sine wave inputs and is known as an equivalent or effective bit (EB) test [3]. This test provides a relation between the effective resolution of the recorder and the frequency of the input signal. In this test, pure sine wave signals are digitized and the digital record is fit with the best-fit sine wave function. The difference between the digitized value and best-fit sine function value is taken as the instantaneous error from which rms error is calculated. Then the rms error is evaluated for an ideal digitizer, i.e., a digitizer where the only error considered is due to quantization. The number of equivalent bits (EB) is calculated from the relation:

$$EB = N - \log_2 RATIO \qquad (11.3)$$

where RATIO is the rms error of the real digitizer divided by the rms error of an ideal digitizer. The shape of EB versus frequency characteristic for recorder looks somewhat similar to the frequency response of an oscilloscope as shown in Figure 11.2 [3]. The frequency response gives an exact, well-defined and fully accurate account of error introduced in the signal when recording it with a particular analog instrument. However, unlike the frequency response of an analog oscilloscope, the EB characteristic only provides statistical representation of the random errors that may be introduced by the digital recorder. Thus, when the same signal is measured repeatedly with the same digital recorder, there will be some differences between the successive records. The EB characteristics sheds some light on the possible magnitudes of such errors. On the other hand, when the same signal is measured repeatedly using the same analog oscilloscope, the records obtained will always be the same and will have the same differences from the true signal. Technical procedures and computer models have been proposed to assess the error bounds of such recorders using input signals such as ramps, etc. [5,6]. Techniques for calibration of digital recorders, dynamic testing of such recorders and methods used for evaluation of amplitude, time parameters and overshoot for impulses and the associated errors in such measurements have been discussed in literature [7,8].

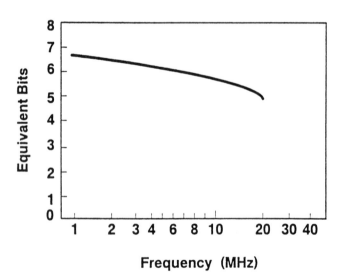

Figure 11.2 Equivalent bit characteristic for a 100 MHz, 8 bit recorder. (Data from Ref. 3.)

11.2.4 Applications

Digital recorders are used for impulse voltage and current measurements during high voltage testing of insulation. For standard lightning impulse tests, the recorders can be separated into the following two groups based on the performance requirements [8]:

1. For testing self-restoring insulation where only the evaluation of impulse amplitude is required with a resolution of 0.4% and a time uncertainty of 1%, an 8 bit recorder with a minimum sampling rate of 40 MSa/s is sufficient.
2. For tests on equipment containing non–self-restoring insulation such as transformers, records of two impulses and their resulting neutral currents are compared. For such cases full and chopped impulses including tests with front-chopped waves are sometimes required. Therefore, a 10 bit recorder (which will allow an amplitude resolution of 0.1%) with 60 MSa/s sampling rate is required.

Standard switching impulses can be accurately measured by using sampling rates of as low as 120 kSa/s with recorder of 10 bits. Digital recorders can also be used to record switching surges on power system networks [9]. The rise time of such surges can vary from 50 ns to hundreds of microseconds depending on system components. An 8 bit, 200 MSa/s digitizer can record such surges with an acceptable accuracy. A restrike during switching operations can also be recorded provided that either enough memory is available or the digitizer's clock speed is dynamically changed during the record [9].

Partial discharges having rise and fall times of a few nanoseconds can be digitally recorded if the sampling rate is either high enough (greater than 5 GSa/s) or the PD pulses are slowed down by filtering [3]. Digital recorders with high resolution (12 bits) and high sampling speeds (10 GSa/s) are now available and can be used for measurements associated with short circuit testing as well [7].

11.3 DIGITAL TECHNIQUES IN HV TESTS

In addition to the use of digital recorders for impulse and other measurements, discussed in the previous section, there are several other areas in which digital techniques, involving both hardware and software, are being applied successfully. Some of these applications and their advantages are briefly outlined here.

11.3.1 Improving Measurement Accuracy by Deconvolution

It is known from the convolution theorem that output, y(t), from a measuring system is related to its input, x(t), by the impulse response, h(t), of the system as follows:

$$y(t) = \int_{-\infty}^{\infty} x(\tau)h(t - \tau)d\tau \qquad (11.4)$$

If y(t) and h(t) were known analytic functions, deconvolution could be used to determine x(t). This allows for improving the measurement accuracy by determining the errors caused by the nonideal characteristics of voltage dividers and measuring circuitry, etc. Therefore, the method of deconvolution has been used by several researchers to improve the accuracy of measuring systems with nonideal impulse responses. However, since y(t) and h(t) can only be measured with limited resolution and may also be contaminated with noise, the deconvolution technique should be used with great care and must not be applied to compensate for an unsuitable measuring system in standard impulse measurements [7]. However, in routine measurements, deconvolution represents a useful tool to reduce errors caused by measuring system limitations.

11.3.2 Impulse Testing of Power Transformers

Impulse testing of high voltage transformers is based on the comparison of the voltage and the neutral (or ground current) transients when a standard lightning impulse is applied to the transformer's HV winding at the basic insulation level (BIL) and at a reduced voltage level (usually 50 to 75% of BIL) (for details see chapter 12). This is a nondestructive test in which it is assumed that the test and measuring circuit remains essentially the same during the reduced and the full wave tests and, therefore, any differences in voltage and current records are caused only by partial or complete insulation failure. In the conventional impulse tests, oscillographic records of transients are generally obtained by camera-equipped oscilloscopes, and in some cases it is difficult to conclude if the insulation has failed or passed the test, due to limited resolution of the oscillographic records of the transients. Using digital recorders, these quantities can be recorded in a digital format and, consequently, the differences between voltages and currents at reduced and at full voltage levels can be calculated, compared, magnified and displayed for a better judgement and evaluation of the state of the tested insulation. Thus, the use of a digital recorder in the time domain mode improves the transformer's insulation diagnostics.

In addition to the time domain applications, digital data of the applied voltage and the resulting current records can also be manipulated to determine the transfer function of the tested winding in the frequency domain. The transfer function, TF(ω), is defined as:

$$\text{TF}(\omega) = \frac{\text{I}(\omega)}{\text{V}(\omega)} \tag{11.5}$$

where V(ω) and I(ω) = input line voltage and the output neutral current functions in the frequency domain (ω). Since voltage and current values are sampled in a digital format, such data can be converted to V(ω) and I(ω) by fast fourier transform (FFT) computations with a digital processor. The state of the transformer winding's insulation is determined by comparing the transfer function obtained at full and at reduced test voltages. TF(ω) versus ω plots have been found to reveal that the local breakdowns in the winding insulation between two turns can shift the resonance frequencies in the transfer function thereby causing a shift in the transfer function poles (Figure 11.3b). However, partial discharge between two windings causes reductions in the peak value of the transfer function corresponding to some resonance frequency (Figure 11.3a). It has been further observed that small changes in the applied impulse waveform or chopping times do not significantly influence the shape of the transfer function. Thus, the transfer function method permits an unambiguous acceptance or rejection of the transformer. Therefore, along with time domain analysis, this technique is also finding increasing acceptance by utilities, transformer manufacturers and test laboratories, thus resulting in improved diagnostics of the winding insulation [7,8,10,11].

11.3.3 Partial Discharge Measurements

The occurrence of partial discharges (PDs) inside high voltage components (see also chapters 8 and 12) aids the degradation of insulation leading to its premature failure. Therefore, most power apparati have traditionally been tested for PD using a particular type of measuring circuit and conventional analog instruments [12]. The quantities usually measured involve discharge inception voltage (DIV), discharge extinction voltage (DEV) and apparent magnitude of the PD pulses. These quantities can also be determined using digital recorder as mentioned earlier in section 11.2.4.

In today's environment of inexpensive high speed ADC and microcomputers, it is possible to quantify different parameters of PD pulses with relative ease. A tremendous amount and variety of PD data are now being gathered by many researchers and new methods for the diagnostics of the

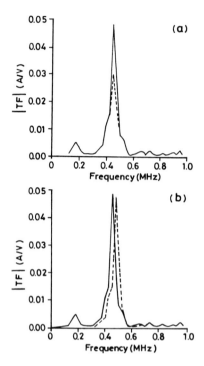

Figure 11.3 Changes in transfer function due to (a) a discharge between two windings and (b) a breakdown between two adjacent turns. (From Ref. 11 © IEEE, 1994.)

insulation integrity are being proposed. A comprehensive bibliography and a brief review of the applications of digital techniques for partial discharge measurements has been prepared by an IEEE working group [13].

Using digital techniques, pulse-height distributions, pulse-phase distributions and pulse-time distributions can be obtained to statistically quantify the partial discharge parameters [13,14]. In recent years, many advances have been made in developing sophisticated hardware and software for better quantification of partial discharges which were previously performed only partly by using multichannel analyzers [15]. Consequently, in order to assist engineers in the assessment of the quality and the condition of the insulation, digital techniques have helped to improve PD measurement sensitivity and to capture the stochastic properties of PD pulses. Some of the areas where digital techniques are providing useful data include on-line PD measurements in real-life, noisy environments through the use of digital filtering, pattern-recognition and noise-reduction techniques, PD source identification and PD site location [8,13]. In addition, digital analysis of calibration pulses for PD measurements have demonstrated that the calculated charges agree with the nominal measured values from 5 pC to 500 pC with $\pm 10\%$ error and that variations of the test circuits influence such values by less than 5% [7]. Despite all such advances, the persisting general lack of knowledge on the aging rates of various insulating materials exposed to partial discharges has largely prevented the use of computer-controlled pulse-distribution methods on a routine test basis for electrical apparatus [16].

11.4 TESTING AUTOMATION

High voltage testing is essential to ensure the insulation quality of the manufactured component. High voltage tests can be separated into three main parts (see chapter 12 for details): routine tests during the manufacturing process for quality assurance, type test as part of research and development quality assurance, and on-site tests to ensure that proper installation and commissioning steps are followed. In addition, sometimes high voltage tests are also required for periodic assessment of the insulation quality.

Routine tests, which are the most frequent, offer the greatest potential for testing automation. Computer-controlled test procedures and voltage sources lead to a reduction in staff, testing time and error probability. In addition, test results can be stored in digital form, a test report can be prepared and printed automatically, and the results can also be transmitted to a host computer which collects all the relevant data for a given customer order, so that high voltage tests can become part of the computer integrated manufacturing [8]. In automated testing, the operator can input the data for the required test and the computer will select the test program, perform the required calculations (e.g., for air density and humidity correction factors) and adjust the generator parameters. These adjustments can then be confirmed by the operator and the test sequence is completed along with the recording and printout of the test results. Computer-controlled impulse testing systems are available from high voltage test system manufacturers [17].

Other applications where computer-controlled testing and data-acquisition systems are useful include long-term testing and monitoring of insulation in multistress aging tests which are conducted to assess the suitability of new materials and to estimate the expected life of high voltage components.

11.5 ELECTRIC FIELD MEASUREMENTS

Electric field strength (E) is one of the most important parameters in the insulation design of high voltage apparati. A knowledge of E is also a prerequisite to assess the biological effects of electromagnetic radiation and the electromagnetic compatibility of electronic devices. The capacitive probe, flux meter, and dipole antenna all have been successfully used for measuring electric fields. Some of these sensors are metallic which limits their use to the field measurement at the surface of metallic electrodes only. In addition, the measured field value may be affected by the presence of the probe in the interelectrode area. In recent years, new field sensors based on electro-optic effects have been developed which overcome most of these disadvantages and such sensors will be discussed in the forthcoming sections. This section outlines the basic principle of a commercially available, capacitive probe–based, electric field measurement system [8,18].

The measuring system consists of a spherical (80 mm in diameter) field sensor, two transmission cables, a receiver and a recording instrument. The probe measures electric field components in two orthogonal directions, which are transmitted by two fiber optic cables and then converted to an electrical signal by the receiver. A capacitive probe usually consists of a sensing electrode of area A immersed in a medium of absolute permittivity ε. The probe is connected to a reference electrode through a measuring resistor R_m. It can be represented by the equivalent circuit of Figure 11.4, where C is the probe capacitance and V_m is the measuring voltage. For this circuit [8]:

$$i_o(t) = \varepsilon \int_A \frac{dE}{dt} dA = C \frac{dV_m}{dt} + \frac{V_m}{R_m} = i_c + i_R \qquad (11.6)$$

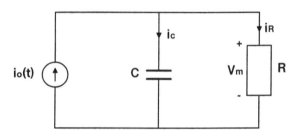

Figure 11.4 Equivalent circuit of an electric field measuring probe.

If the sensor is terminated with the characteristic impedance of the measuring cable, then $i_c \ll i_R$, and therefore:

$$V_m = R_m \varepsilon A \frac{dE}{dt} \qquad (11.7)$$

Since the measuring voltage is proportional to the derivative of the applied field, E can be measured by integrating the measuring signal V_m. On the other hand, if the sensor is terminated with a high impedance (i.e., $R_m > 1\ M\Omega$), $i_R \ll i_C$, the measuring signal is proportional to E since equation (11.6) can be expressed as:

$$V_m = \frac{\varepsilon E A}{C} \qquad (11.8)$$

Both principles are used in practical applications and several designs of such probes exist.

If the distance between a high voltage electrode and the field probe is larger than the probe diameter, electric field can be measured near the electrode with good accuracy as long as no corona occurs at the high voltage electrode. Using such a probe, high voltage can also be measured if the probe is calibrated at a fixed position. Capacitive probes are useful for measuring AC and impulse fields and have been successfully used for such purposes [8].

11.6 ELECTRO-OPTIC SENSORS

Due to rapid advances in optical telecommunications technology, practical applications of electro-optic transducers are increasing for the measurements of voltages, currents as well as electric and magnetic fields associated with high voltage power networks. Consequently, a variety of electro-optic measuring sensors have been developed. For the measurement of high voltages and electric fields, such sensors have the following advantages [19]:

1. They directly measure the electric field in space
2. They contain no electronic circuit and no power sources
3. They respond to changes in the electric field in the frequency range from DC to GHz
4. The field distribution is disturbed only marginally by the sensors itself, thereby resulting in better accuracy

Furthermore, such sensors are very small, and therefore easy to transport and install [20]. There is an increasing interest to integrate electro-optic and magneto-optic transducers for monitoring electrical quantities in power apparati such as circuit breakers, transformers, GIS, etc. Fiber optic sensors can also be used to measure temperature, pressure, vibrations, gas density, etc., and are being used for power equipment monitoring as well [8]. Electro-optic sensors for electric field intensity and voltage measurements can broadly be classified into the following two main types: (1) sensors based on the Pockels effect and (2) sensors based on the Kerr effect. A brief introduction about the principle, characteristics and applications of these sensors is provided next.

11.6.1 Electro-Optic Effect

The velocity of light depends on the refractive index of the material through which the light passes. The refractive index of some materials has different values for two mutually orthogonal polarizations of the light wave. This property, whereby a material possesses two refractive indices, is known as birefringence. In some materials, birefringence occurs when an electric field is applied across the material. The birefringence, therefore, causes the orthogonal components of polarization of the light to travel at different velocities, which causes a phase shift between the waves. This is the electro-optic effect. The refractive index n for each orthogonal component of the optical polarization is related to the applied electric field E by [19]:

$$n = n_o + aE + bE^2 + ... \qquad (11.9)$$

where n_o = normal refractive index in the absence of any applied field and a and b = coefficients for the electro-optic effects [19]. The term aE shows that n varies linearly with E, i.e., a linear electro-optic effect commonly known as the Pockels effect. However, the term bE^2 shows a quadratic electro-optic effect which is referred to as the Kerr effect. The higher-order terms including and above E^3 normally contribute very little to the refractive index changes and, therefore, can be usually ignored. Generally, in a given material, the Pockels and Kerr effects do not appear simultaneously, but one effect becomes dominant depending on the symmetry and the electrical-polarization structure of the material. The Pockels effect is expected to appear in solid dielectric materials and the Kerr effect mainly in liquid insulants [19].

11.6.2 Pockels Sensors

Consider the light wave e which is linearly polarized and therefore has two orthogonal components e_x and e_y, and propagating along the z-axis, as shown in Figure 11.5. The phase velocities of two components will vary in inverse proportion to the refractive indices n_x and n_y for each polarization component of the light. So, when birefringence occurs, $n_x \neq n_y$, and therefore the two components will travel at different velocities. After the light passes through the electro-optic material, i.e., the Pockels crystal, the two components are out of phase due to different velocities and have a phase shift $\Delta\theta$ induced by the electric field. When the Kerr effect and other higher order terms are neglected, this phase shift is given as [19]:

$$\Delta\theta = \Delta\theta_o + AE \quad (11.10)$$

where $\Delta\theta_o$ = phase shift corresponding to the natural birefringence in the absence of any electric field ($\Delta\theta_o$ is usually $\approx$ zero for most crystals) and A = Pockels cell's sensitivity coefficient, which is given as:

$$A = \left(\frac{2\pi}{\lambda}\right) n_o^3 r_p L \quad (11.11)$$

where λ = wavelength of the incident light, r_p = Pockels coefficient and

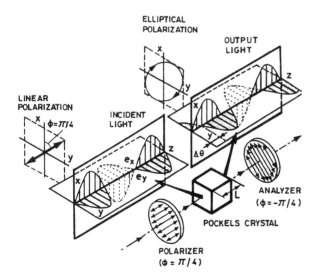

Figure 11.5 Principle of electro-optic effect. (From Ref. 19 © IEEE, 1996.)

L = length of Pockels crystal along the z-axis. The phase shift $\Delta\theta$ causes varying degrees of elliptical polarization of light, and consequently the linear polarization of the incident light results in an elliptically polarized state at the output of the Pockels crystal, as shown in Figure 11.5. If the output light passes through a polarizer oriented orthogonally to the input polarizer as in Figure 11.5, the output light intensity P_o varies nonlinearly with E. However, if an optical bias of $\pi/2$ radians is added to the original phase shift of equation (11.10), P_o is related to the input light intensity P_i by:

$$P_o = \frac{P_i}{2}(1 + \sin \Delta\theta) \qquad (11.12)$$

When $\Delta\theta$ is relatively small, $\sin \Delta\theta \approx \Delta\theta \approx \Delta\theta_o + AE \approx AE$, since $\Delta\theta_o$ is normally zero in most cases. Thus, P_o varies linearly with applied electric field E. The crystal producing Pockels effect is called a Pockels crystal, a Pockels device or a Pockels cell. Table 11.1 shows some important properties of a few Pockels crystals [19]. Crystals having a large Pockels effect (A values) are desired to sense electric fields. In addition, the Pockels crystal should have low dielectric constant, high resistivity, small piezoelectric constant and no natural birefringence in order to minimize measurement errors.

A practical Pockels sensor system consists of a coherent light source, an optical conversion device for producing circular polarization, a Pockels crystal, a polarizing plate and a photodetector. Figure 11.6 shows two typical Pockel sensor systems, whereas Figure 11.7 shows the relationship

Table 11.1 Selected Properties of Some Pockels Crystals

Material	Sensitivity coefficient A/L ($\times 10^{-5}$/V)	Relative permittivity	Resistivity (Ω cm)
$LiNbO_3$	40–100	50[a], 100[a]	10^{16}
$LiTaO_3$			
ADP, KDP	30	20[a], 50[a]	10^{10}–10^{14}
Quartz	1	4	$>10^{16}$
$Bi_{12}SiO_{20}$	40–70	50	10^{14}
$Bi_{12}GeO_{20}$			
ZnS, ZnTe	30–80	7–10	10^7–10^8
$Bi_4Ge_3O_{12}$	10	16	10^{14}

[a]These values are for two orthogonal directions.
Source: Ref. 19.

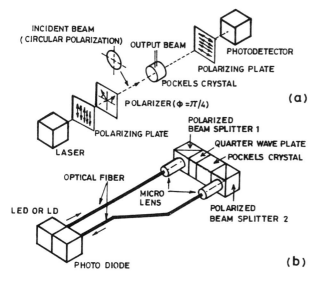

Figure 11.6 Two typical Pockels sensor systems employing (a) free space transmission of laser and (b) optical fiber sensor. (From Ref. 19 © IEEE, 1996.)

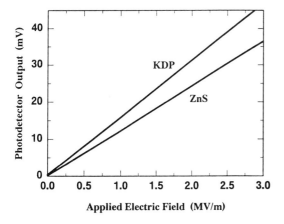

Figure 11.7 Variation of optical sensor output with applied electric field. (Data from Ref. 19.)

between optical output of the sensor and the applied electric field for two Pockels crystals [19]. A laser, a light-emitting diode (LED) or a laser diode (LD) is usually used as a source of light. All optical devices in which polarization of light is modified are assembled into a single unit. After passing through this unit, the light has an intensity that depends on the applied electric field, and thus the electric field can be measured by measuring the light intensity. Pockels sensors for measuring electric field or voltage have been developed and used by many investigators [19–23]. Some of such sensors are commercially available as well. For a Pockels sensor to be used in the measurement of voltage, the sensor should be calibrated by a standard voltage transformer or a calibrated voltage divider. The factors which should be considered in the choice of a Pockels sensor are response nonlinearities due to Kerr and higher-order effects, influence of temperature and vibrations on sensor performance, frequency response, and oscillations introduced in the output signal of the Pockels crystal by the piezoelectric effect, etc.

In many countries, Pockels voltage sensors have been incorporated into electric power networks and power apparati such as GIS. Pockels field sensors are applied to the measurement of not only the electrostatic field, but also the space charge modified fields in electrical discharges due to DC, AC, lightning impulse or switching impulse voltages. This has helped in a better physical understanding of the space charge and discharge phenomena in gas insulation [19]. The sensors are being developed for measuring field vectors and are being miniaturized with the introduction of optical waveguide technology and are showing improved performance. It is anticipated that the applications of Pockels sensors in high voltage components and insulations systems will increase.

11.6.3 Kerr Sensors

Kerr-effect–based electro-optic sensors are also used for electric field and voltage measurements and work on the same electro-optic principle as the Pockels sensor. However, in this case the phase shift $\Delta\theta$ between the two orthogonal components of a light wave after it travels through an electro-optic Kerr cell is proportional to E^2. If we assume no Pockels effect and no natural birefringence (i.e., in the absence of an applied field $\Delta\theta_o = 0$), then $\Delta\theta$ is given as:

$$\Delta\theta = 2\pi r_k L E^2 = B E^2 \qquad (11.13)$$

where r_k = Kerr constant and L = effective length of the optical path in the Kerr cell. The Kerr cell sensitivity coefficient B is equal to $2\pi r_k L$, as opposed to Pockels cell sensitivity coefficient A given by equation (11.11).

Since it is difficult to measure $\Delta\theta$ directly, it is calculated from the measured relative transmitted light intensity ratio i.e. P_o/P_i, where P_o is the transmitted light intensity and P_i the incident light intensity.

Liquid dielectrics usually exhibit Kerr effect because their molecules have different polarizabilities in at least two different perpendicular molecular axis and are, therefore, used in Kerr cells. Table 11.2 gives values of Kerr constant r_k and breakdown strength for a few dielectrics which have large Kerr constants [24]. Kerr sensors are used to measure either high voltages or to measure electric field values in stressed liquid dielectrics. For voltage measurements, application of linearly polarized light at an angle of $\pi/4$ radians to the direction of the applied electric field gives equal polarization components parallel to and perpendicular to the externally applied field. This light is elliptically polarized by the Kerr cell. After exiting from the cell, the light is passed through a second polarizer which is oriented orthogonally to the input polarizer. Then output optical intensity is given as:

$$P_o = P_i \sin^2\left(\frac{\Delta\theta}{2}\right) = P_i \sin^2\left[\pi L r_k \frac{V^2}{d^2}\right] \quad (11.14)$$

where V = applied voltage between the cell electrodes, and d = effective distance across which V is applied. This type of arrangement uses crossed polarizers. Other optical arrangements are also used for E field measurements. Equation (11.14) can also be used to measure unknown voltage V. Kerr cells employing nitrobenzene are used for high voltage measurements and as standard voltage dividers [25–27].

Kerr electro-optical transducers have also been used to measure the spatial field distribution in stressed dielectric liquids in order to understand the role of space charge in their breakdown. The measurement range of the electric field magnitudes is limited by the Kerr constant (r_k) of the liquid dielectric and the length L of the electrode for lower limit, and by the breakdown strength E_B of the dielectric medium for the upper limit. For such measurements, the applied field magnitude E should also be much higher than the field magnitude E_m necessary to reach the first light maxima to produce some bright and dark isochromatic fringes. Table 11.2 also shows E_m values for a fixed value of L for some liquid dielectrics. Figure 11.8 shows optic measurement systems using linearly and circularly polarized light [24]. For both arrangements, either cross polarizers or aligned polarizers can be used resulting in different forms of the P_o/P_i relationships [24]. In recent years, the field measurement sensitivity has been greatly improved by the use of electric field modulation and elliptically polarized incident light [28]. Moreover, the dynamic behavior of two-dimensional electric field distributions has been observed because of advanced tech-

Table 11.2 Kerr Constant r_k, Field Magnitude E_m and Breakdown Strength E_B of Some Dielectric Liquids

Liquid specimen	r_k (m/V²)	E_m [V/m] (L = 3 cm)	E_B [V/m]
Nitrobenzene	20×10^{-12}	2.3×10^6	8.5×10^6
Purified water	3.4×10^{-14}	7.7×10^6	14.0×10^6
Transformer oil	3.5×10^{-15}	55.1×10^6	20.0×10^6

Source: Ref. 24.

niques such as two-dimensional lock-in amplifier, high quality optic devices, highly sensitive charge coupled device cameras and computer image processing techniques [24,29]. A method to measure three-dimensional electric field distributions in liquid dielectrics using the Kerr effect combined with AC field modulation and circularly polarized light has been

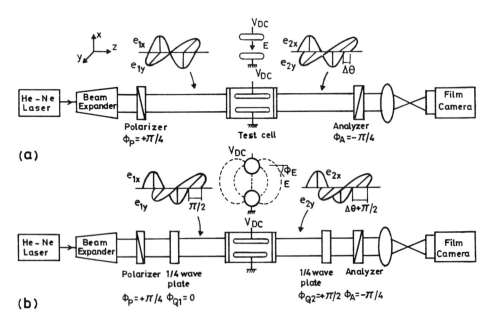

Figure 11.8 Optic measurement system for nonuniform electric field in the plane perpendicular to the light propagation: (a) using linearly polarized incident light and (b) using circulary polarized incident light. (From Ref. 24 © IEEE, 1996.)

discussed by Shimizu et al. [30]. The applications of Kerr effect for diagnostics of liquid dielectrics are expected to grow in the future.

11.7 MAGNETO-OPTIC SENSORS

Magneto-optic sensors are used for measuring current and magnetic field intensity in high voltage power networks. Current measurement is required during high power testing in addition to its use for metering and relaying purposes. Magnetic field measurements are also essential because of increasing concern about the possible health hazards of power frequency electro-magnetic fields. Conventional methods of current measurements employing bulky current transformers are being questioned by a large number of electric power utilities due to cost, reliability and safety considerations. Consequently, interest is growing in the development and applications of integrated electro-optic and magneto-optic sensors. Such integrated electro- as well as magneto-optic sensors offer several advantages over their conventional counterparts, i.e., potential and current transformers (PTs and CTs). For example integrated optic sensors are nonintrusive, inherently insulating, corrosion resistant, compact, lightweight, flexible and immune to electromagnetic interference. The output signals of these sensors are compatible with the signal levels required for digital protection systems. Therefore, such magneto-optic transducers are briefly described next.

11.7.1 Faraday Effect

The magnetic field and current measuring optical sensors are usually based on magneto-optic effect which was first observed by Michael Faraday 150 years ago and is commonly known as the Faraday effect. According to this effect, when a linearly polarized light ray propagates through a magneto-optic medium in the presence of an external magnetic field, it undergoes a rotation of the plane of polarization proportional to the strength of the magnetic field (H). The rotation angle $\Delta\theta$ (radians) is related to the magnetic field intensity H (ampere-turn/m) and interaction length L (m) by the following equation:

$$\Delta\theta = \mu K_v \int \vec{H} \cdot \vec{dL} \quad (11.15)$$

where K_v = Verdet constant of the material (radians/ampere-turn), μ = relative permeability, and $\vec{H}$ and $\vec{dL}$ = components in the direction of light propagation. The Verdet constant K_v varies with light wavelength and is material as well as temperature dependent. When a linearly polarized light encircles a current carrying conductor, by using Ampere's law, equation (11.15) can be rewritten as:

$$\Delta\theta = \mu K_v N_t I \tag{11.16}$$

where N_t = number of turns of the optical path around the current carrying conductor. Thus, from equation (11.16), the angle of rotation $\Delta\theta$ is directly proportional to the enclosed electric current I. This method rejects the magnetic field signals due to external currents which are also quite strong in power systems. Equation (11.15) can also form the basis of magnetic field measurements using magneto-optical effects. Table 11.3 shows Verdet constant values for a few magneto-optic materials.

Usually Faraday crystals show sensitivity to ambient temperature since the Verdet constant K_v is sensitive to the crystal temperature. In addition, some of the magneto-optic materials also exhibit electro-optic (Pockels) effect. Therefore, in the choice of a suitable Faraday material, these factors have to be considered. Usually materials with large K_v constant but having less sensitivity to the temperature and smaller r_p (Pockels coefficient) coefficients are preferable. Thus, several factors such as frequency response, temperature effect, effect of noise and vibrations on the sensors, and errors caused by neighboring conductors on the measured values are some of the important considerations in the selection and use of magneto-optic transducers [22,23,31,32].

11.7.2 Magnetic Field Sensors

A practical magnetic field sensor will have polarized light going through a Faraday cell which is subjected to the magnetic field to be measured. The optical output is then passed on to an analyzer. A linear relation between sensor output voltage and the applied magnetic field can be produced

Table 11.3 Verdet Constant K_v for Some Magneto-Optic Materials

Material	K_v (rad/A)
Optical glass (SF-57)	3.32×10^{-5}
Bismuth silicate (BSO) $Bi_{12}SiO_{20}$	3.66×10^{-5}
ZnSe	5.48×10^{-5}
Terbium gallium garnet (TGG)	8.2×10^{-5}
$Y_3Fe_5O_{12}$ (YIG)	3×10^{-3}

as clear from equation (11.15). Figure 11.9 shows the structure of a practical magnetic field sensor which utilizes two Faraday crystals, i.e., a BSO crystal with a Verdet constant of K_{v1} and a ZnSe crystal with a Verdet constant of K_{v2} [22]. The overall rotation angle $\Delta\theta$ in this case will be given by:

$$\Delta\theta = K_{v1}L_1H_1 + K_{v2}L_2H_2 \qquad (11.17)$$

where L_1, L_2 = lengths of the two crystals and H_1 and H_2 = associated magnetic field values. This sensor provides a linear output with applied magnetic field over the range of $0 \leq H \leq 8$ kA/m with a sensor output voltage of ≈ 0.6 V per 1 kA/m of the applied magnetic field. Since BSO crystal is also a Pockels sensor, such an arrangement can lead to development of an integrated voltage and current measurement system.

11.7.3 Optical Current Transducers

Optical current transducers (OCTs) are finding extensive applications in power systems because of their several advantages over the conventional current transformers (CTs). There is a large variety in the available OCT designs, ranging from conventional CT with optical readout to the all-optical sensors. The IEEE Working Groups on Emerging Technologies and Fiber Optic Sensors [33] provide a comprehensive review about the various available instruments and their experience with their use in the power system. It has also provided a list of key references in OCT technology.

There is a growing interest to develop a single optical transducer for both voltage and current measurements in high voltage networks [22,23]. In future, it is expected that such transducers will be used extensively in key high voltage components for monitoring and diagnostic purposes.

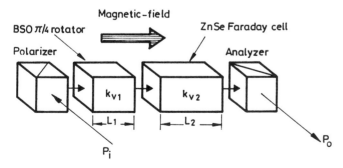

Figure 11.9 Structure of a magnetic field sensor. (From Ref. 22 © IEEE, 1987.)

11.8 MEASUREMENTS OF VERY FAST TRANSIENTS IN GIS

Electrical transients in gas-insulated switchgear (GIS) result from either breakdown or partial discharge. Phase-to-ground or phase-to-phase breakdown can occur during the operation of disconnect switches or circuit breakers. The breakdown generated transients have rise times of the orders of 5–10 ns. Due to the coaxial construction and low loss attenuation characteristic of a GIS (10 dB/km at 1 GHz), these internal surges can propagate around the GIS installation with very little attenuation [34]. Hence, various reflections of the transients can occur which can lead to overvoltages. In addition, at the station terminals, the transients can couple into the GIS enclosure ground system, secondary wiring as well as the control circuitry. Therefore, there is a strong interest to measure and investigate such transients. Partial discharges in GIS can result from a number of sources such as free conducting particles, solid dielectric components, floating components and conductors with surface defects, all of which generate transients with ≤1 ns rise times. Since partial discharge induced transients are indicative of incipient faults, their early detection and location is of considerable importance.

For the measurement of voltage induced transients, external voltage dividers are usually inadequate and specially designed probes have to be included in the metal enclosure of the GIS. Such a probe is normally a capacitive voltage divider which is placed close to the point where the transient voltage is to be measured, since the voltage may differ along the length of the GIS due to propagation effects [8]. The probe normally consists of a plane metal electrode which is placed close to, but is insulated from, the grounded enclosure. The stray capacitance of ≈1 pF between the line conductor and the probe forms the high voltage arm of a voltage divider. The low voltage arm of the divider consists of some tens of nanofarads capacitance arranged coaxially between the probe and the enclosure and can be obtained by different methods. The probe is placed in a region of low electric stress to maintain the dielectric integrity of the GIS and is connected to the recording instrument by a coaxial cable matched at both ends. The cable length is kept as small as possible [8]. It is properly calibrated to ensure divider ratio is accurately known at high frequencies as well [35].

For partial discharge measurement, coaxial probes or coaxial capacitive voltage dividers, with flat response from 10 Hz to more than 1 GHz, have been successfully designed and built with a wide dynamic range (divider ratios ranging from 3:1 to 2000:1) in order to investigate various sources of partial discharges in GIS [36,37]. Such dividers can also be used to record breakdown induced transients of modest magnitudes.

11.9 SPACE CHARGE MEASUREMENT TECHNIQUES

Space charge is formed in stressed dielectrics and plays an important role in the breakdown and DC leakage current behavior in polymers. Furthermore, the formation of space charge is believed to have a strong influence on electrical trees initiation under various conditions such as DC voltage application, polarity reversal and short circuits. The formation and accumulation of space charge in XLPE power cables may distort the electrical stress distribution in the insulation, thereby causing excessive stresses that may result in breakdown under some adverse conditions. Therefore, the effect of space charge must be considered in the design and testing of high voltage cables and other solid insulation. For this reason, many efforts have been made during the last decade to develop nondestructive techniques of space charge measurements. The aim of such measurements is to understand the processes that take place in the dielectric under study and make it possible to select materials and interfaces that minimize the risk of breakdown in HV applications, thereby improving the performance and reliability of the insulation system.

Previously, thermally stimulated currents (TSCs) were measured to gain an insight into the space charge accumulation in dielectrics. The TSC technique involves the polarization of the dielectric by a static field at a high temperature. The sample is then cooled quickly to a much lower temperature in the presence of the external field. The external field is then removed, the electrodes of the sample are short circuited to allow it to discharge isothermally for a suitable period of time, and then the sample is warmed at a constant rate. A current corresponding to dipole relaxation is recorded as a function of temperature [38]. A plot of this current with temperature is the thermogram which may contain current peaks related to molecular relaxation process and delocalization of charges injected in the dielectric during the polarization process (Figure 11.10). The TSC spectra thus obtained is then used to estimate the space charge distribution. Though useful, TSC method is not sufficient to determine space charge distribution fully.

To overcome the limitations of the TSC method, pulse methods for space charge measurements were developed. Pulse methods of measuring space charge are useful to find the spatial and temporal variation of space charge in a stressed dielectric and have seen significant advances in recent years. These methods can be broadly classified as follows [39]:

1. Pressure wave propagation (PWP) methods. The methods based on this principle include piezoelectrically induced pressure step (PIPS) method and laser induced pressure pulse (LIPP) method

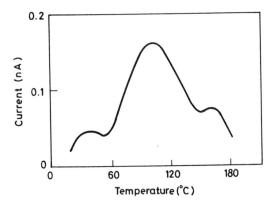

Figure 11.10 Thermally stimulated discharge current thermogram.

2. Pulsed electro-acoustic (PEA) method
3. Laser intensity modulation method (LIMM)
4. Thermal pulse (TP) and thermal step pulse (TSP) methods

Some of these techniques will be briefly discussed next.

11.9.1 Piezoelectric Transducers

Piezoelectric transducers (or electrets) used for space charge measurements should have the piezoelectric characteristics such that when exposed to an electric field, acoustic waves are generated in the transducers; by contrast, when the pressure is applied, surface charges are induced on the transducer. Thus, when an electric field e(t) is applied to the transducer of Figure 11.11a, the resulting acoustic wave p(t) is generated which is given as:

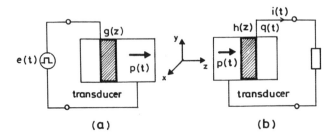

Figure 11.11 Piezoelectric transducers.

$$p(t) = K \int_0^t g(z = v\tau)e(t - \tau)d\tau \qquad (11.18)$$

where g(z) = electrostriction constant of the transducer, v = acoustic velocity of the pulse and K = a constant. When a pulsed acoustic wave p(t) is propagated through the transducer of Figure 11.11b, surface charges q(t) are induced in the transducer which are given as:

$$q(t) = K \int_0^t h(z = v\tau)p(t - \tau)d\tau \qquad (11.19)$$

where h(z) = piezoelectric constant of the transducer, i.e., transfer function of the transducer.

11.9.2 Charged Dielectrics

Polymer insulating materials possess piezo strain and stress characteristics, though only with very small values. However, when a space charge is accumulated in polymeric materials, the piezo strain and stress characteristics of the charged material becomes large, as in an electret transducer. Therefore, when a dielectric with trapped space charge $\rho(z)$ is exposed to an electric field e(t) in Figure 11.12a, the acoustic pulse p(t) thus generated becomes [40]:

$$p(t) = K \int_0^t \rho(z = v\tau)e(t - \tau)d\tau \qquad (11.20)$$

where v = acoustic velocity of the pulse p(t).

Similarly, when a charged dielectric is exposed to an acoustic pulse p(t), the position where the space charge is accumulated is compressed, and a current i(t) flows through the external circuit of Figure 11.12b. This current is given by [40]:

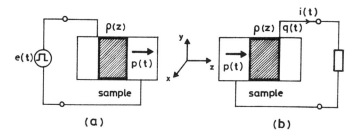

Figure 11.12 Piezoelectric effects in charged dielectrics.

$$i(t) = K \int_0^t \rho(z = v\tau)p(t - \tau)d\tau \qquad (11.21)$$

If the duration of the acoustic wave is much narrower than the travel time of the acoustic wave through the sample, i(t) is proportional to the accumulated space charge. Thus, the current waveform shows the space charge distribution.

11.9.3 Pressure Wave Propagation Technique

In this method, a pressure pulse p(t) is impressed upon the charged dielectric, and the resulting current i(t) is measured as in Figure 11.12b. The techniques used for generation of p(t) differ between the PIPS and the LIPP methods. In the PIPS method, the pressure pulse is generated by stressing a piezoelectric transducer with a narrow voltage pulse, as shown in Figure 11.11a. However, in the LIPP method, a laser pulse of subnanosecond duration is irradiated to the target electrode of the dielectric sample. The evaporation of the electrode material under laser irradiation generates a pulsed acoustic wave which is used for charge measurements.

Using the PIPS method, the space charge distributions in dielectric films of 50 ~ 200 μm can be directly obtained if the duration of the applied voltage pulse is about 1.25 ~ 5 ns. Thus, nanosecond-duration high voltage pulses are required for such measurements and circuits have been built to generate such pulses [39,40] (see also chapter 10). The LIPP method is applicable to both thin (10–100 μm) and thick (1 to ~20 mm) specimens. The time scale of the space charge measurements is related to the acoustic wave transit time across the specimen. The duration of the pressure wave (or that of the laser pulse) determines the resolution of spatial charge distribution. The main drawbacks of the LIPP method are the size and the cost of the laser system required for pressure pulse generation.

11.9.4 Pulsed Electroacoustic Method

The pulsed electroacoustic (PEA) method has been used by many researchers in recent years and involves the applications of a pulsed electric field across the dielectric with accumulated space charge. This causes a pulsed force that generates a pulsed acoustic wave that is detected by using a piezoelectric transducer. The induced charge q(t) on the surface of the transducer is proportional to the space charge distribution in the dielectric provided the duration of applied pulsed electric field Δt is much shorter than the propagation time of the acoustic wave through the sample.

The space resolution, Δl, in the PEA method is given as:

$$\Delta l = v\Delta t \qquad (11.22)$$

where Δt = duration of the externally applied field, and v = velocity of propagation of the acoustic wave in the sample and is ≈ 2000 m/s for low density polyethylene (LDPE). For example, if a voltage pulse of 30 ns duration is used in LDPE, the theoretical resolution of space charge measurements is ~60 μm. In practice, since the acoustic wave broadens as it propagates towards the transducer, the measurement resolution is ~100 μm [41]. Thus, for better resolution a smaller value of Δt is required. Furthermore, in order to obtain q(t), which is linearly proportional to p(t), the propagation time Δt for the detection transducer should also be much shorter than the wave propagation time through the sample. The acoustic wave can be detected using a polyvinylidene fluoride film (PVDF) transducer having 10 ~ 30 μm thickness.

The PEA method has been used to study the formation of space charges in dielectric films as well as in coaxial cable samples [39–42]. Figure 11.13 shows an example of output signal waveform showing spatial charge distribution in LDPE obtained by the PEA method [40]. The positive space charge is distributed at the interfaces of the electrodes and the sample film (thickness is 600 μm). These results were obtained from a gamma-ray–irradiated sample after a DC voltage of 15 kV was applied. As the negative charge is swept out, only the positive charge is left in the sample. A 1 kV pulse voltage of 15 ns duration was also applied and the output signal includes both the surfaces charges induced by the applied DC voltage and the space charges [40]. Figure 11.14 shows the effect of polarity of DC voltage on the space charge distribution in XLPE cable

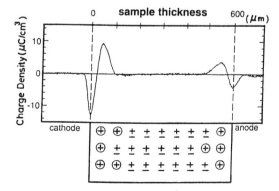

Figure 11.13 Signal waveform showing spatial charge distribution in low density PE. (From Ref. 40 © IEEE, 1994.)

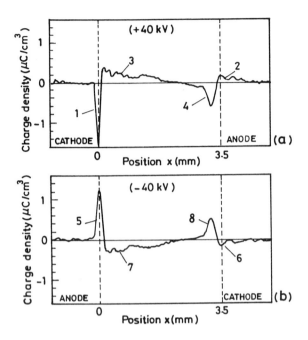

Figure 11.14 Polarity effect of DC voltage on space charge accumulation in XLPE cable insulation. (From Ref. 41 © IEEE, 1995.)

sample [41]. In Figure 11.14a, 1 is the negative surface charge induced on the cathode, 2 is positive surface charge induced on the anode, whereas 3 and 4 are positive and negative space charges accumulating in the XLPE. When the voltage polarity is reversed, all these charges reverse polarities as shown by the numbers 5, 6, 7 and 8 (marked on Figure 11.14b).

11.9.5 Other Techniques

Other techniques such as the laser intensity modulation method (LIMM) or thermal methods (TP and TSP methods) can also be used to measure space charge or electric field distribution in polarized dielectrics. TP and TSP methods utilize the thermal expansion induced by illumination of a laser or a flash lamp of one of the electrode surfaces. This expansion moves through the dielectric specimen and induces a charge signal which is measured to determine the field distribution in the sample, the charges induced on the electrodes and the total space charge [39]. The signals for thermal methods are inherently slower than the acoustic-based method discussed earlier and, therefore, such methods are not suitable for studying transient phenomena. Furthermore, these methods have less spatial resolution.

LIMM [43] is a useful method that has been successfully employed to determine the spatial distribution of polarization and space charges in PVDF electret and feroelectric ceramics but is more complicated as compared to the PEA method.

11.10 ELECTRO-OPTICAL IMAGING TECHNIQUES

Photonic emissions associated with the initiation and propagation of electrical discharges are often used to investigate the breakdown and prebreakdown mechanisms in various dielectrics. The electro-optical devices commonly used for such applications include photomultipliers, image intensifiers, and high speed image converter and streak cameras. The use of such advanced technologies in discharge studies have improved our understanding of the breakdown phenomena in long air gaps, compressed gases, vacuum and liquid dielectrics. Laser-Schleiren and Shadowgraphic techniques (see chapter 5) along with image intensifiers and high speed cameras have been used to understand the development of streamers in liquid dielectrics. In gaseous and vacuum dielectrics, use of such electro-optic devices is helping researchers to investigate fast breakdown and flashover processes. Some recent examples of the uses of such electro-optic imaging techniques are provided in the literature [44–48]. Figure 11.15 shows an experimental arrangement for photography of fast (ns) discharges with speeds of $\sim 10^8$ cm/s on spacer surfaces in gases [46]. An example of the use of photomultipliers in the early detection of discharges in SF_6 insulated equipment is provided in Binns et al. [47]. Electro-optical and commercial imaging equipment (camera, video recorders and monitors) have been applied to study the growth of electrical trees in XLPE under mechanical stress and strain [48]. A brief introduction to some of the electro-optical imaging devices commonly used in the investigations of dielectrics is provided next.

11.10.1 Photomultipliers

The photomultiplier is a very sensitive detector of radiant energy in the ultraviolet, visible and near infrared regions of the electromagnetic spectrum. Figure 11.16 shows a schematic representation of a typical photomultiplier tube (PMT) and its operation. The basic radiation sensor is the photocathode, which is located inside a vacuum envelope. On the incidence of light at photocathode, the photo-electric emission occurs at the photo-

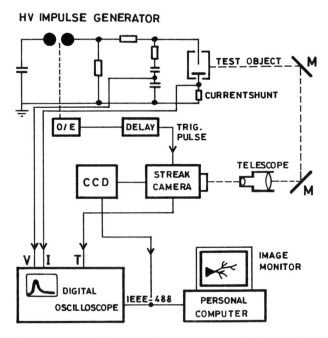

Figure 11.15 Schematic diagram for streak photography of spacer flashover.

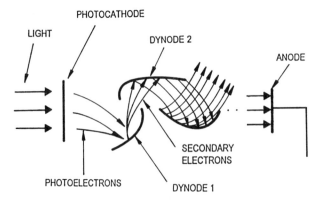

Figure 11.16 Schematic representation of a photomultiplier tube.

cathode resulting in the production of photoelectrons. These photoelectrons are directed by an electric field to an electrode or dynode within the PMT envelope. Dynode provides current amplification by emitting a number of secondary electrons for each impinging primary photoelectron. These secondary electrons, in turn, are directed to a second dynode and so on, until a final gain of $\sim 10^6$ is achieved. The electrons from the last dynode are collected by an anode that provides the current signal which is then measured. Amplification ranging from 10^3 to 10^8 with extremely fast time response, with rise times as short as a fraction of a nanosecond, makes photomultiplier a useful detector. Important parameters of a PMT are its spectral response, luminous responsivity (given in mA/lumen), gain, rise time, maximum average anode current and anode dark current. For applications in the studies of breakdown and prebreakdown events, a broad spectral response and a fast rise time (≤ 1 ns) are desirable. Photomultipliers are generally most suitable for detecting photon emission associated with corona and prebreakdown types of partial discharges.

11.10.2 Image Intensifiers

An image intensifier directly intensifies electronically, images at extremely low light levels. It is made up of a photocathode, an electron lens, microchannel plates (MCP) and a phosphor screen. When an optical image is focused onto the photocathode (Figure 11.17), it emits electrons in accordance with the intensity of the input optical image, thereby converting the input optical image into an *electronic image*. The electronic image is then focused on the MCP where it is intensified, and then strikes the phosphor screen of the intensifier where an intensified optical image is reproduced. The result of this process is $\sim 10^4$ to 10^5 intensification of the original incident light. The MCP normally used acts as a secondary electron

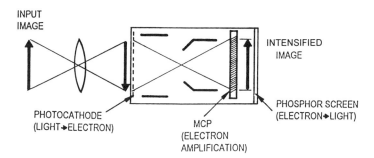

Figure 11.17 Operation of an image intensifier.

multiplier and consists of an array of millions of glass capillaries fused into the form of a disk. When an electron enters and hits the capillary wall, secondary electrons are produced from the wall. These electrons are then accelerated by an electric field and strike the opposite wall to produce additional electrons, thereby providing the overall image amplification. Besides gain, spectral response and photocathode luminous sensitivity, the other parameters that influence the performance of an image intensifier are phosphor screen material, which influences phosphor's spectral emission and decay characteristics, image magnification, limiting resolution and distortion. For investigations of fast electrical discharge phenomena or to image weak intensity discharges, image intensifiers play an important role.

11.10.3 Image Converters

An image converter is a device for converting an image formed by infrared or ultraviolet radiation, which are invisible to the human eye, into a visible light image. It employs a photocathode that is sensitive to infrared or ultraviolet radiation, and a phosphor screen that emits visible light. The spectral response of the photocathode determines whether the image converter is suitable for infrared or for ultraviolet radiation. If the photocathode has a broad frequency spectral response, it may image cumulative effects of ultraviolet, visible as well as infrared radiations. Image converters can be used for a wide variety of diagnostics such as monitoring of power equipment to locate hot spots, detecting ultraviolet radiations and studying gas discharges.

11.10.4 High Speed Cameras

The streak camera is a device that measures ultrafast light phenomena and delivers light intensity variation with time as well as with position and is extremely useful for the investigation of discharge phenomena. Thus when high speed cameras are used in combination with proper optics, it is possible to measure time variation of the incident light with respect to the position of light, thereby enabling the measurements of leader and streamer discharge speeds in liquids, gases or vacuum insulants. Such cameras usually offer streak and multiframing functions at high framing speeds.

In the streak mode, pictures are possible with a spatial resolution of ~ 0.5 ns/15 mm. In multiframing mode, $\sim 10^7$ frames per second allow investigations of discharges with temporal resolution of the order of some tens of nanoseconds. Both streaking as well as multiframing modes are used in insulation related studies.

In recent years there has been increasing tendency to use high speed cameras in combination with charged coupled device (CCD) cameras along with digital video and personal computers, as shown in Figure 11.15, which allow digital image storage and retrieval. The use of digital imaging techniques is expected to grow in the coming years for monitoring as well as investigations of dielectrics.

REFERENCES

1. IEC Publication 1083-1, "Digital recorder measurements in high voltage impulse tests, Part 1: Requirements for digital recorders," 1991.
2. IEEE/ANSI Standard 1122-1987, "Digital recorders for high voltage impulse measurements," 1987.
3. G. C. Stone and J. Kuffel, IEEE Elect. Insul. Magazine, Vol. 5, No. 3, pp. 9–17, 1989.
4. T. R. McComb, J. Kuffel and R. Malewski, IEEE Trans. on PWRD, Vol. 2, No. 3, pp. 661–670, 1987.
5. T. R. McComb, J. Kuffel, R. Malewski and K. Schon, IEEE Trans. on PWRD, Vol. 5, No. 3, pp. 1256–1265, 1990.
6. J. Kuffel, R. Malewski and R. G. van Heeswijk, IEEE Trans. on PWRD, Vol. 6, No. 2, pp. 507–515, 1991.
7. T. R. McComb, C. Fenimore, E. Gockenbach, J. Kuffel, R. Malewski, K. Schon, L. Van der Sluis, B. Ward and Y. X. Zhang, IEEE Trans. on PWRD, Vol. 7, No. 4, pp. 1800–1804, 1992.
8. H. M. Ryan (ed.), *High voltage engineering and testing*, Peter Peregrinus Ltd., London, England, 1994.
9. B. A. Lloyd, B. K. Gupta, S. R. Compbell and D. K. Sharma, IEEE Trans. on PWRD, Vol. 4, No. 2, pp. 932–936, 1989.
10. R. Malewski and B. Poulin, IEEE Trans. on PWRD, Vol. 3, No. 2, pp. 476–489, 1988.
11. E. Hanique, IEEE Trans. on PWRD, Vol. 9, No. 3, pp. 1261–1266, 1994.
12. F. H. Kreuger, *Partial discharge detection in high-voltage equipment*, Butterworths & Co. Publishers, London, England, 1989.
13. B. H. Ward, IEEE Trans. on PWRD, Vol. 7, No. 2, pp. 469–479, 1992.
14. R. J. Van Brunt, IEEE Trans. on Elect. Insul., Vol. 26, No. 5, pp. 902–948, 1991.
15. N. H. Malik and A. A. Al-Arainy, IEEE Trans. on Elect. Insul., Vol. 22, No. 6, pp. 825–829, 1987.
16. R. Bartnikas, Discussion published in IEEE Trans. on PWRD, Vol. 7, No. 2, pp. 477–478, 1992.
17. "Computer aided control system for impulse test plant," Bulletin E137, Haefely, Basel, Switzerland.
18. "Field measuring systems," Bulletin E142, Haefely, Basel, Switzerland.

19. K. Hidaka, IEEE Elect. Insul. Magazine, Vol. 12, No. 1, pp. 17–28, 1996.
20. N. A. Jaeger and F. Rahmatian, IEEE Trans. on PWRD, Vol. 10, No. 1, pp. 127–134, 1995.
21. S. J. Huang and D. C. Erickson, IEEE Trans. on PWRD, Vol. 4, No. 3, pp. 1579–1585, 1989.
22. T. Metsui, K. Hosoe, H. Usami and S. Miyamoto, IEEE Trans. on PWRD, Vol. 2, No. 1, pp. 87–93, 1987.
23. A. Cruden, Z. J. Richardson, J. R. McDonald and I. Andonovic, IEEE Trans. on PWRD, Vol. 10, No. 3, pp. 1217–1223, 1995.
24. T. Takada, Y. Zhu and T. Maeno, IEEE Elect. Insul. Magazine, Vol. 12, No. 2, pp. 8–20, 1996.
25. T. J. Englert, B. H. Chowdhury and E. Grigsby, IEEE Trans. on PWRD, Vol. 6, No. 3, pp. 979–985, 1991.
26. T. R. McComb, F. A. Chagas, R. C. Hughes, G. Rizzi and K. Schon, Electra, No. 161, pp. 105–119, 1995.
27. R. E. Hebner, R. A. Malewski and E. C. Cassidy, Proc. of IEEE, Vol. 65, No. 11, pp. 1524–1548, 1977.
28. T. Maeno and T. Takada, IEEE Trans. on Elect. Insul., Vol. 22, No. 4, pp. 503–508, 1987.
29. M. Zahn, IEEE Trans. on Dielectrics and Elect. Insul., Vol. 1, No. 2, pp. 235–246, 1994.
30. R. Shimizu, M. Matsuoka, K. Kato, N. Hayakawa, M. Hikita and H. Okubo, IEEE Trans. on Dielectrics and Elect. Insul., Vol. 3, No. 2, pp. 191–196, 1996.
31. J. Song, P. G. McLaren, D. J. Thomson and R. L. Middleton, IEEE Trans. on PWRD, Vol. 10, No. 4, pp. 1764–1771, 1995.
32. Y. Yamagata, T. Oshi, H. Katsukawa, S. Kato and Y. Sakurai, IEEE Trans. on PWRD, Vol. 8, No. 3, pp. 866–873, 1993.
33. IEEE Working Groups on Emerging Technologies and Fiber Optic Sensors, IEEE Trans. on PWRD, Vol. 9, No. 4, pp. 1778–1788, 1994.
34. S. M. Ghufran Ali and W. D. Goodwin, IEE Power Engineering Journal, Vol. 2, No. 1, pp. 17–26, 1988.
35. "Monograph on GIS very fast transients," CIGRE, Paris, France, 1989.
36. S. A. Boggs, G. L. Ford and R. C. Madge, "Measurements of transients potentials in coaxial transmission lines using coaxial dividers," in *Measurement of electrical quantities in pulse power systems*, NBS, Washington, DC, 1981.
37. S. A. Boggs, G. L. Ford, and R. C. Madge, IEEE PES Winter Meeting Paper 81 WM 139-5, IEEE, New York, 1981.
38. C. Lavergene and C. Lacabanne, IEEE Elect. Insul. Mag., Vol. 9, No. 5, pp. 5–21, 1993.
39. T. Mizutani, IEEE Trans. on Dielectrics and Elect. Insul., Vol. 1, No. 5, pp. 923–933, 1994.
40. Y. Li and T. Takada, IEEE Elect. Insul. Mag., Vol. 10, No. 5, pp. 16–28, 1994.

41. X. Wang, D. Tu, Y. Yanaka, T. Muronaka, T. Takada, C. Shinoda and T. Hashizumi, IEEE Trans. on Dielectrics and Elect. Insul., Vol. 2, No. 3, pp. 467–474, 1995.
42. Y. Li, M. Yasuda and T. Takada, IEEE Trans. on Dielectrics and Elect. Insul., Vol. 1, No. 2, pp. 188–195, 1994.
43. D. K. Das Gupta and J. S. Hornsby, IEEE Trans. on Elect. Insul., Vol. 26, No. 1, pp. 63–68, 1991.
44. B. M. Coaker, N. S. Xu, R. V. Latham and F. J. Jones, IEEE Trans. on Dielectrics and Elect. Insul., Vol. 2, No. 2, pp. 210–217, 1995.
45. O. Linhjell, L. Lundgaard and G. Berg, IEEE Trans. on Dielectrics and Elect. Insul., Vol. 1, No. 3, pp. 447–458, 1994.
46. S. Y. Li, K. D. Srivastava and G. D. Theophilus, IEEE Trans. on Dielectrics and Elect. Insul., Vol. 2, No. 1, pp. 114–120, 1995.
47. D. F. Binns, A. H. Mufti and N. H. Malik, IEEE Trans. on Elect. Insul., Vol. 25, No. 2, pp. 405–414, 1990.
48. E. David, J. L. Parpal and J. P. Crine, IEEE Trans. on Dielectrics and Elect. Insul., Vol. 3, No. 2, pp. 248–257, 1996.

12
Insulation Testing

12.1 OBJECTIVES OF TESTING

Electrical insulating materials are used in various forms to provide insulation for high voltage power networks. These materials should possess good insulating properties over a wide range of operating parameters since the reliability of the system depends on their insulation quality. High voltage tests are applied to determine the ability of the insulation to meet its design requirements. High voltage tests can be destructive or nondestructive. The destructive tests are used to measure the dielectric strength of insulation and are made on sample pieces of the material or apparatus. Nondestructive tests are carried out to ensure that the quality of insulation is adequate for the required service conditions. In this case, the insulation is not exposed to excessive test voltages. The customer views the testing as an examination of the ability of the system or apparatus to meet the guaranteed specifications of dielectric properties. Through the insulation testing, a manufacturer wants to check the design and to ensure the suitability of the selected materials. A researcher may use these tests and others to develop better insulating materials and superior equipment design.

Normal operating voltages overstress the system insulation only under special circumstances, such as contamination of external insulation. Power system equipment has to withstand its operating voltage in addition to occasional transient overvoltages. These transients may have external or internal origins. External overvoltages are associated with lightning discharges and are not dependent on the system's operating voltage level. As a result, the importance of stresses produced by lightning decreases as the operating voltage increases. Internal overvoltages are generated by switching operations and due to occurrence of faults on the power system network. These are called switching surges, and their magnitude depend on the system rated voltage. Basic insulation level (BIL) and basic switching level (BSL) depend on the levels of lightning and switching surges, respectively. The designed values of BIL and BSL, which depend on the system operating voltage, determine the insulation dimensions of the equipment.

In the insulation design, the areas of specific importance are (1) determination of the voltage stresses that the insulation must withstand and (2) determination of the response of the insulation when subjected to such voltages. The relationship between the electric stresses imposed on the insulation and the dielectric strength forms the framework of insulation coordination. Another requirement for the satisfactory performance of insulation is the minimization of internal voids, as their presence may lead to internal discharges and an ultimate breakdown. Partial discharge (PD) tests are carried out to help in a proper design and selection of void-free insulating components for power system applications.

Sometime tests are also required to ensure the safety of operating personnel in areas where a buildup of hazardous high voltages is possible. Potential examples of such cases are found in oil industry and as streaming electrification in oil-filled transformers.

12.2 HV TEST CLASSIFICATION

There are several ways of classifying the high voltage tests. The tests may be destructive or nondestructive. They may be named in association with type of the HV used, i.e., alternating voltage, direct voltage or impulse voltage tests. These may be further classified according to the type of measurement, such as a volt-time characteristic. Some of the tests are required during the production of insulating materials or equipment, whereas the others are carried out after equipment production and/or installation. These include the routine, type and special tests. Figure 12.1 shows a schematic diagram of the main HV test classifications.

Destructive testing is associated with non–self-restoring solid insulating media. The nondestructive tests can be performed to find the insulation quality of any material, even after it forms part of an equipment. Such tests are mainly carried out to assess the electrical properties, such as DC resistivity, the dielectric constant and loss factor over a wide range of operating parameters. The quality of insulation is usually assessed by mea-

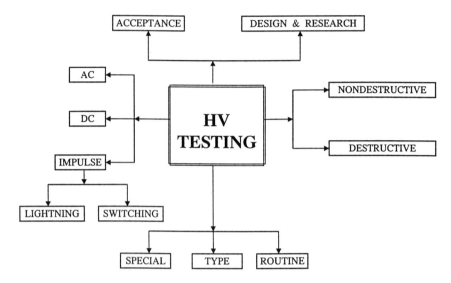

Figure 12.1 Main high voltage (HV) test classifications.

suring the loss factor at high voltages and also by conducting partial discharge tests to detect any deterioration or internal faults in the insulation of the apparatus.

Standard tests can be classified as follows:

1. Routine test: This test is carried out by the manufacturer on every unit produced. It can be part of the manufacturing process. An example is the power frequency overvoltage test for insulation, i.e., 1 min power frequency voltage withstand test.
2. Type test: This test is performed by the manufacturer for a few samples to check if the new product meets certain specifications. Users may ask for such a test on selected samples. An example is the lightning impulse voltage test on oil-filled distribution transformers.
3. Special test: As the name implies, this is a special test which is performed to check insulation behavior under certain extreme conditions, e.g., "measurement of acoustic sound level" for oil-filled distribution transformer.

The range of high voltage tests depends on the nature of the equipment or system being tested, but generally they might involve the following:

1. Short-time dry/wet withstand/flashover test

Insulation Testing

2. Induced overvoltage withstand test
3. Full/chopped wave impulse voltage withstand test
4. tan δ (or loss angle) test
5. Temperature rise test
6. Electrochemical test
7. Electromechanical test
8. Mechanical test
9. Puncture test
10. Porosity test
11. Aging test
12. Pollution test
13. Radio influence voltage (RIV) measurement or corona related tests (e.g., radio interference, corona loss and audible noise)

Usually standard tests and test procedures are defined by national and international standards organizations for each equipment or component of the HV network.

12.3 TEST VOLTAGES

The test voltages normally used can be divided into three main groups: (1) direct voltages, (2) power frequency or low frequency alternating voltages, and (3) impulse voltage, which are further divided into lightning impulses and switching impulses. The generation and measurement techniques of such test voltages were presented earlier in chapter 10. Tables 12.1 and 12.2 list the recommended test voltages adopted for testing equipment having rated AC voltage ranging between 1 and 765 kV [1]. The system voltage levels can be broadly classified as shown in Table 1.1. The performance of equipment rated at MHV, HV and EHV are verified by the following types of tests:

1. The system behavior under AC operating voltages, temporary overvoltages and switching overvoltages is generally checked by a short duration (usually 1 min) power frequency voltage withstand test.
2. For insulation aging and external insulation contamination, the performance is generally checked by long-duration power frequency voltage test.
3. The performance under lightning impulses is checked by lightning overvoltage tests of equipment rated up to 300 kV. For equipment with $V_n > 300$ kV, such a test is complimented with or replaced by the switching impulse voltage test.

Table 12.1 Standard Insulation Levels for $1\text{ kV} < V_n < 300\text{ kV}$.

Highest voltage for equipment V_n (r.m.s) kV	Rated lightning impulse withstand voltage (peak) kV	Rated power-frequency short duration withstand voltage kV
3.6	20	10
	40	
7.2	60	20
4.4*		19
12	75	28
17.5	95	38
(13.2, 13.97, 14.52)*	110	34
24	125	50
26.4*	150	50
36	145	70
	170	
36.5*	200	70

* based on current practice in the USA and Canada (the high value of LI is used when kVA>500). The other voltage levels based on current practice in Europe and many other countries. For these voltage levels the choice between the values of LI depends on the degrees of lightning exposure, type of neutral earthing and the type of overvoltage protection.

52	250	95
72.5	325	140
123	450	185
	550	230
145	650	275
170	750	325
245	850	360
	950	395
	1050	460

The choice between the values of LI depends on many factors such as neutral earthing conditions, and the existence of protective devices and their location and characteristics.

Source: Ref. 1.

Insulation Testing

Table 12.2 Standard Insulation Levels for $V_n > 300$ kV

Highest voltage for equipment V_n (r.m.s) (kV)	Rated switching impulse (SI) withstand voltage (peak) (kV)	Rated lighning impulse (LI) withstand voltage (peak) (kV)
	750	850
300	850	950
362	950	1050
420	1050	1175
525	1175	1300
		1425
	1300	1550
765	1425	1800
	1550	1950
		2100
		2400

Source: Ref. 1.

12.3.1 Tests with Direct Voltages

For direct voltage tests, the test voltage should not contain AC components corresponding to a ripple factor of more than 3% at rated current. It is required that the rate of voltage rise above 75% of the estimated final test voltage value, be about 2% per second [2]. The value of the test voltage should be maintained within ± 1% of the specified level thorough the test if duration is ≤ 60 sec, otherwise the tolerances are ± 3%.

12.3.2 Tests with Alternating Voltages

The alternating test voltage should be sinusoidal with a frequency normally in the range 45–60 Hz. Its peak/$\sqrt{2}$ value should not differ from its true rms value by more than 5%. Partial discharge should not significantly reduce the test voltage. This is usually achieved if the total HV circuit capacitance is of the order of 500 to 1,000 pF and the current, with the test object short circuited, is at 1A or higher [2]. For dry tests on small samples of solid insulation or insulating liquids, a short-circuit current of 0.1A can be sufficient. However, for tests under artificial pollution, the required short-circuit current will be higher (as high as 15A) and depends on the ratio of series resistance R_S to the steady-state reactance X_S of the voltage source, including the generator or supply network, at the test frequency [2].

12.3.3 Tests with Impulse Voltages

The standard lightning and switching impulse voltages were defined in chapter 10. According to IEC-60 [2], the rated impulse voltage withstand tests are as follows:

1. For tests on non–self-restoring insulation, three impulses of the specified polarity are applied at the rated withstand voltage level. The requirements of the test are satisfied if no failure occurs.
2. For withstand tests on self-restoring insulation, the following two methods are applicable:

    Fifteen impulses of specified shape and polarity having peak value equal to the rated withstand voltage are applied. The requirements of the test are satisfied if not more than two failures occur.
    The 50% breakdown voltage (V_{50}) is determined. The test requirements are satisfied if V_{50} is not less than $1/(1 - 1.3\sigma)$ time the rated impulse withstand voltage, where σ is the per unit standard deviation of the breakdown voltage. The value of the V_{50} and σ could be determined using either probability or up and down methods [2]. Test procedures of these methods are described next.

Probability Method

In this method, n impulses (e.g., n = 20) of a fixed peak value are applied at each test voltage level. The value of V_{50} is obtained from a curve of breakdown probability versus corresponding applied voltage. Several voltage levels should be selected to cover the probability range and to ensure acceptable accuracy which increases with an increase in the number of voltage applications at each level.

Up and Down Method

One impulse voltage having an amplitude V close to the estimated V_{50} is applied. If this impulse cause a breakdown, the next applied impulse should have the voltage $V - \Delta V$, where ΔV is the voltage step and is approximately equal to 3% of V. If there is no breakdown at the level V, the next impulse voltage should have a value of $V + \Delta V$. This procedure is continued until 20–30 impulse voltages have been applied. If there are a total of

N applications with voltage V_i for ith application, where i = 1, 2, 3, . . . N, then:

$$V_{50} = \left(\frac{\sum_{i=1}^{N} V_i}{N}\right) \quad (12.1)$$

and for normal distribution:

$$\sigma = \left[\sum_{i=1}^{N} (V_i - V_{50})^2/(n-1)\right]^{1/2} \quad (12.2)$$

If σ cannot be calculated, it can be assumed as 0.03 and 0.06 p.u. for lightning and switching impulses, respectively.

12.4 TEST PROCEDURES AND STANDARDS

In the early days of the power industry, only national standards existed. However, as time passed, it became necessary to establish international standards to regulate the dealings with such apparati. Setting international agreements on acceptance tests based on similar requirements can considerably simplify approval of new equipment and enhance international co-operation leading to mutual benefits to the industry and users in different countries:

1. The manufacturer can produce HV apparatus in accordance with international standards, and then they can easily sell their products in any country
2. Equipment can be produced in larger quantities having unified standard, and hence cost per piece is reduced
3. Users in any country can purchase equipment conforming to international standards at competitive prices in the open market

In order to perform a certain test on a device or a component, a comprehensive standard procedure should be followed in order to ascertain the reliability of the test. This includes:

Choosing the sample
Preparing the sample for test
Specifying the condition at which the test should be performed and the correction procedure if different conditions prevail

The detailed steps of executing the test
Data collection
Results, analysis and reporting

The above procedures are described in general standards and specific apparatus standards. Several national and international standards exist for HV testing. International standards are more general, because they try to accommodate most of the possible cases and leave some of details to the local or national standards. For example, the international standard may leave the specifications of minimum and maximum ambient temperature to the local conditions. The ambient temperature may range from -40 to $+30°C$ in Sweden, and from -7 to $+55°C$ in Saudi Arabia.

The following are some of the standard organizations which deal with high voltage apparatus and systems.

1. International Electrotechnical Commission (IEC). This is the most recognized international body dealing with standards in electrotechnology. It publishes many standards on different electrical apparati, and devices. IEC may have, for each HV apparatus, several standards outlining its specifications and the applicable methods for testing. In the next section, some of these standards will be highlighted when discussing examples of HV testing of power apparati. IEC, like any standards organization, updates its standards over the years. There are several special committees within IEC dealing with special subjects. One example of such committees is the International Special Committee on Radio Interference (CISPR).

2. American National Standard Institute (ANSI). This is an American body dealing with standards. In some cases, other countries may also have to deal with ANSI standards due to their importance and common use of the American manufactured products. In the range of medium and high voltages, there are some common areas and some differences between ANSI and IEC standards, and for this reason individual apparatus standards should be consulated. On the other hand, there are commonalities between the two organizations regarding EHV and UHV equipment standards, since:

> Such equipment and techniques were developed recently when the Americans and the Europeans started to cooperate in these matters
> The cost of manufacturing at these voltage levels is very high, therefore such equipment manufacturing is limited to a few companies world wide, and hence it becomes easier to correlate their activities

3. Institute of Electrical and Electronics Engineers (IEEE). The IEEE was formed in the United States more than 100 years ago to enhance the development of electrical engineering. Besides publishing technical jour-

Insulation Testing 351

nals and magazines, it also publishes several standards concerning electrical apparatis. Since it is an international organization located in the United States, it tries to adopt a "compromise approach" between ANSI and IEC.

4. British Standard Institute (BSI). This may be the oldest standard organization that still exists. With the advent of IEC, which is European in specific and international in general, the publicity and importance of British standards have started to decline—like that of other individual European standards.

5. Other standards-making institutes:

Verband Deutscher Elektrotechniker e.v. (VDE), Germany
Association Francaise de Normalisation (AFNOR), France
Canadian Standards Association (CSA), Canada
Japanese Industrial Standards (JIS), Japan

Table 12.3 lists the latest editions of some IEC standards dealing with HV equipment.

12.5 TESTING OF HV MEASURING DEVICES

When performing various tests, a measuring system is employed to quantify the measured items. A common practice for HV measurements is to use a potential divider, connecting cable and LV measuring instrument (see section 10.7 for details). The complete measuring system includes all the components and their interconnections and any auxiliary equipment connected with the measuring system.

The LV meter must reproduce faithfully the required HV parameters that are being measured. This is achieved if the potential divider is frequency independent, the cable has low attenuation over a wide frequency range, the LV meter has a suitable bandwidth and there exists good matching between these three components. In order to check the suitability of the measuring system, a test may be performed in order to determine the following:

1. The maximum impulse amplitude and steepness, in addition to the impulse duration for which the system is suitable
2. The accuracy with which the system can measure the impulse amplitude, its front time and its tail time; in addition to the ability of the system to detect and measure any oscillations of the impulse
3. The influence of the proximity of objects on the accuracy of the measured values

Table 12.3 Some IEC Standards Relevant to Testing of Power System Components

HV apparatus	Standard no. and year of latest edition	Title	Impression (Imp.) or amendment (Amd.)
1. General	60	High voltage test technique	
	60.1 (1989)	General definition and test requirement	2nd Imp. (1991)
	60.2 (1994)	Test procedure	
	60.3 (1994)	Measuring devices	
	60.4 (1977)	Application guide for measuring devices	2nd Imp. (1987)
2. Transformers	76	Power transformers	
	76.1 (1993)	General	
	76.2 (1993)	Temperature rise	
	76.3 (1980)	Insulation level and dielectric tests	2nd imp. (1991)
	76.3.1 (1987)	External clearances in air	3rd Imp. (1991)
	76.4 (1976)	Tapping and connections	3rd Imp. (1990), Amd. 1 (1979)
	76.5 (1976)	Ability to withstand short circuit	Amd. 2 (1994)
3. Cables	55	Paper insulated cable	
	55-1 (1978)	General	Amd. 1 (1989)
	55-2 (1981)	General and construction requirement	Amd. 1 (1989)
	141	Tests on oil-filled and gas pressure cables and their accessories	
	141.1 (1993)	Oil filled with V_n up to 400 kV	Amd. 1 (1995)
	141-2 (1963)	Internal gas pressure cables with V_n up to 257 kV	Amd. 1 (1967)
	141-3 (1963)	External gas pressure cables with V_n up to 275 kV	Amd. 1 (1967)
	141-4 (1980)	Oil, high pressure cables with V_n up to 400 kV	Amd. 1 (1990)
	183 (1984)	Guide to the selection of HV cables	Amd. 1 (1990)
	230 (1966)	Impulse tests on cables	
	502 (1994)	Extruded solid dielectric insulated power cables (1–30 kV)	
	840 (1988)	Tests for power cables with extruded insulation	Amd. 1 (1993)

Insulation Testing

4. Circuit Breakers	56 (1987)	HV AC circuit breaker	Amd. 1 (1989), Amd. 2 (1995)
5. Switchgear	427 (1989)	Synthetic testing of HV AC circuit breakers	Amd. 1 (1992), Amd. 2 (1995)
	694 (1980)	Common clauses for HV switchgear and control gear standards	Amd. 1 (1985), Amd. 2 (1993), Amd. 3 (1995)
6. Switches	265	HV switches	
	265-1 (1983)	HV switches for $1\ \text{kV} < V_n < 52\ \text{kV}$	Amd. 1 (1984), Amd. 2 (1994)
	265-2 (1988)	HV switches for $V_n \geq 52\ \text{kV}$	Amd. 1 (1994)
	420 (1990)	HV AC switch fuse combination	
7. Contactors	470 (1974)	HV AC contactors	Amd. 1 (1975)
8. Surge Arrestors (SA)	99	Surge arrestor (SA)	
	99-1 (1991)	Nonlinear resistor type gapped SA for AC system	
	99-2 (1962)	Explosion type lightning arrestor	
	99-3 (1990)	Artificial pollution testing of SA	
	99-4 (1991)	Metal-oxide SA without gap for AC system	
9. Insulators	383 (1983)	Tests on insulator of ceramic or glass for overhead lines with $V_n > 1\ \text{kV}$	Amd. 1 (1993)
	506 (1975)	Switching impulse tests on HV insulators	
	507 (1991)	Artificial pollution tests on HV insulators to be used on AC system	
10. Bushings	137 (1995)	Bushings for AC with $V_n > 1\ \text{kV}$	
11. Fuses	282	HV fuses	
	282-1 (1994)	Current limiting fuse	Amd. 1 (1995)
	282-2 (1995)	Explosion fuse	
12. HV capacitor	143 (1992)	Series capacitors for power systems	
	143-2 (1994)	Protective equipment for series capacitor banks	
13. Digital recorders for measurement in HV impulse test	1083	Digital recorders for measurements in HV impulse tests	
	1083-1 (1991)	Requirements for digital recorders	

V_n = rated systems voltage.

Most of the factors affecting the performance of a measuring system can be evaluated from the step response of the system. Step response is the output of the system as a function of time when a step voltage is applied between its high voltage and ground terminals. Once the step response of a measuring system is known, the output of the system for any other shape of input voltage can be determined.

The step generator must have zero impedance (Z_{int}) to obtain the true step response of the measuring system. This is impossible because of the large distance between the two input terminals. To minimize Z_{int}, the two input terminals are brought close together as much as possible and then the step response is taken. This is not the true step response of the measuring system, but in most cases will give a response very close to the true one.

The quality of the measuring system is checked by the value of the response time. When the response time is short, the system can measure fast impulses accurately. From step response it is possible to determine both the error in the measurement of the impulse peak amplitude and the error in the measurement of the chopping time. The response time (T) is essentially the area under the unit step minus the area under the normalized unit step response. It has the dimension of time, since the X-axis is time and the Y-axis is simply a relative numerical value. The response time can be either positive or negative, depending on the shape of the step response. If the step response is oscillatory or has an overshoot above the unit amplitude line, the response time will be negative if the sum of the negative areas is greater than the sum of the positive areas. In the case of damped type, the response time is simply the area between the step response and the step input. Figure 12.2 shows damped and oscillatory types of step responses. The algebraic sum of all of the areas in case of oscillatory type is the response time, which in this case is negative. A negative response time means that the measurement of a front-chopped impulse or a chopped ramp will show a higher peak voltage and higher chopping time than the actual values of these parameters.

If the rate of rise of impulse voltage is denoted by S and the chopping time by T_c, the true peak amplitude is $S \cdot T_c$. Provided that the chopping of the impulse occurs in the region where the response is rising at the same rate as the input, the measured value of the amplitude will be [$S(T_c - T)$], since ($T_c - T$) is the measured time to chop. This quantity is greater than T_c when the response time is negative, as in Figure 12.3 and therefore the measured value of the peak amplitude will be larger than the true value. The error in the measured value is clearly [$S(-T)$].

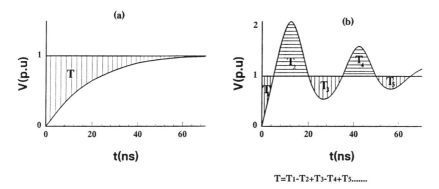

Figure 12.2 Evaluating the response time from a step response oscillogram: (a) damped and (b) oscillatory response.

12.6 PARTIAL DISCHARGE TEST

Before discussing testing of various power apparati, a brief description will be first presented for the two main nondestructive testing techniques, i.e., partial discharge (PD) and dielectric loss or tan δ measurements. The basic mechanism of PD and PD measuring schemes will be discussed here, and tan δ will be described in the next section. These nondestructive tests are usually used to assess the electrical properties of dielectric materials over a wide range of operating and environmental conditions. The term *partial discharge* means the discharge is not complete between the two conducting electrodes. If such a discharge occurs in the air, it is called a corona. Many

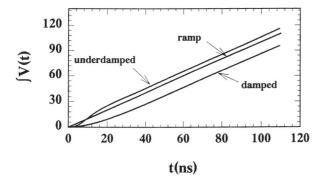

Figure 12.3 Integrals of the step response and unit step input of Figure 12.2.

solid insulating materials contain gas voids (cavities) which may be bridged by local discharges causing a PD. PD is considered one of the major reasons behind the aging and eventual failure of solid insulating materials; albeit there is no direct relation available between PD level and the expected life of the insulation as yet.

12.6.1 Partial Discharge Modeling

Development of partial discharge in a cavity enclosed in a dielectric slab was explained earlier in section (8.4) with the help of Figure 8.5 and Figure 8.6. It was illustrated that when V is the applied voltage across the dielectric, the voltage V_c across the cavity will be given as:

$$V_c = V \frac{C_b}{C_b + C_c} \qquad (12.3)$$

This voltage may cause void breakdown depending on the instantaneous value of V_c and the dielectric strength of air inside the void, since the stress in the void (E_C) may become much higher than the average stress in the dielectric. The energy released in this discharge will cause deterioration of the dielectric.

12.6.2 Partial Discharge Detection System

The observed effects of a PD—which include the energy loss, electric current pulse, sound and electromagnetic radiation and changes in the material properties—can be used for PD detection. Each effect is useful in certain applications depending on the apparatus and the type of information needed. In this section, details will be presented concerning the subject of electric pulse detection since it is the most commonly used method for the PD detection. For others the reader can refer to Krueger [3].

The electric pulse generated from a PD can be detected easily using electrical detection arrangements. Figure 12.4 shows the two basic detection circuits which are termed as straight and balanced methods. The balanced method is advantageous since in this case the external interference is eliminated as it produces identical signals at the input of the differential amplifier. The specimen capacitance C_S in Figure 12.4 is assumed to be lumped, and this assumption is valid for most of the power apparati. However, in cases like long cable and transformers, C_S is a distributed parameter and a special measure should be considered.

The detection system can be calibrated by injecting a pulse of known charge magnitude to simulate the actual charge. The linear relation between

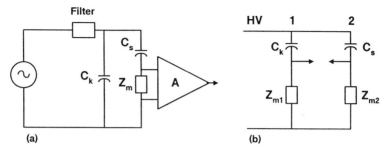

Figure 12.4 Partial discharge detection circuits: (a) straight detection circuit and (b) balanced method of discharge detection. C_k = coupling capacitor and Z_m, Z_{m1} and Z_{m2} are measuring impedances.

this known charge and the measured one is the calibration factor for the instrument. If a square wave generator with an output V_o is coupled to the detection system through capacitor C, then the calibrating charge q_o is given as:

$$q_o = CV_o \qquad (12.4)$$

There are other ways of calibrating the detection system, like measuring well-known signals such as charges from point-plane corona or charges in an artificial cavity [3].

Traditionally, the value of apparent charge (q_a) is used as an indicator for the PD level. q_a is measurable and it is related to the real charge q_c by [3]:

$$q_a = q_c \frac{C_b}{C_b + C_c} \qquad (12.5)$$

Recently, the use of q_a as an accurate indicator for a quantitative evaluation of PD has been questioned. Instead it is proposed that the discharge energy be used as an appropriate measurable quantity to characterize PD inside the dielectrics [4].

12.6.3 Partial Discharge Classification

In addition to PD detection, its source (e.g., cavity discharge, surface discharge, corona) can be known through the PD classification. Research has been carried out on various PD sources in order to identify the PD source through the evaluation of typically measured discharge parameters. These include pulse height distributions, PD frequency and location on the AC cycle and statistical nature. From such measurements, PD classification can be attempted. In earlier studies, classification was performed by studying the discharge patterns on the oscilloscope [5]. Table 12.4 shows typical PD discharges, their sources, and the influence of time and voltage.

In recent years, digital signal processing techniques have been advantageously used to automate the PD pattern recognition [6–8]. Krivda [8] used the "fingerprints" derived from various PD data to discriminate between different PD sources, and then to classify these PD sources accordingly. Commercial applications of such PD classification techniques are expected in the future.

12.7 DIELECTRIC LOSS TEST

When a dielectric material is placed between two electrodes, a capacitor is formed, and when a voltage is applied across this dielectric, a certain amount of current (I) flows. Thus, the dielectric is represented by an RC equivalent circuit. If a parallel RC equivalent circuit is used, $\tan \delta = 1/\omega C_p R_p$ (ideally, $R_p = \infty$). On the other hand, if a series equivalent circuit is used, $\tan \delta = \omega C_s R_s$ (ideally, $R_s = 0$). The energy dissipated as heat in the dielectric is proportional to $\tan \delta$, and thus the value of $\tan \delta$ is an important measure of the insulation quality. The maximum value of $\tan \delta$ is normally given in component or system specifications. The variation of $\tan \delta$ with the applied voltage is an important relation which gives useful information about the quality of the insulation and if an imperfection exists in the insulation or not. As the voltage across the dielectric increases, the value of $\tan \delta$ is roughly constant in the beginning; then, at some specific voltage, it starts to increase appreciably with voltage indicating the inception of a PD. Similarly, the properties of the dielectric material change with time and $\tan \delta$ measurements give some measure of this change. There are many causes of changes in solid insulation, including changes due to variations in temperature, mechanical stress, chemical reactions, water absorption and surface state deterioration. Similarly, changes in oil properties take place due to gas and water absorption, chemical changes and introduction of impurities. Periodic measurements of $\tan \delta$ can provide an indication of the status of the insulation quality in such cases.

Capacitance and $\tan \delta$ are measured using bridges having two fixed arms, consisting of the test object and a standard capacitance, and two adjustable arms. The sensitivity of the measuring bridge is the most important and perhaps the limiting factor in low-frequency applications. If high-frequency AC is used, the sensitivity is limited by the frequency-dependent error of the resistance elements and the stray capacitance. Schering bridge,

Table 12.4 Partial Discharge Display Patterns, Minimum Detectable Discharge (MDD), Discharge Extinction Voltage (DEV) and Discharge Inception Voltage (DIV)

Discharge source	Discharge display	Pulse amplitude and inception voltage characteristics	Effect of increasing the voltage	Effect of time
1. Corona from sharp point in air		Pulses have fixed amplitude and are uniformly distributed around −ve voltage peak	No. of pulses increase, then same will appear around the +ve peak.	No effect.
2. Internal discharge in a solid dielectric bounded cavity		DIV is well defined and above MDD, pulses have similar amplitude although differences of 3:1 from one side to the other are normal.	Little or no effect.	The time of voltage application has little effect on the discharge pattern during single test. Pulse amplitude and location have some randomness.
3.a. Several internal cavities b. Discharge between two dielectric surfaces c. Surface discharge		Pulses are the same for +ve and −ve half cycles, 1:3 differences in magnitudes is normal. DIV is distinct and above MDD, DEV≤DIV.	Pulses increase till becoming unresolved.	No effect (at least up to 10 min).
4. Luminar cavity		Pulses are the same in +ve and −ve half cycles.	Pulses are resolved at inception becoming rapidly unresolved with increase in voltage. The magnitude increases steadily with increase in voltage.	If voltage is high, pulses magnitude gradually increases but becomes stable after ~10 min.
5. Gas bubbles in an insulating liquid in contact with moist cellulose (e.g., oil impregnated paper)		These can be random in magnitude and location on the waveform. The bubbles are generated by electric field in moist cellulose. They increase by discharges but dissolve in liquid and disappear upon removing stress for a period of time.	Well-defined DIV with steep rise in magnitude as voltage is increased.	If voltage is fixed above DIV the magnitude of pulses increase by 100 times in few minutes. With extinction and reinception DIV is about 3 times less than the original value. If restarting after 1 day, high DIV will be restored.

Table 12.4 Continued

6. Cavity between metal or carbon and dielectric		Pulses are asymmetrical, i.e., unequal in number and magnitude in the two half cycles. When the metal face at the cavity is grounded, a small number of pulses with large magnitudes at $-$ve peak and large number with low magnitudes at $+$ve peak occur. When the metal is at HV, the situation will be reversed.	DIV is well defined and the magnitude remains unaltered as voltage is increased.	The time application (up to 10 min) has little effect on the display.
7. Gap-type discharge in nearby objects		Symmetrical pulses in magnitude and number.	The pulses cover larger portion of the waveform and no change in magnitude. DIV is well defined above the noise level and DIV = DEV	No effect.
8. Tree growth	(A) (B)	Symmetrical pulses, the location can be erratic as can be the magnitude, DIV is variable and higher than DEV. Large and unrepeatable changes in magnitude may occur particularly at high voltage.	The pattern can be resolved at low voltage and become unresolved at high voltage. Pulse magnitude increases rapidly with voltage.	If display is similar to A the magnitude is likely to be stable for minutes or longer. If it is similar to B then the pattern is likely to be erratic with rapid and substantial changes in magnitudes in few minutes.

Source: Ref. 5.

shown in Figure 12.5, is the most widely used bridge for tan δ measurements. From Figure 12.5, it can be deduced that at balance:

$$C_x = C_s \frac{R_4}{R_3} \qquad (12.6)$$

$$\tan \delta = \omega C_4 R_4 \qquad (12.7)$$

A variety of bridge circuits and accessories are available to suit the various measuring conditions. To overcome the stray capacitance of the high voltage arm, a Wagner earth device is used in connection with the bridge [26]. In recent years, the measurements using the Schering bridge have been automated to save time for the repetitive kind of measurements.

12.8 TESTING OF HV APPARATUS

In this section, the main HV tests performed on some of the power apparati as per IEC standards will be outlined. First, the various tests for the individual apparatus will be mentioned and then a brief description of some of these tests will be provided.

12.8.1 Distribution Transformers

The following tests are required on most power transformers, as recommended by relevant IEC publications [9,10]:

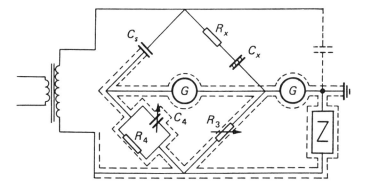

Figure 12.5 Schering bridge with Wagner earth arrangement to eliminate stray capacitance. G = galvanometer, Z = additional arm impedence.

1. Routine tests
   a. Measurement of winding resistance
   b. Measurement of voltage ratio and check of voltage vector relationship
   c. Measurement of impedance voltage, short-circuit impedance and load loss
   d. Measurement of no-load loss and current
   e. Test on tap changers
2. Type tests
   a. Dielectric tests
   b. Temperature rise tests

Some special tests such as a short-circuit test, measurement of zero-sequence impedance on three-phase transformers and measurement of acoustic sound level may also be required. The two type tests mentioned above will be discussed.

Insulation Level and Dielectric Tests

A transformer has both internal as well as external insulation. A failure in the non–self-restoring internal insulation is catastrophic and normally leads to the transformer being withdrawn from service for a long period. If the transformer's rated voltage is less than 300 kV, then only lightning and AC voltage tests are required. However, switching voltage tests are also required when the system voltage is higher than 300 kV.

Impulse Testing of Oil-Filled Transformers

The purpose of the impulse tests is to determine the ability of the transformer insulation to withstand the transient voltages due to lightning. Since the transients are impulses of short rise time, the voltage distribution along the transformer's windings is not uniform. The equivalent circuit of a transformer winding for impulses is shown in Figure 12.6. If an impulse wave is applied to such a network, the voltage distribution along the element will be uneven due to the existence of the capacitances C_g and C_s. The voltage stress across the winding is characterized by the following equation if the neutral is isolated [11]:

$$V(x) = V \, \frac{\cosh\left(\dfrac{\alpha x}{L}\right)}{\cosh \alpha} \tag{12.8}$$

where $\alpha = \sqrt{C_g/C_s}$. Thus the voltage distribution along the transformer is not uniform, and the initial part of the winding (close to the line side) will

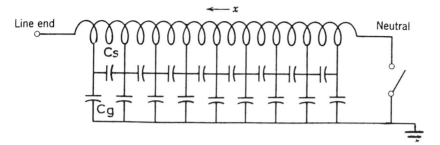

Figure 12.6 Equivalent circuit of transformer winding under fast impulses.

have higher stress and thus it will be the first to breakdown in case of high voltage surges.

Impulse test of an oil-filled transformer is usually performed using both the full-wave and the chopped-wave impulses with chopping time from 2 to 6 μs. To prevent large overvoltage from being induced in the windings that are not under test, they are short circuited and connected to ground through low impedance.

The schematic diagram showing connections for the impulse testing of a three phase Δ/Y distribution transformer is given in Figure 12.7. Here, winding UW is under test with full impulse voltage whereas windings UV and VW are subjected only to half of the applied voltage. In the case of Y/Y winding connections, each HV winding is tested separately. In transformer testing, it is essential to record the waveforms of the applied voltage and the resulting current through the winding under test. Sometimes, the transferred voltage in the secondary winding and/or the neutral current is also recorded.

Following sequence of impulse voltage applications is used for oil-filled transformers:

1. One full wave impulse voltage of reduced (50–75% of BIL) magnitude
2. One full wave impulse of 100% BIL
3. One or more reduced chopped impulses
4. Two chopped impulses at 100% BIL
5. Two full wave impulses of 100% BIL

The fault in winding insulation is detected by general observations of noise, smoke, etc., during voltage application. The inspection of voltage and current oscillograms give more accurate indication of the failure, especially

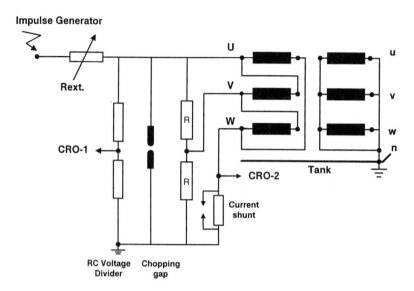

Figure 12.7 Arranagement of transformer for impulse voltage test on terminal U and winding UW.

the partial one. A partial or complete failure of winding appears as a partial or complete collapse of the applied impulse voltage. However, the impulse voltage may not show a small partial failure since the sensitivity of the voltage waveform method is low and this method does not detect faults which occur on less than 5% of the total winding. The failure detection is enhanced by a current oscillogram, which usually shows a record of the impulse current flowing through a resistive shunt connected between the neutral and the ground or between the low voltage winding and the ground. The current oscillogram usually consists of a high frequency oscillation, a low frequency disturbance and a current rise due to reflections from the ground end of the windings. When a major fault such as breakdown between turns or between one turn and the ground occurs, high frequency pulses are observed in the current oscillogram and the waveshape changes. For local failure such as a partial discharge, only high frequency oscillations are observed without a change of waveshape. To detect any failure, voltage and current oscillographs for the full wave impulses are compared with the initial reduced wave records. In addition, chopped impulses are compared as well. The IEC test criterion is met if there are no significant differences between the oscillograms corresponding to reduced and full voltage applications. Figure 12.8 shows examples of the voltage and current oscillograms corresponding to reduced and full voltage impulses. In this figure, the three possible cases are displayed, i.e., no failure, partial winding failure and complete failure (complete winding, external flashover or breakdown between winding and the transformer tank).

Power Frequency Test

A voltage of specific value (see Table 12.1 and 12.2) is applied across each of the transformer windings. The transformer is considered to pass the test if no breakdown takes place during the test.

Temperature Rise Test

The temperature rise values for the windings, core and oil of transformers designed for operation at normal altitude (≤ 1000 m) are specified by IEC according to the cooling medium employed [8]. A reduced temperature rise may be considered for transformers located at altitudes higher than 1000 m. The temperature rises are checked by subjecting the transformer to simulated normal load condition for a specific time period. Details of the measurement procedures and interpretation of the test results are specified in relevant standards [9].

12.8.2 HV Cables

Cables are subjected to electrical and thermal stresses while in service. During their transportation, installation and repair, they are usually subjected to mechanical stresses also which may cause insulation cracking or produce voids, which in turn can act as sites of PD that may lead finally to the breakdown of the cable insulation. Therefore, HV cables are subjected to mechanical and electrical tests [12]. The following are the main electrical tests as recommended in IEC standard.

1. Routine tests
   a. Electrical resistance of the conductor
   b. AC voltage withstand test
   c. Partial discharge tests

2. Type tests. The following electrical tests are specified by IEC in the sequence indicated:
   a. Insulation resistance (R_{in}) at room temperature
   b. PD test
   c. Bending test plus PD test (to check if the bending creates void or not)

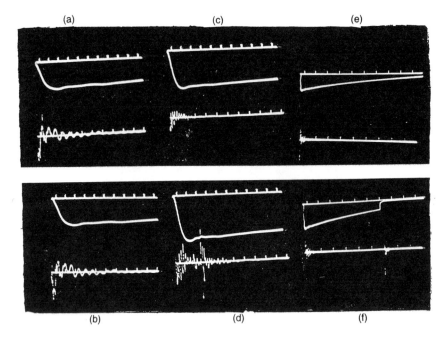

Figure 12.8 Oscillograms of transformer windings subjected to impulse testing. Upper traces represent applied voltage pulse, whereas lower traces are for current flow. Cases (a) and (b) represent no failure; (c) and (d) indicate partial failure whereas in (e) and (f) winding insulation has totally collapsed. Oscillograms (a), (c) and (e) are for reduced full wave; whereas (b), (d) and (f) are for full wave. Time scale = 1 μs/div in (a)–(d) and 10 μs/div in (e) and (f).

    d. Tan δ as function of voltage and capacitance measurements.
    e. Tan δ as function of temperature
    f. Insulation resistance at operating temperature
    g. Heating cycle test plus PD test
    h. Impulse voltage withstand test followed by AC voltage test
    i. HV alternating current test

In addition to the above, there are several nonelectrical type tests as well.

    3. Special tests. There are several special tests such as conductor examination, check of dimensions and hot set test.

For PD measurements and voltage withstand tests, samples have to be prepared and terminated carefully. Otherwise, excessive leakage, corona or end flashover may occur during testing. The normal length of the cable

Insulation Testing

samples used varies from 5 to 10 m. The terminations are usually made by shielding the end conductor with stress shields to relieve the excessive high electric stresses which would occur otherwise.

12.8.3 Circuit Breakers

Circuit breakers are tested to evaluate their constructional and operational characteristics, in addition to check their performance in making and breaking the expected high currents. The main properties of a circuit breaker are:

1. Electrical properties. These include the arc characteristics, the current chopping characteristics and the shunting effects in interruption.
2. Nonelectrical properties. These include the type of media in which the arc is extinguished, operation time, the number of operations and the size of the arcing chamber.

In addition, the circuit performance characteristics should also be specified. These include the degree of electrical loading, rated voltage, the type of fault in the system which the breaker has to clear, the circuit time constant, the time of interruption, the rate of rise of recovery voltage and the restriking voltage.

The following are the main tests specified for the circuit breakers [13]:

1. Routine tests
   a. AC dry test of the main circuit
   b. Voltage tests on control and auxiliary circuits
   c. Measurement of the resistance of the main circuit
   d. Mechanical operation tests
   e. Design and visual checks
2. Type tests
   a. Dielectric tests
   b. Radio influence voltage (RIV) test
   c. Temperature rise test
   d. Measurement of the resistance of the main circuit
   e. Mechanical and environmental tests
   f. Making and breaking tests
   g. Short circuit tests

Testing of circuit breakers requires highly equipped laboratory and sophisticated testing procedures. A brief description is presented here for some of the circuit breaker tests.

Mechanical Test

More than 80% of the total failures in circuit breakers are associated with mechanical failures. To check its mechanical performance, the circuit breaker is subjected to 2000 operating cycles with no voltage or current in its main circuit. During the test, replacement of any part and mechanical adjustments are not permissible, though lubrication of the mechanical parts is allowed according to the manufacturer instructions.

Temperature Rise Test

When a circuit breaker is in service, the temperature of any part of the breaker should not exceed the specified limits of temperature rise. These limits vary for individual parts and for circuit breaker types. This test is carried out in conditions similar to the normal operating conditions where the normal rated current flows and the mounting is under normal service conditions. If the maximum observed temperature rise is within the specified limits, the breaker is deemed to have passed the test.

Dielectric Test

This test is done to ensure circuit breaker can withstand the expected overvoltages within the power system. AC and impulse voltages are used for the tests. The breaker under test should withstand the specified test voltages without flashover or puncture. The specified AC voltage is applied for 1 min to indoor and outdoor breakers under dry or wet conditions. Positive and negative polarity impulse voltages of specified amplitudes are applied 15 times. The test criterion is satisfied if the number of flashovers at either polarity does not exceed 2.

Short-Circuit Tests

Short-circuit tests are usually carried out according to test duties that specify the test current, the percentage of DC component, and the transient and the power frequency recovery voltages. The purpose of the test is to ensure that the circuit breaker is capable of making and breaking the circuit under short-circuit conditions without damage to its components. This test is subdivided into many tests, including the following:

1. Breaking and making current tests
2. Short-time current test
3. Operating sequence test
4. Single-phase short-circuit test

Insulation Testing

5. Out-of-phase switching test
6. Short-line fault test
7. Capacitor charging current breaking test
8. Small inductive current breaking test

The short-circuit tests need high-power testing plants since the circuit breaker needs to break high current while the high voltage is applied. Such high-power test facilities are very costly. An alternative solution is to use synthetic method of circuit-breaker testing. It basically consists of two independent voltage and current sources (Figure 12.9). The idea is to synchronize momentarily the injection of high current and application of high voltage. The current source injects the current into the circuit breaker under test at a relatively reduced voltage. The voltage source injects a high voltage across the circuit-breaker contacts at the moment it is made to interrupt the high current. The high voltage source usually contains a large capacitor bank. The actual synthetic testing circuits would also include elaborate control and instrumentation schemes, in addition to the protection arrangements in case of a breaker failure.

12.8.4 Switchgear

Switchgear normally contains circuit breakers, switches, busbars and other metering and monitoring instruments. In addition to the tests performed on

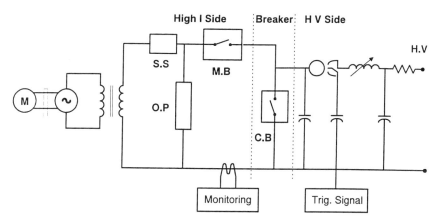

Figure 12.9 Synthetic testing circuit for circuit breakers: C.B. = circuit breaker under test, M.B. = main breaker, S.S. = synchronizing switch and O.P. = overvoltage protection.

the circuit breaker, there are several other tests concerning the performance of the whole switchgear. The following are the main tests specified for the switchgear [14].

1. Routine tests
   a. AC dry test
   b. Voltage tests on control and auxiliary circuit
   c. Measurement of the resistance of the main circuit
2. Type tests
   a. Dielectric tests, which include tests with lightning and switching impulse voltages, AC voltage, artificial pollution, PD and AC tests on auxiliary and control circuits
   b. Radio influence voltage (RIV) test
   c. Temperature rise test
   d. Measurement of the resistance of the main circuit
   e. Short-time withstand current and peak withstand current test.

A brief description of some of the tests for switchgear is provided next.

Radio Interference (or Influence) Voltage (RIV) Test

A voltage of $1.1V/\sqrt{3}$ shall be applied to the switchgear and maintained for at least 5 min (V is the rated voltage). The voltage shall then be decreased by steps down to $0.3V/\sqrt{3}$. Then the voltage is raised again by steps to the initial value and finally decreased by steps to $0.3V/\sqrt{3}$. At each step, RIV measurements are made and then RIV is plotted against the applied voltage. The equipment meets the test criterion if RIV at $1.1V/\sqrt{3}$ does not exceed 2500 μV.

Dielectric Tests

Tests are made on switchgear and controlgear completely assembled, as in service and mounted with minimum clearances and height as specified by the manufacturer. Lightning impulse, switching impulse (if rated voltage is more than 300 kV) and power frequency voltages are applied in a method similar to that mentioned earlier for the circuit breakers.

12.8.5 Surge Arresters

Surge diverters or lightning arresters are used to protect the power system components against transient overvoltage due to lightning and switching surges. To ensure their working performance, surge arresters must be tested using standard test procedures. Ideally, lightning arresters should have in-

Insulation Testing

finite resistance for operating power frequency voltages and zero resistance for transient overvoltage, thus discharging the heavy transient current, and then recovering its insulation after the surge has finished without allowing the power frequency current to continue. Surge arresters are identified by their rated voltage, rated frequency, normal discharge current and class of long-duration discharge [15]. The following are the main tests specified for surge arresters.

1. Routine test: Dry power frequency sparkover voltage test
2. Type tests
   a. Power frequency sparkover voltage test
   b. Lightning (and switching) impulse sparkover voltage test
   c. Volt-time curve for lightning impulses
   d. Residual voltage test
   e. Current impulse withstand test
   f. Operating duty test
   g. Test of arrester disconnectors

The standard acceptance tests are dry AC voltage and lightning impulse tests. In some cases, residual voltage test is added. These tests should be made on a number of samples equal to the higher whole number to the cube root of the number of arresters to be supplied. Some of the above tests are described briefly below.

Power-Frequency Sparkover Tests

Dry and wet tests are made on complete arresters. The voltage initially applied to the arrester should be of a low value in order to avoid sparkover of its series gaps. The permissible time in which the applied voltage may exceed the rated voltage of the arrester is in the range 2–5 sec. After sparkover, the test voltage is switched off by automatic tripping within 0.5 sec. The power-frequency sparkover voltage is the average of five test results.

Impulse Sparkover Tests

These tests are performed on the same samples as those used for power-frequency sparkover tests. The procedure of test for lightning impulses is the same as for circuit breakers discussed earlier.

Impulse Current Withstand Tests

The rated voltage of these tests must be in the range of 3–6 kV. Each sample is subjected to two current impulses of the standard shape with peak impulse values depending on the arrester class. Before each test, the samples must be at approximately the ambient temperature. Subsequently, the power-frequency sparkover voltage is determined. It should not change by more than 10%. Also, no evidence of puncture or flashover of the nonlinear resistors or significant damage to the series gaps or grading circuit should occur.

12.8.6 Ceramic and Glass Insulators

Insulators are named as type A if the puncture path is at least half of the creepage length, i.e., the shortest flashover path through air on insulator surface. Otherwise, these are called type B insulators. Insulators used on overhead transmission lines and in substations are subjected to routine tests, type tests and sample tests, as described next [16].

1. Routine tests
    a. Thermal shock test
    b. Visual examination
    c. Mechanical test
    d. Electrical test (on some insulator types)
2. Type tests
    a. Dry lightning impulse voltage withstand test
    b. Wet switching impulse voltage withstand test
    c. Wet AC voltage withstand test
3. Sample tests
    a. Verification of locking system and dimensions
    b. Mechanical and electromechanical failing load test
    c. Thermal shock test
    d. Puncture test
    e. Porosity test
    f. Galvanizing test

Some of the above tests are described briefly below.

Dry Lightning Impulses Withstand Test

The dry insulator is tested under positive and negative lightning impulses. It shall be adequate to test with one polarity if it is evident that this particular polarity will give lower flashover voltage. Two test procedures are in common use for the lightning impulse withstand test, i.e., the withstand procedure with 15 impulses and the 50% flashover voltage procedure discussed earlier in section 12.3.3.

Insulation Testing 373

Wet Switching Impulse Withstand Test

The insulator is tested under positive and negative 250/2500 μs switching impulse voltage waves. The wetting procedure is specified in the standard. The 50% flashover procedure is normally used. If it is not possible to use this procedure, the withstand procedure can be used.

Wet Power-Frequency Withstand Test

This test is applicable only to insulators for outdoor use. With insulators under wet conditions, a voltage of about 75% of the test voltage shall be applied and then increased gradually with a rate of rise of about 2% of this voltage per second. The test voltage is maintained at this value for one minute. If no flashover occurs, the insulator is deemed to meet the test requirements. For wetting, the spray is arranged such that the water drops fall approximately at an angle of 45° to the vertical. The test object is sprayed for 15 min including the adjustment time before the voltage application and the spray is continued following the voltage application. The characteristics of the precipitation are:

Vertical and horizontal components = 1 to 1.5 mm/min
Limits for individual measurements = 0.5 to 2.0 mm/min
Temperature of collected water = ambient temperature ± 15°C
Conductivity of water corrected to 20°C = 100 ± 15 μS/cm

12.8.7 Bushings

A bushing is a structure carrying one or several conductors through a partition such as wall or tank and insulating it (them) therefrom, incorporating the means of attachment to the partition. The bushing insulation is solid, liquid, gas or composite. The following are the main tests specified for HV bushings [17]:

1. Routine tests
   a. Tan δ and capacitance measurements
   b. Dry AC test
   c. PD test
   d. Tape insulation test
   e. Pressure test for gas type bushings
   f. Pressure test for gas or liquid type bushings
   g. Tightness at fixing devices
2. Type tests
   a. Wet AC test

b. Dry lightning impulse test
  c. Dry or wet switching impulse test
  d. Thermal stability test
  e. Temperature rise test
  f. Thermal short-time current withstand test
  g. Dynamic current withstand test
  h. Cantilever load withstand test
  i. Tightness test for liquid type

Wet AC test, dry lightning impulse test and switching impulse tests are performed in manners similar to those discussed earlier for insulators.

12.8.8 High Voltage Fuses

A high voltage fuse is introduced in the circuit to interrupt the current in case of a fault similar to the LV fuse. However, it contains HV insulation to isolate the thread of the fuse from the other parts of the circuit. There are several standard tests to be carried out to ensure the proper operation and the quality of insulation of HV fuses. The following are the type tests recommended by the relevant IEC standard [18]:

1. Dielectric test
2. Temperature rise test
3. Breaking test
4. Tests for time/current characteristics

The main HV test is the dielectric test where the standard test voltages are applied successively with one terminal of the output of the impulse generator or one point of the power frequency source connected to the earth. The voltage is applied:

1. Between the terminals and all earthable metal parts
   a. With the fuse including the fuse link and its fuse carrier completely assembled and ready for service
   b. With the fuse link and its fuse carrier removed
2. Between terminals
   a. For drop out fuses, the fuse carrier should be in the drop out position
   b. For disconnector fuses, the fuse carrier or the fuse link should be removed from the fuse base

The tests required are lightning impulse tests under dry conditions and power frequency tests under dry and wet conditions. The procedures for these tests are similar to those described for outdoor insulators.

12.9 ELECTROSTATIC HAZARDS

The problem of electrostatic charge buildup concerns a wide variety of human activity ranging from large-scale industries to domestic activities. The effects of such a buildup can cause major hazards with potential loss of life and money. After describing how the electrostatic charges are generated and measured, some practical examples of their hazards will be given in conjunction with the means of controlling and preventing such hazards. Some typical cases of the electrostatic buildup in power system components will be mentioned as well.

12.9.1 Electrostatic Charge Generation and Measurement

Electrostatic charge buildup and its potential hazards can occur if the following three conditions prevail:

1. Charging of the material or nearby structures occurs
2. The leakage of such charge is so slow that charge accumulation takes place
3. The conditions for explosion or shock are favorable

The presence of high humidity reduces the surface resistivity, and hence the relaxation time constant τ_c and consequently the danger of electrostatic hazards decrease with humidity. A static charge is due to an excess or a deficiency of electrons. Static electricity is generated by the contact and separation of different materials. When materials are in contact, electrons from one material can move across the interface to the surface of the other material where they will align themselves since the first material will now have a slight positive charge due to the loss of the electrons. In order for the electrostatic charge buildup to take place, one or both materials must be insulators. Materials differ widely in their ability to give or receive electrons on contact with other objects. Typical triboelectric series is shown in Table 12.5, which serves as a useful rule of thumb to predict the behavior of one substance when rubbed with another. When any two of the materials on the list are rubbed together, the one higher in the list becomes positive with respect to the one lower on the list. However, it cannot be used indiscriminately [19]. The nature of the charges produced on a surface depend on the environmental conditions, the physical conditions of the surface and the presence of impurities as well. The value of the discharge will depend, in addition to the above, on the area of contact and speed of separation.

Table 12.5 Triboelectric Series

Positive End
Asbestos
Glass
Mica
Wool
Cat fur
Lead
Silk
Aluminum
Paper
Cotton
Wood
Sealing wax
Ebonite
Ni, Cu, Ag, brass
Sulphur
India rubber
Negative End

Source: Ref. 19.

The mechanism by which the materials retain the charge depend on the value of its relaxation time τ_c, which depends on the resistivity ρ and the dielectric constant ϵ_r of the material:

$$\tau_c = \epsilon_o \epsilon_r \rho \qquad (12.9)$$

The value of τ_c is determined by the leakage paths available to the charge. The leakage path through air depends on the presence of ions and the possible presence of ionizing materials. The leakage path through powders involves the resistivity of the powders, which is dependent on humidity and packing. After the electrostatic charge is generated and retained, a possible spark and thus fire and/or explosion can take place.

The amount of charge or the value of the voltage developed on the surface of an insulating body can be measured by using special meters. The most common devices in this regard are electrostatic voltmeter, electroscope and the different kinds of tube-based meters. The main common property of these meters is the extreme high value of internal impedance, which is necessary to avoid charge leakage during the measurement.

12.9.2 Examples of Electrostatic Hazards

The following are some examples of hazards which exist in industry [19].

1. *Aviation Industry.* Static charges are developed on an aircraft due to the physical contact of the aircraft in flight with airborne dust and water particles. Several accidents have occurred during fuelling operations because of static discharges resulting from inadequate bonding and grounding.
2. *Flour and Grain Industry.* Material movement by means of conveyor belts and elevators can be responsible for charge accumulation. Fine particles of grain dust suspended in the air can act as sources of explosions.
3. *Gas Industry.* Movement of a gas that is contaminated with metallic oxides, or liquid particles can produce electrostatic charge.
4. *Paper and Printing Industries.* The electrostatic charges can build up by the movement of the paper itself over the various rolls and the machinery of manufacture. In the presence of flammable inks and solvents used in the printing process, electrostatic charge buildups have caused many fires and an occasional explosion.
5. *Refining Oil Industry.* Electrostatic charge has caused in the past many hazardous fires in refineries. Extensive precautions against this hazard are important to avoid its occurrences.

12.9.3 Electrostatic Charge Control by Grounding and Bonding

Many "static" problems can be solved by bonding the various parts of the equipment together and grounding the entire system. However grounding in some cases is not enough. If the material being processed is rather bulky and has high resistivity, the charge on the upper portion of the material will be effectively insulated from ground and may result in a spark discharge. Generally the area of contact with ground is an important parameter in this regard. The following methods are used to mitigate the problem of electrostatic charge buildup.

1. Humidity control. When the humidity is increased the electrostatic charge will not be retained since charge leakage will be increased as explained earlier. However, the humidity should be increased only to the level where it does not harm the concerned process.
2. Static collectors via metallic combs.
3. Neutralizing the electrostatic charges by their opposite charges produced by special devices called neutralizers.

When using one of the above methods, care must be exercised by maintaining the relevant instrumentation and connections through proper preventive maintenance schedule. Failing to perform such a maintenance may lead to unpleasant consequences. At present there are some standards such as IEC 801-2 [20] which outline the procedures for electrostatic discharge requirements of industrial processes.

12.9.4 Electrostatic Buildup in Power Systems

In high voltage power systems, electrostatic charge buildup problems are involved in some apparati such as switchgear and large power transformers. For example, IEC 255-22-2 deals with electrostatic discharge test of electronic relay [21].

The electrostatic charge can build up in flowing insulating liquids as described earlier in section 5.4. In the electrical power industry, these charges can be found in large power transformers with forced oil circulation. The need for higher power ratings led to the application of new insulating materials and increased oil flow rates and hence more electrostatic buildup. Takagi et al. [22] reported this problem in 1978 for the first time, while Crofts [23] examined the transformers failures caused by electrostatic charge buildup due to oil circulation. Kedzia [24] used the electrostatic charge buildup as an indicator for the level of oil aging. More research should be directed to the proper measurement of the electrostatic charge buildup and control and mitigation in large power transformers in particular and power apparati in general. Peyraque et al. [25] suggested the use of the leakage current values on the tank and in the windings as a means of measuring the level of electrostatic charge buildup on transformers. Hence the dangerous level of the charges can be avoided by properly monitoring the leakage current. They show that this current is strongly influenced by temperature, applied voltage and oil flow rate.

REFERENCES

1. IEC 71, "Insulation Coordination, Part 1: Terms, Definitions, Principles and Rules and Part 2: Application Guide", IEC, Geneva, Switzerland, 1976.
2. IEC 60, "High Voltage Testing Techniques", Part 1: General Definitions and Test Requirement, Part 2: Test Procedures", IEC, Geneva, Switzerland, 1989 and 1994.
3. F. Kreuger, *Partial Discharge Detection in High Voltage Equipment*, Butterworths, London, England, 1989.
4. J. Sletbak, IEEE Trans. on Dielectrics and Elect. Insul., Vol. 3, No. 1, pp. 126–130, 1996.

5. D. Nattrass, IEEE Elect. Insul. Mag., Vol. 4, No. 3, pp. 10–23, 1988.
6. F. Kreuger, E. Gulski and A. Krivda, IEEE Trans. on Elect. Insul., Vol. 28, No. 6, pp. 917–931, 1993.
7. R. Van Brunt, IEEE Trans. on Dielectrics and Elect. Insul., Vol. 1, No. 5, pp. 761–784, 1994.
8. A. Krivda, "Recognition of Discharges, Discrimination and Classification", Ph.D Thesis, Delft University, Delft, The Netherlands, 1995.
9. IEC 76 Power Transformers, Part 1: General, Part 2: Temperature Rise, Part 3: Insulation and Dielectric Tests, and Part 5: Ability to Withstand Short Circuit, 1993, 1993, 1980, and 1976.
10. IEC 76-3-1, "Power Transformers: External Clearances in Air", 1987.
11. A. Greenwood, *Electrical Transients in Power System*, Wiley Interscience, New York, 1971.
12. IEC 502, "Extruded Solid Dielectric Insulated Power Cables (1-30 kV)", IEC, Geneva, Switzerland, 1994.
13. IEC 56-4, "High Voltage Alternating Current Circuit Breakers," IEC, Geneva, Switzerland, 1987.
14. IEC 694, "Common Clauses for High Voltage Switchgear and Controlgear Standards", IEC, Geneva, Switzerland, 1980.
15. IEC 99-1, "Non Linear Resistor Type Gapped Surge Arrester for AC Systems", IEC, Geneva, Switzerland, 1991.
16. IEC 383, "Tests on Insulators of Ceramic Material or Glass for Overhead Lines with Nominal Voltage Greater than 1000 V", IEC, Geneva, Switzerland, 1983.
17. IEC 137, "Bushings for Alternating Voltages above 1000V", IEC, Geneva, Switzerland, 1995.
18. IEC 282, "High Voltage Fuses, Part 1: Current Limiting Fuse, and Part 2: Expulsion and Similar Fuses", IEC, Geneva, Switzerland, 1994 and 1995.
19. IEEE STD 142-1972, "IEEE Recommended Practice for Grounding", 1972.
20. IEC 80-1, "Electromagnetic Compatibility for Industrial Process, Measurement and Control Equipment, Part 1: Electrostatic Discharge Requirement", IEC, Geneva, Switzerland, 1991.
21. IEC 255-22-2, "Current Relays, Part 22: Electrical Disturbances Tests for Measuring Relays and Protection Equipment, Section Two: Electrostatic Discharge Tests", IEC, Geneva, Switzerland, 1989.
22. T. Takagi, I. Ishii, T. Okada, K. Kurita, R. Tamura and H. Murata, "Reliability Improvement of 500 kV Large Capacity Power Transformer", CIGRE Paper No. 12-02, Paris, France, 1978.
23. D. Crofts, "The Static Electrification Phenomena in Power Transformers", Proceedings of CEIDP, Claymont, Delaware, pp. 192–199, 1986.
24. J. Kedzia, IEEE Trans. on Elect. Insul., Vol. 24, No. 2, pp. 175–178, 1989.
25. L. Peyraque, C. Boisdon, A. Beroeul and F. Buret, IEEE Trans. on Dielectrics and Elec. Insul., Vol. 2, No. 1, pp. 40–49, 1995.
26. B. Hague, *Alternating-Current Bridge Methods,* 5th edition, Pitman and Sons, London, England, 1959.

Index

Accelerated aging, 263–266
    frequency, 263
    multifactor stress, 265
    thermal, 265
    voltage, 263
Acceptance tests, 344
AFNOR, 351
Aging of insulation, 263–265
Air clearances, 69
    conductor-conductor, 50, 56
    conductor-cross arm, 56
    conductor-ground, 50, 56
    conductor-rod, 56
    conductor-rope, 56
    conductor-window, 56
Air insulation, 49–82
    applications, 5, 9, 10, 49, 50
    modeling, 49, 50
Air gaps, 9, 10
Alkyl benzene, 113, 116, 124
ANSI, 96, 351–352
Antioxidant additives, 121
    DBP, 121
    DBPC, 121

Arc, 30, 71, 72
    industrial applications, 71
    interruption, 72, 96–98, 198
    reignition, 72, 96–98
Arc interruption in SF_6, 96–98
Arc interruption in vacuum, 197–202
Arc in vacuum, 198, 199
    anode spots, 198
    cathode spots, 198, 199
    diffused arc discharge, 198
    vacuum gap after the arc, 199
Arresters (*see* Surge arresters)
Askerels (PCBs), 6, 114
Atmospheric conditions, effects on discharges, 57–61(*see also* Gas parameters)
Attachment coefficient, 24, 83, 85, 95
Audible noise, 41, 73, 74
Avalanche, 30–34, 37, 44, 86, 87
Avogadro number, 22
Basic impulse insulation levels (BIL, BSL), 106, 107, 343

Bohr's theories, 25
Boiling point, 47
Boltzmann constant, 3, 22
Boyle's law, 21
Breakdown of composite
  dielectrics, 93, 94, 213–220
  cavity breakdown, 214
  edge breakdown, 213
  surface erosion, 217
  tracking, 217
Breakdown field strength, 5–7,
  46–48, 86, 95
Breakdown in gases, 29–46, 85–90
  in long air gaps, 53–59
  at minimum voltage, 35, 36
  under nanoseconds pulses, 43,
    44, 288, 289
  in nonuniform fields, 35–42,
    53–57, 293
  in uniform fields, 29, 34, 35,
    86, 87, 289–292
Breakdown in liquid dielectrics,
  129–143
  breakdown process at anode,
    142, 143
  breakdown process at cathode,
    138–141
  bubble theory of breakdown,
    133
  cavitation breakdown theory,
    132
  electronic breakdown theory,
    131
  post breakdown events in
    liquids, 143
  suspended particle breakdown
    theory, 131
  with optical techniques,
    135–138
  without optical techniques,
    131–134
Breakdown probability, 51–53

Breakdown of SF_6, 85–95
  breakdown voltage calculation,
    86, 90, 91
  discharge parameters, 85, 86,
  factors effecting breakdown
    voltage, 91–95
  nonuniform field breakdown,
    87–91
  uniform field breakdown, 86, 87
Breakdown in solid dielectrics,
  173–186
  avalanche breakdown, 175
  electromechanical breakdown,
    174, 181
  electronic breakdown, 174,
  filamentary thermal breakdown,
    180
  impulse thermal breakdown,
    178, 180
  intrinsic breakdown, 174,
  steady state thermal breakdown,
    179
  thermal breakdown, 177–181
Breakdown strength, 5–7, 46–48,
  86, 95
Breakdown in vacuum, 192–198
  factors effecting breakdown,
    192
  vacuum breakdown theories,
    195
    anode theory, 197
    clump theory, 195
    cathodic theory, 196
    interaction theory, 196
Breakdown voltage, 21, 32, 35, 41,
  42, 51–66, 69, 87, 88, 91–95,
  114, 115, 289–293
  with 0% probability, 51
  with 10% probability, 51
  with 50% probability, 51, 62–66
  with 100% probability, 52
Breakers (*see* Circuit breakers)

Bridges, 357, 360
Bruce breakdown formula, 291
BSI, 350
Bubble theory, 133
Bundled conductors, 13, 71
Burst pulse corona, 70
Busbars, 99, 103, 104, 106
Bushings, 9, 100, 104, 106, 229, 269, 352
   testing of, 372, 373

Cables, 6–8, 241–273
   accelerated aging, 263–266
   accessories, 266–270
   armors, 244
   belted, 244–246, 248, 249, 253, 263, 264
   capacitance, 246, 248
   compressed gas, 6, 8, 107, 242, 243, 266, 267, 271
   constants, 245–249
   construction, 241–244
   cross-bonding, 250, 251
   current-carrying capacity (ampacity), 251–253
   DC testing, 273
   electric fields in, 249, 250
   inductance, 248, 249
   insulation, 242, 243
   insulation resistance, 249
   internal gas pressure, 351
   jointing, 266–268
   life (see Accelerated aging)
   locating faults, 270–272
   losses, 250–253
   materials, 241–244
   oil-filled, paper-insulated, 241–244, 250, 255, 262, 266–267, 270–272, 351
   resistance, 245, 246, 250
   screens and jackets, 243
   selection, 351
   sheath grounding, 251

[Cables]
   sheath phenomena, 250, 251
   superconducting, 242, 273
   synthetic insulation, 242, 243
   terminations, 269, 270
   testing of, 351, 365, 366
   types, 244
Calculation of electric fields, 11, 12, 36
Capacitors, 7, 353
Capacitive dividers, 293, 294
Capacitive voltage transformer, 103
Cascaded rectifier circuits (see High voltage DC generators)
Cascaded transformers, 279, 280
Cathode processes
   in gases, 27
   in vacuum, 189, 190, 197
Cavities in insulation, 16, 17, 105, 214, 253, 254, 258
Ceramic insulating materials, 151
Charged dielectrics, 331–332
Charged particles generation, 23, 24
Charge suppressant, 124
   benzotriazole (BTA), 124
Chemical reactions in oils (see Insulating oils)
Chlorinated hydrocarbons, 6, 114
Circuit breakers, 8, 71, 100–102, 351
   arc interruption in, 96–98
   RRRV (rate of rise of restriking voltage), 96–97, 200
   SF_6, 101–102
   short-line faults in, 97
   terminal faults in, 97
   testing of, 105, 351, 367–369
   types of, 8, 101, 102
CIGRE, 76
CISPR, 76, 350
Coefficient of attachment, 24, 83, 85, 95

Coefficient of diffusion (*see* Electron diffusion)
Coefficient of ionization, 30–32, 85, 86, 90
Coefficient of variation, 52
Collision cross-section, 23
Composite dielectrics, 7, 15–17, 209–236
  chemical deterioration, 218
  dielectric constant of composites, 211
  electrochemical deterioration, 218
  interfacial polarization, 212
  materials, 223
  multiple layers, 226
  properties, 210
  tracking, 217
  tracking index, 218
Compressed-gas cables (*see* Cables)
Conduction in liquids, 125
  conduction under high fields, 127
  conduction under low fields, 126, 127
Conductivity (*see* Properties of dielectrics)
Conductor bundles, 13, 71
Contactors, 352
Contamination of air gaps (*see* Sand/dust particles influence)
Corona, 35, 69, 70, 72, 289, 358
  AC, 37–38
  advantages, 41
  applications in industry, 41, 289
  current measurement, 301, 302
  generation, 301, 302
  glow, 37, 38, 40
  inception, 36
  nanosecond pulsed, 44, 288, 289
  negative DC, 37, 39, 74
  onset streamer, 37–39
  positive DC, 37, 39, 40, 74

[Corona]
  prebreakdown streamer, 37, 38, 40
  in SF_6, 87–90
  trichel pulses, 37–39
  water purification, 41
Corona audible noise, 41, 73, 74
Corona current pulse, 37, 74
Corona-free breakdown, 87
Corona onset voltage, 36, 87, 88, 90
Corona power loss, 41, 73
Corona radio interference, 41, 73–80
  average level, 75
  peak level, 75
  quasi-peak level, 75–80
  rms level, 75
CSA, 350
Corona stabilized breakdown, 88, 106
Cross-linking techniques, 155
Curing techniques, 155
  catalyst curing, 155
  hardener curing, 155
  radiation curing, 155
Current pulses in liquid dielectrics, 129, 130
Current transducers, 327
Current transformers, 103, 327
Current-carrying capacity of cables (*see* Cables)

Damped capacitive divider, 294
De-excitation time, 44
Deionization of gases, 28
  by electron attachment, 83
  by recombination, 28
Deltatron circuit, 279
Diagnosis of transformer lifetime
  by degree of polymerization, 231
  by dissolved gas analysis, 232
  by measurement of furfural, 233

Index 385

Dielectric
  constant (or permittivity), 3
  paper and board, 149
  liquids, 111 (*see also* Liquid
    dielectrics)
  loss factor, 4, 6, 7
  relaxation time, 375, 376
Dielectric gases, 21–48, 83, 84
  characteristics, 6–7, 83, 84
  choice of, 5, 6, 46–48
Dielectric losses in solids, 3, 4,
  167, 170, 250–253, 357
Dielectric loss test, 358, 361
Dielectric strength, 5–7, 46–48, 86,
  95
Digital recorders, 306–311
  applications, 310, 311
  measurement errors, 308, 309
  parameters, 307, 308
  technical assessment, 309–310
Digital techniques in HV tests,
  311–315
  deconvolution, 311, 312
  partial discharge measurements,
    314, 315
  transfer function method, 312–
    314
Dipole polarization, 169
Discharges in voids, 359, 360 (*see
  also* Partial discharge)
Disconnectors, 99–103
Dissipation factors, 3, 4, 250, 265
Distribution transformer testing, 8,
  312–314, 361–366
Dividers, 293, 294
Dust and sand effect on breakdown
  (*see* Sand/dust particles)

Earthing switch, 101, 102
Elastomers, 161
  ethylene-propylene diene
    monomer (EPDM), 161, 162,
    164, 242

[Elastomers]
  ethylene-propylene monomer
    (EPM), 161
  ethylene-propylene rubber
    (EPR), 161, 163, 171, 242,
    268, 272
  HTV rubber, 162
  RTV rubber, 162
  silicone rubber, 162, 164
Electric field, 10–18
  in cable insulation, 249, 250
  in cavities, 16, 17, 215–217,
    253
  computation, 11, 12, 36
  control and optimization of, 18, 93
  enhancement factor, 12–14, 41,
    190
  estimation, 12
  at free particles, 17, 18
  intensification at protrusions,
    14, 15, 88, 91–93, 249, 258
  at interfaces, 15, 16
  measurements, 315–323
  in multidielectric media, 15, 93
  nonuniform, 5, 11, 239
  types, 11
  uniform, 5, 11, 289–292
  utilization factor, 14, 86, 87, 90
Electric conduction in liquids,
  125–129
Electric polarization, 3–5 (*see also*
  Polarization)
Electric stress control, 18, 93
Electrical treeing, 254–258
Electrochemical treeing, 261
Electrode effects on breakdown,
  54–58, 92, 93
Electrode optimization, 18, 107
Electrolytic tank, 12
Electromagnetic compatibility,
  106, 316
Electromagnetic interference, 41,
  45, 46, 73–80, 243

Electron
    attachment, 24, 83, 85, 95
    attachment cross-section, 47
    avalanche, 30–34, 37, 44, 86, 87
    critical avalanche length, 43, 86
    detachment, 26
    diffusion, 28
    drift time, 44
    drift velocity, 44
    emission, 24, 27, 128, 131
    energy, 25
    excitation, 25
    free path, 22
Electron polarizability, 169
Electronegative gases, 32, 83 (see also Sulfur hexafluoride gas)
Electro-optical high voltage measurements, 317, 318, 322
Electro-optical imaging techniques, 335–339
Electro-optic effect, 318
Electro-optic sensors, 317, 318, 324
    Kerr sensors, 322–324
    Pockels sensors, 318–324
Electrostatic generators (see High voltage DC generators)
Electrostatic hazards, 343, 375–378
    buildup in power system, 378
    control of, 377–378
    generation of, 375
    in industry, 377
    measurement of, 375
Electrostatic precipitators, 1, 41
Electrostatic voltmeters, 295, 296
Emission (see Electron emission)
Epoxy resins, 7–9, 100, 149, 164, 166
EPRI, 73, 74
Erosion of dielectrics, 253, 254 (see also Partial discharges)
Estimation of minimum discharge voltage in SF_6, 90

Esters, 115, 116
External insulation (see Outdoor insulators)
Extra high voltages (EHV), 1, 2, 45, 46, 80, 105, 106, 243, 254, 272, 350

Faraday effect, 324–326
Field emission of electrons, 24, 27, 128, 131
Field factor, 12–14, 41, 190
Field inside a cavity, 16, 17, 215–217, 253
Field measurement, 316–324
Flexible laminates, 234–236
    classification, 234
    components, 235
    designation, 235
    properties, 235
    selection, 236
Formative time lag, 43
Front time of impulse, 282–287
Fuse, 352, 374
Full-wave rectifiers (see High voltage DC generators)

Gap factor, 54
Gap-type discharge, 44–46, 302, 360
    current waveform of, 46
    interference from, 79, 80
    measurement, 302
    sources, 45
Gas bubbles, 359 (see also Voids in insulation)
Gas density monitor, 98, 100, 105
Gas dielectric (see Dielectric gases)
Gas insulated switchgear (GIS), 6, 7, 9, 98–107, 316, 327, 328
    components, 99–103
    conductor systems, 99–100, 104
    dimensions, 98
    earthing switches, 101, 103

[Gas insulated switchgear]
  effect of moisture on, 84–85, 93–94, 98, 102, 108
  enclosure configurations, 99, 100
  gas system, 100, 101, 104
  insulation coordination, 105–107
  layout, 99, 100
  particle contamination, 91,92
  pressure effect, 87–93, 97, 98, 100–102
  spacers, 7, 93, 94, 100
  testing, 105
  very fast transient, 327, 328 (*see also* High frequency transients)
Gas mixtures, 6, 94, 95
Gas parameters, 21–23
  effects on breakdown, 87–90, 94, 95 (*see also* Atmospheric conditions)
Gay Lussac's law, 21
Generation of high voltages (*see* High voltage impulse generators)
Generating voltmeter, 297, 299–301
Glass, 151
  electric, 173
Glow discharge
  negative, 38
  positive, 40

High frequency transients, 105–107, 282, 327, 328
High molecular weight hydrocarbon oil, 116–118
High speed cameras, 338, 339
High voltage AC generators, 279–282 (*see also* Testing transformers)
High voltage applications, 1, 2

High voltage bushings, 9, 100, 104, 106, 229, 269, 352
High voltage DC generators, 276–279
  cascaded multiplier circuits, 278
  deltatron circuit, 279
  electrostatic generator, 276
High voltage dividers, 293, 294
High voltage equipment design parameters, 18, 19
High voltage fuse, 352
  testing, 374
High voltage impulse generators, 282–288
  circuits, 283, 285, 287
  control of, 283, 284
  for lightning impulses, 285, 286
  modeling, 282
  multistage circuits, 285, 286
  for switching impulses, 285–288
High voltage measurements, 352
  deconvolution, 311–312
  by digital recorders, 306, 307, 308, 311, 353
  by electro-optical methods, 323–324
  by electrostatic voltmeter, 295, 296
  by generating voltmeter, 297, 299–301
  by peak voltmeter, 295, 296, 298, 299
  by rod gaps, 293, 296
  by series impedance, 297
  by sphere gaps, 289–292, 296
  by voltage dividers, 293, 294, 297
  by voltage transformers, 297
High voltage power capacitors, 7, 353
  impregnants, 228, 230
High voltage test techniques, 352
Humidity effect
  on breakdown, 58–61

[Humidity effect]
  correction factor, 60, 61
  electrostatic build up, 375, 377
Hydrocarbon oils, 7
  high molecular weight, 116–118

IEC, 76, 252, 307, 361–374
IEEE, 307, 314, 350
Image converters, 338
Image intensifiers, 337, 338
Impact ionization, 24–26, 85
Impregnated insulation, 7–11, 224, 242, 243, 267
Impregnated paper properties, 159
Impulse breakdown probability, 51–53
Impulse chopping time, 355
Impulse front (or crest) time, 282–287
  critical, 56, 57, 70, 71
Impulse generators (*see* High voltage impulse generators)
Impulse recorders (*see* Digital recorders)
Impulse sparkover, 56–59
Impulse tail time, 57, 282–287
Impulse voltage dividers, 293, 294
Impulse withstand voltages, 106, 107, 343
Insulating liquids (*see* Insulating oils)
Insulating materials applications, 8–10, 98–108
Insulating materials classification, 5–7
Insulating oils, 6, 112, 117–126
  characteristics, 119–122
  chemical stability, 121, 122
  classification, 112
  constitutes, 6
  natural, 111
  petroleum, 112
  properties, 119
  reconditioning, 124, 125

[Insulating oils]
  service aged, 121, 124
  synthetic, 113, 114
  thermal heat transfer, 120
Insulation coordination, 66–69
Insulation level, 106, 107, 343
Insulation losses, 3, 4, 167, 170, 250–253, 357
Insulation resistance, 3, 366
Insulator pollution, 222, 223
Insulators (*see* Outdoor insulators)
Interfacial polarization, 171, 212
Internal discharges, 359, 360
Internal overvoltages, 42, 69, 96–97, 105–106, 282, 241, 343
Intrinsic breakdown, 174
Ion-pair production, 24
Ionization, 25, 30, 70
  by electron detachment, 26
  by electron impact, 24–26, 85
  by photons, 24, 26, 85
  coefficient, 30–32, 85, 86, 90
  thermal, 24, 26, 85

JSI, 351

Kerr effect, 318, 322–324
Kinetic energy, 22
Kinetic theory of gases, 21–23

Laplace's equation, 11, 14
Layout of GIS, 99, 100
Leader, 30, 69–71
Leader channel, 70, 88, 89
Leakage current, 273
Lightning impulse, 51, 285, 286, 310, 311
Liquid dielectrics, 6, 111–128 (*see also* Insulating oils)
  electrohydrodynamic motion, 128
  nonpolar liquids, 125–126
  polar liquids, 125, 126
  streaming electrification, 123, 124

Index

Live line maintenance, 71
Loss angle, 3, 4
Losses in dielectrics, 3, 4, 167, 170, 250–253, 357

Magnetic field sensors, 326, 327
Magneto-optic sensors, 324–327
Marx generator (*see* High voltage impulse generators)
Mean free path, 23, 24
Measurement
    by digital methods, 306–312
    by electro-optic methods, 317–324
    of electric fields, 316–324
    of high voltages (*see* High voltage measurements)
    of magnetic fields, 324–327
    by magneto-optic methods, 324–327
    of very fast transients, 327, 328
Measurement system testing, 351, 354, 355
Medium high voltage, 1, 2
Metal oxide arresters (*see* Surge arresters)
Methods to suppress trees, 261–263
Mica and its products, 150
Mineral oils (*see* Insulating oils)
Moisture effect on breakdown (*see* Humidity effect on breakdown)
Molecular velocities, 22
Multiframe photography, 137, 334, 338
Multiple rod gaps, 67–69
Multiplier circuit (*see* High voltage DC generators)

Nanoseconds pulse generation, 288, 289
Negative corona (*see* Corona)
Negative ion, 23, 24, 38, 88

Non-polar liquids (*see* Liquid dielectrics)
Nonuniform fields, 5, 11, 239
Nonuniform field gaps, 5, 11, 35–42, 239

Oil, insulating (*see* Insulating oils)
Oil-impregnated insulation, 224
Oil-filled cables (*see* Cables)
Onset voltage, 36, 87, 88, 90
Optical current transducers (OCT), 327
Orientation polarization (*see* Polarization)
Oscillatory switching impulses (*see* High voltage impulse generators)
Outdoor insulators, 217–224, 353
    composite insulators, 217
    polymeric insulators, 217, 221
    hydrophobicity, 223
    testing, 353, 372, 373
Overvoltage, 260 (*see also* Surge)
    lightning, 42, 282, 342–343
    protective devices, 10, 53, 63-65, 68, 69, 106
    switching, 42, 69, 282, 342, 343 (*see also* Internal overvoltages)

Partial discharge, 35, 216, 253–258, 260, 262, 263, 273, 306, 311, 342, 355–358
    apparent charge, 357
    classification, 357, 358
    degradation caused by, 217, 255–258
    detection system, 356, 357
    extinction voltage, 359, 360
    inception voltage, 253–254, 359, 360
    minimum detectable discharge, 359, 360

[Partial discharge]
  measurements of, 105, 254, 306, 311, 314, 315
  modeling, 353
  origin, 253–254
  recognition, 358
  significance, 253, 254
Photomultipliers, 335–337
Piezoelectric transducers, 330, 331
Particles in liquids (*see* Insulating oils)
Paschen's law, 34, 35, 87
Peak voltage measurement (*see* High voltage measurements)
Peak voltmeter, 295, 296, 298, 299
Perfluorocarbons, 47
Perfluoro-n butane, 48
Perfluoropolyether, 118, 119
Permittivity
  complex, 3–4
  relative, 3–4, 15–17, 84, 246, 250, 253
Petroleum oils, 112
  additives, 112, 121, 124
  aromatics, 112
  naphthenes, 112
  paraffins, 112
  pour point, 112
Phase to ground insulation, 49, 50, 53, 70
Phase to phase breakdown, 69–71
Phase to phase insulation, 49, 50, 53
Photoionization, 24, 26, 85
Photocopying machine, 41
Photoelectric emission, 24, 27, 128, 131
Plancks constant, 25
Point-plane gaps (*see* Rod-plane gaps)
Pockels effect, 318
Poisson's equation, 11, 14
Polar liquids, 125, 126

Polarization, 3–5, 167
  dipole, 169
  electronic, 167
  interfacial, 171, 212
  orientation, 169
Pollution, 222, 223
Poly-butylenes, 113, 114
Polyvinylchloride (PVC), 7–9, 160, 166, 242, 244, 263, 264
Polyethylene (PE), 7–9, 153, 156, 242, 252, 257, 265, 268, 333, 334
  cross-linked (XLPE), 79, 154, 158, 171, 172, 242–244, 250, 252, 254, 263–267, 272, 273
  high density (HDPE), 154, 157, 242
  linear low density (LLDPE), 157, 243
  low density (LDPE), 153, 263
  medium density (MDPE), 157
Polymerization, 153
Polymers, 153–167
  homopolymers, 153
  thermoplastic, 7, 8, 156
  thermoset, 155
Polymethylsiloxane, 114, 115, 117, 120, 121
Polypropylene, 244
Potential dividers, 293, 294
Potential transformers, 103, 277, 352, 378
Porcelain, 152, 166
Prebreakdown streamer
  negative, 37, 38
  positive, 37, 40
Properties of composites, 7, 209, 210
Properties of gases, 6, 46–48, 83–85
Properties of liquids, 6, 114, 115, 117
Properties of solids, 7

Index 391

Properties of vacuums, 6
Precipitators, 1, 41
Properties of dielectrics, 2–8
 complex permittivity, 3, 4
 DC conductivity, 3
 dielectric permittivity, 3
 dissipation factor, 3
 insulation resistance, 3, 365
 loss angle, 3
 resistivity, 3, 6, 7, 249, 250
Protective gaps (*see* Rod-rod gaps)
Pulse generators (*see* High voltage impulse generators)
Quasi-uniform field, 61, 62

Radio influence voltage (RIV), 345
Radio interference (*see* Corona radio interference)
Rate of rise of impulse voltage, 354
Rain effect on discharge, 61
Raleigh Taylor instability in liquids, 141
Rate of rise of restriking voltage (*see* Arc interruption)
Recombination of ions, 28, 29
Relative air density (RAD), 36, 58–61
 correction factor, 59, 61, 290, 291, 293
Relaxation time constant, 375, 376
Resistive divider, 293, 294
Resistive capacitive divider, 294
Resonant transformers, 281
Response time of measuring system, 351, 354, 355
Ripple
 factor, 279, 280
 voltage, 279
Rod-plane gaps, 13, 37, 50, 53, 56, 61–63, 88
Rod-rod gaps, 10, 50, 53, 56, 58, 63–69, 293
Rod-structure, 55

Sand/dust particles influence
 on air breakdown voltage, 62–66
 on measuring spheres, 62, 65, 66, 291
 on protective gaps, 62–65, 68, 69
 on transmission line design, 62, 63
Schering bridge, 357, 360
Schlieren optical system, 136, 137, 335
Secondary avalanches, 33
Secondary electrons, 31, 32
Semiconductor switching for nanosecond pulse generators, 289
SF_6 equipment (*see* Gas insulated switchgear)
SF_6 insulation, 83–110, 243, 269
SF_6 gas (*see* Sulfur hexafluoride)
Shadowgraphic optical system, 136, 137, 335
Silicone oils, 114, 117, 120, 121
Single-stage generator (*see* High voltage impulse generator)
Skin effect, 245, 246
Solid dielectrics, 7, 147–186
 alumina, 152
 ceramic insulating material, 7, 151
 classification, 148
 dielectric board, 149
 dielectric paper, 149, 231
 glass, 7, 151, 173
 inorganic insulating materials, 7, 148
 mica products, 150
 organic insulating materials, 7, 148
 porcelain, 152
 steatite, 152
 thermal classes, 236
 treeing in, 255–263, 360

Space charge measurements, 328–335
  laser intensity modulation method, 330, 334
  pulsed electroacoustic method, 330–334
  pressure wave propagation method, 330, 332
  thermal pulse method, 329, 330, 334
  thermal step pulse method, 330, 334
Space charge field, 38, 41, 69, 322-324, 333, 334
Space charge in polymers, 328, 329, 332-334
Spacers (*see* Gas insulated switchgear)
Spark discharge, 30, 71
Spark gaps (*see also* Sphere gaps), 288, 289
Sphere gaps, 52, 65, 66, 289–292, 297
Sphere to plane gap, 37, 41, 42, 50, 63
Standards, 345, 349–351
Standard deviation, 52, 348
Statistical time lag (*see* Time to breakdown)
Step response, 351, 354, 355
Streamer, 30, 33, 34
Streamer breakdown criterion, 34, 86, 90, 91
Streamer mechanism of breakdown, 32–34, 86–88
Streaming electrification, 123–124,
Sulfur hexafluoride (SF_6), 6, 8–10, 16, 46, 83–109, 243, 269
  breakdown characteristics, 87, 88
  breakdown processes, 85, 86
  decomposition of, 84, 85, 94, 102, 108
  properties of, 83–85

[Sulfur hexafluoride]
  storage and handling of, 108, 109
  toxicity of, 84, 85
Surface discharge, 358
Surface irregularity factor, 36
Surge arresters, 10, 103, 104, 106, 107, 352
  metal oxide, 10, 103, 106, 107
  testing, 353, 370-372
Surge
  lighting, 42, 282, 310, 311, 362, 370
  switching, 42, 69, 282, 287, 302, 311, 370
  with double peak, 71
  with long duration, 287
  with multiple peaks, 71
  with oscillations on tail, 287
Surge voltage, 42, 306, 310, 311
Switches, 353
Switchgear, 353, 358 (*see also* Gas insulated switchgear)
Switchgear testing, 105, 369, 370
Switching impulses, 51, 282–288, 311
Synthetic hydrocarbons, 113 (*see also* Liquid dielectrics)
Synthetic insulating materials, 149
Synthetic testing of circuit breakers, 105, 352, 369

Tail time, 57, 282–287
Tape insulation, 373
Television interference (TVI), 48, 50, 302
Temporary overvoltage, 51, 260
Test
  aging, 345
  acceptance, 3
  alternating current (AC), 273, 306, 343, 345–348, 366
  bending, 366

Index 393

[Test]
    breaking, 372
    cantilever load withstand, 374
    current impulse, 371
    direct current (DC), 273, 306, 343, 345, 347
    dielectric, 352, 362, 367, 368, 370, 374
    dry, 344, 367, 372-374
    dynamic current withstand, 374
    electromechanical, 345, 372
    galvanizing, 372
    impulse, 345–349, 353, 362–366, 371
    induced overvoltage, 345
    mechanical, 345, 367, 368, 372
    operating duty cycle, 371
    partial discharge, 253, 254, 366, 373
    pollution, 345
    porosity, 345, 372
    power frequency, 365, 366, 371
    pressure, 373
    procedures, 349–351, 352
    puncture, 345, 372
    radio influence voltage (RIV), 367, 370
    residual voltage, 371
    routine, 343, 344, 361, 362, 366, 367, 370–373
    sample, 343, 372
    short circuit, 367–369, 374
    special, 344, 362, 366
    standards, 349–351
    tan δ, 345, 366, 373
    tape insulation, 373
    temperature rise, 345, 352, 365–368, 374
    thermal shock, 372
    thermal stability, 372
    tightness, 373, 374
    time/current, 374

[Test]
    types, 343, 344, 362, 366, 367, 370–374
    wet withstand, 344, 372–374
Test voltages
    generation, 276-288
    measurement, 289-301
Testing, 2, 105
    automation, 315
    classification, 342–345
    destructive, 342, 343, 348
    nondestructive, 342–344, 348, 355
    probability method of, 348
    under impulse voltages, 343
    up-and-down method of, 348, 349
Testing transformers
    cascaded, 279, 280
    resonant, 281
    single-unit, 279
Tetrachloroethylene, 117, 118
Thermal breakdown (*see* Breakdown in solid dielectrics)
Thermal classes of insulating materials (*see* Solid dielectrics)
Thermal energy, 22
Thermal ionization, 24, 26, 85
Thermal rating of cables, 251–253
Thermal resistance, 251–253, 268
Thermal step methods, 329–330, 334
Thermally stimulated currents, 329
Time to breakdown, 43, 64, 70, 71, 291
    formative, 43
    statistical, 43
Townsend breakdown criterion, 32
Townsend first ionization coefficient, 31, 41
Townsend second ionization coefficient, 31, 41
Townsend theory of breakdown, 30
Tracking, 217

Transformer testing, 8, 312–314, 361–366
Transient current measurements, 301, 302
Transient overvoltage, 42, 306, 310, 311
Transient recovery voltage, 72, 96–97
Treeing (*see* Solid dielectrics)
Triboelectric series, 374–375
Trichel pulses (*see* Corona)

U curve, 56, 59
Ultra-high voltages (UHV), 1, 2, 54, 56, 69, 73, 80, 243, 350
Ultraviolet light, 66
Uniform field, 5, 11, 289–292
Uniform field breakdown, 29, 34, 35, 86, 87, 289–292
Uniform field gaps, 11, 13, 29–35, 86
Universal gas constant, 22
Up-and-down method (*see* Testing)

V_{50}, 348
Vacuum circuit breakers (VCB), 199, 202–206, 292
  construction, 203
  contact material, 201
  current chopping, 201
  limitations, 204
  RRRV in VCB, 200
  merits and demerits, 205
  spiral contacts, 200

Vacuum dielectrics, 6, 188–207
  microdischarge, 191
  nonmetallic emission, 189, 190
  metallic emission, 189
  prebreakdown emission, 189
Van de Graaff generator (see under High voltage DC generators electrostatic)
VDE, 351
Voids in insulation, 16, 17, 105, 214, 253, 254, 258
Volt-time characteristics, 43, 66–69, 343, 371
Voltage classification, 1, 2
Voltage dividers, 293, 294
Voltage drop, 277–279
Voltage gradient (*see* Electric field)
Voltage multiplier circuits (*see* High voltage DC generators)
Voltage ripple, 279
Voltage stresses, 51
Voltage transformers, 103, 277, 352, 378
Wagner earth device, 361
Water in insulating liquids, 119, 124, 227
Water treeing, 258–263, 273
Waveshape of impulse, 282–285, 287
Wet withstand test (*see* Test)
Wind effect on discharge, 61